LIBRARY

LIBRARY

Please return this book or have the
loan renewed on or before the date
stamped below

GLASGOW

CALEDONIAN
UNIVERSITY

7 FEB 2001   APR 2004

8 DEC 2001

10 JAN 200   0 APR 2004.

30 APR 2004

- 1 OCT 2003

AN 1992

BRE Digests   **Building Construction**

BRE Digests

Building Construction

Essential information from the Building Research Establishment

THE CONSTRUCTION PRESS
LANCASTER   LONDON   NEW YORK

Published in 1977
by
The Construction Press Ltd
a company within the Longman Group.

Second impression 1978

The Construction Press Ltd
Lunesdale House,
Hornby
Lancaster LA2 8NB
England.

Longman Group Ltd
5 Bentinck Street
London W1M 5RN.

Longman Inc.
New York

690
B86

ISBN 0 904406 42 3

Foreword

This book and its companion volumes bring together in a bound and edited form the monthly publications of the Building Research Establishment known as the BRE Digests. Respected throughout the construction industry, the BRE Digests have earned a reputation for authoritative information presented in a readily understood manner. Each Digest takes a subject of particular building concern and provides an analysis of the most important and useful relevant information available about it.

The first edition of this series of volumes presented the BRE Digests in a format which has proved to be outstandingly useful and successful. They were originally published in 1973 with a somewhat uncertain expectation of likely demand, but three printings were required to fill the orders that started to arrive from almost every country in the world.

Since the publication of the first edition, however, about 40 new Digests have appeared and many of the earlier ones have been revised and updated. This new edition therefore will meet a very real need in the many offices and colleges where it is essential to be up to date. These volumes contain all current Digests up to and including that issued in February 1977, with the exception of No. 149.

The editorial format of the first edition has been retained,with the series contained in four volumes on Construction, Materials, Services, and Defects and Maintenance respectively. Each volume includes a comprehensive index to the whole series, thereby making easy cross-reference from one volume to another.

New BRE Digests are published at monthly intervals by HMSO and these may be purchased either singly or on a subscription basis through any supplier of HMSO publications or direct from their London bookshop:

HMSO
PO Box 569
London, SE1 9NH.

ACKNOWLEDGEMENT

We have pleasure in acknowledging the co-operation of both the Building Research Establishment and Her Majesty's Stationery Office in granting us permission to publish the BRE Digests in this volume.

Contents

1 Foundations

Soils and foundations : 1

This Digest and the next show how soil conditions on site affect the design of foundations and the subsequent behaviour of buildings.

In the past the choice of foundation type has been little influenced by soil conditions but the increasing use of sites with difficult soils has more serious implications for foundation design and calls for a better appreciation of the way soil reacts to applied loads and natural forces.

This Digest therefore considers the principles of soil behaviour and examines the movements which result from the construction of foundations or which can be induced by other factors such as the weather, vegetation and subsidence.

Digest 64 deals with the choice of building sites and the investigations necessary to determine soil conditions. Digest 67 considers the types of foundation most appropriate to these conditions.

Interaction between superstructure, foundation and soil

A building, its foundation and the supporting soil interact with one another in a complex manner, the behaviour of one depending upon, and influencing, that of the others. Foundation design must therefore take into account not only the type of structure to be supported, its function and the constructional materials to be used, but also the soil conditions on site. The basic properties of soils are outlined below.

Soil properties

Soil structure

Most soils consist of solid particles of varying shapes and sizes, with water and, to a lesser extent, air filling the spaces between them. Large particles (sands) are held together mainly by their weight and when loose have very little strength. In fact, strength is related to the closeness of the packing and to the size of external forces; the influence of additional water on strength is only marginal unless it is flowing rapidly (*see* page 7). On the other hand, the amount of water that can be held by fine particles (clays) is very much larger, the water films between the particles being responsible for the characteristic stickiness which binds them together. The strength of clays therefore depends on how much water they contain. The strength increases as the film thickness is reduced, e.g. by the drying action of tree roots.

Shrinkage and swelling

When water is removed from a soil the solid particles tend to move closer together. Such movement is very limited in sands because of the negligible amounts of water held between the particles, but in clays the proportions of solid matter and water are more nearly equal and particle movements can be appreciable. Clays therefore shrink when they are dried, the shrinkage being accompanied by an increase in strength.

Conversely, when clays absorb water the water films thicken and the clays swell and lose strength. In sands the volume changes are almost negligible. Soils of intermediate particle size, such as silts, have properties intermediate between those of sand and clay.

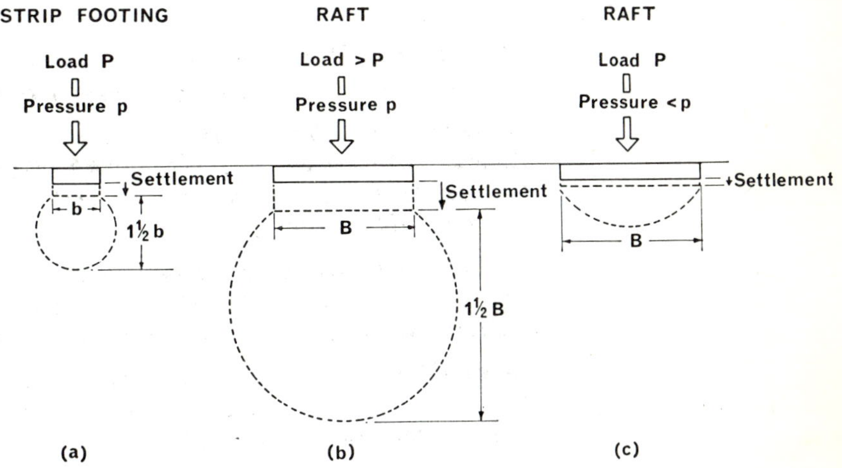

Fig. 1 Stress envelopes and settlements

Consolidation

Increased external pressure on a soil, e.g. that applied by a foundation load, increases both water and soil pressures. Water is squeezed from between the solid particles and driven to areas where the water pressure is less, while the soil particles are forced into closer contact with each other. As the ground is compressed so the foundation settles. This process of consolidation continues until the water pressure has fallen to its original value and the forces between the particles have increased by an amount equal to the newly applied load. If the soil particles and consequently the pore spaces are large, as in sands, the water movements are rapid. Foundations on sands therefore settle quickly once the load is applied and settlement after construction is completed is unlikely to be of any significance.

With clays the story is different; they offer considerable resistance to the expulsion of water and settlement caused by consolidation can continue for years after construction.

If the load on a soil is reduced, e.g. by excavation, the process just described is reversed: water tends to move towards the unloaded areas and swelling of the soil occurs.

Shrinkage (including consolidation) and swelling are thus fundamental to an understanding of soil behaviour. Often they operate simultaneously, as, for example, when the weight of clay removed by excavation approximates to the new structural load. Then long-term swelling may have settlement superimposed upon it.

Movements caused by loading

The extent to which solids deform under pressure depends on properties such as hardness, and the size of the imposed load. Soils behave similarly and a load applied through a foundation always causes settlement. However, not even uniform ground uniformly loaded settles evenly and the complex properties of soil make it difficult to assess the settlement of individual foundations or to predict the distortion of buildings as a whole.

Settlement of shallow foundations

Under normal loading shallow foundations consisting of strip and pad footings and rafts increase the pressure in a uniform soil significantly to a depth and breadth roughly equal to one-and-a-half times the breadth of the foundation. The resulting settlement depends on the increase in pressure and the soil properties in an imaginary envelope bounded by these dimensions (Fig. 1(a)).

For the same bearing *pressure* the settlement of a broad foundation or raft will be greater than for a narrow one (Fig. 1(b)), but the total load carried by the raft will be greater.

It the *load* is the same the pressure on the raft will of course be less than that on the narrow foundation, and the depth to which pressures in the soil are of any importance will be much less than one-and-a-half times the breadth of the foundation (Fig. 1(c)). The settlement of a raft is thus less than that of a strip footing carrying the same load.

In practice this simple appraisal needs modification when firmer or weaker strata pass through the stressed zones as, for example, when sand overlies soft clay.

Settlement of deep foundations

Pile foundations are supported by frictional forces acting at the surface of each shaft and by bearing forces acting at the point or base. Only small movements are necessary for frictional forces to be developed and in clay soils these can be much greater than the bearing forces at the base. In sand or gravel most of the resistance to a pile is provided at its base irrespective of whether this is pointed or flat. The settlement of a single pile under working conditions is related to its breadth or diameter, and is usually very small. In fact, in stiff soils the compression of the pile itself may be comparable to the settlement necessary for the development of supporting forces.

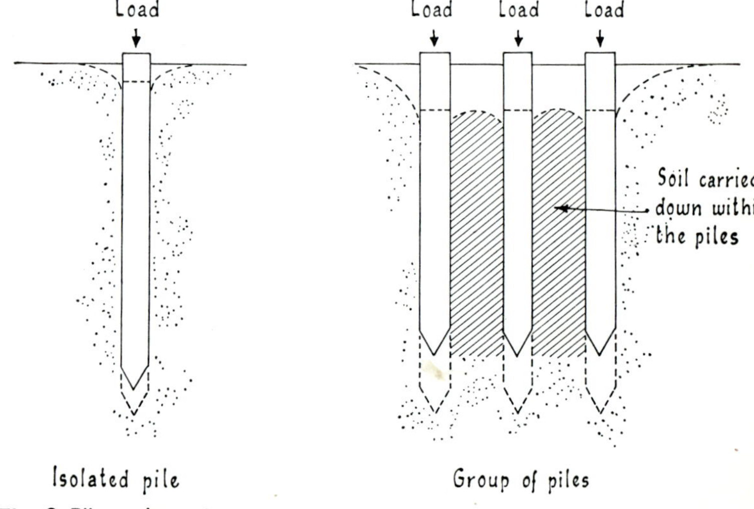

Fig. 2 Pile settlements

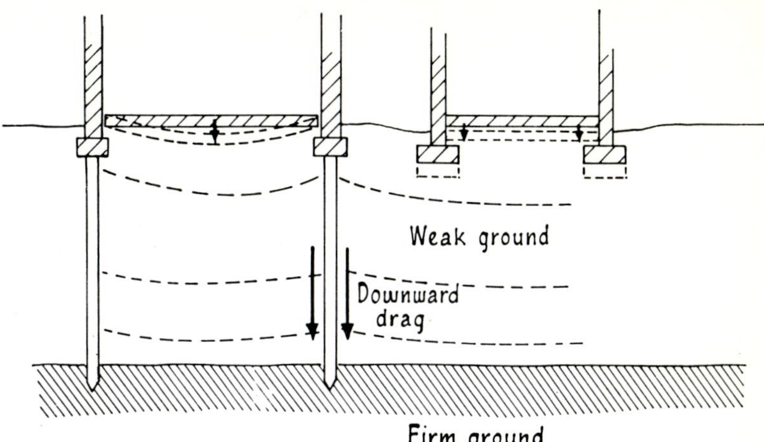

Fig. 3 Weak ground consolidating
and increasing pile load

Piles in a closely spaced group tend to carry the soil within the perimeter of the group down with them as they settle and this soil may be less effective in providing support than the soil outside the group. Thus the strength of a pile group in clay is frequently less than the sum of the strengths of the individual piles. In sand the increased compaction of the soil during pile-driving can often more than compensate for this loss. A closely spaced group behaves very much as a single foundation and the settlement at the design load is likely to be related more to the size of the whole group than to that of a single pile (Fig. 2).

Piles are frequently used to carry loads through soft soils containing a lot of water to a firm layer in which the points are founded. In these circumstances the soft soils may consolidate under the weight of floor slabs or adjacent buildings and exert a downward drag on a pile instead of supporting it; this tends to increase the load which must be carried by the point (Fig. 3).

Basements with a raft floor and retaining walls are a useful form of deep foundation which is less common now than it deserves to be. By excavating for the basement a weight of soil roughly equal to that of the building, the net increase in pressure on the soil—and hence the probable settlement—can be kept very small.

Soil movements from causes other than loading

These movements, which are not necessarily vertical or evenly distributed, are caused, for example, by seasonal weather changes, the growth or removal of vegetation, by earth flows and subsidence and, less frequently, by the particular use of the structure in question. The problems raised by these movements are usually most acute with soils composed of fine particles.

The behaviour of clay soils
Clays which shrink on drying and swell again when wetted are commonly responsible for the movement of shallow foundations. If such clays are firm enough to support buildings of a few storeys they are known as firm shrinkable clays.

(i) *Shrinkage and swelling caused by vegetation and climate.* The roots of plants penetrate soil to considerable depths and dry it when rainfall is low in summer. Beneath large trees and shrubs permanent drying has been detected in the United Kingdom at depths of 5 m or more, and shrinkage of the order of 100 mm has been measured at the ground surface. Beneath grasses shrinkage occurs to depths of about 2 m, but drying to these shallow depths is less likely to be permanent, the water transpired by the grasses in summer usually being replaced by rainfall the following winter. Even so, vertical movements of more than 25 mm have been measured at the surface between April and September. A building on shallow foundations close to vegetation is therefore liable to seasonal movements or long-term settlement, depending on the root growth. To some extent the building protects the clay beneath it from seasonal drying and wetting, and movement is more likely under the outer walls and corners. Shrinkage of clays occurs horizontally as well as vertically and so there is a tendency for walls to be drawn outwards in addition to settling and for cracks to open between the clay and the sides of the foundations. These cracks allow water to enter during the following winter and to soften the clay against or beneath the foundations (Fig. 4).

Fig. 4 also shows typical cracking of the building. During the winter these cracks partially close but, because of the relative displacement of parts of the building and the accumulation of debris within the cracks, closure is seldom complete and the cracks go on widening each dry summer.

To reduce the risk of damage as far as possible it is advisable, *as a rough guide,* not to erect buildings on shallow foundations closer to single trees than their height at maturity. The roots of groups or rows of trees competing for water over a limited area can be much more extensive, and, again *as a rough guide,* one-and-a-half times the mature height of the trees is suggested as the limiting distance. It is of course equally important that young trees should not be planted closer to buildings than these distances.

(ii) *Swelling caused by tree removal.* When trees are felled to clear a site for building, considerable time should be allowed for the clay (which was previously dried by tree roots) to regain water. Otherwise there is a serious risk that as the clay swells it will lift the building.

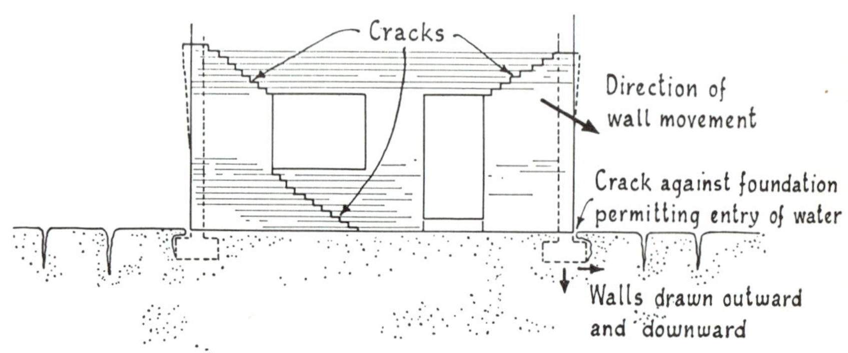

Fig. 4 Cracking associated with shallow foundations on shrinkable clay

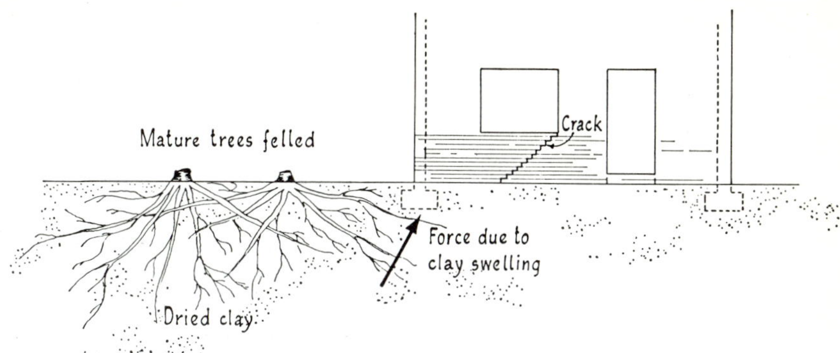

Fig. 5 Consequences of tree felling

And because the swelling is most marked close to the sites of the trees which have been removed, damage is likely from relative movements (Fig. 5). The pressures that dried clays develop when reabsorbing water are often greater than those applied by shallow foundations, and the resulting upward movements can continue for several years. For example, regular measurements made by the Building Research Station on an office block built in 1959 on a site cleared of trees a year previously show an upward movement of about 6 mm per year since construction was completed. Observations suggest that these movements may continue for up to ten years.

(iii) *Shrinkage caused by artificial drying.* Boilers and furnaces inadequately insulated from the clay beneath have been known to dry and shrink the clay, and so cause cracking of the concrete foundation slab through lack of support. Heating appliances not normally provided with adequate air ventilation or water channels between the furnace and the base should have such cooling systems installed as part of the foundation.

The behaviour of sandy soils

Water can move much faster through sands than through fine-grained soils; some foundation problems are directly attributable to this property.

(i) *Loss of ground.* Dense beds of sand are normally excellent foundation soils but occasionally water washes out the finer particles, leaving the coarser material in a less stable condition. This happens, for example, when fine sands and silts are affected by water flowing underground from high ground nearby; these exhibit a 'quick-sand' condition, particularly if an excavation is made, and much of their bearing capacity can be lost. Such situations must be dewatered or avoided completely for building purposes.

(ii) *Frost heave.* During severe winters in the United Kingdom frost may penetrate the soil to a depth of 600 mm or so. If the water table is close to the ground surface and the spaces between the soil particles are of a particular range of sizes—as they are with fine sands, silts and chalk—water can move into the frozen zone and form ice lenses of increasing thickness. As a result the ground surface is lifted in what is known as 'frost heave'. For this reason these materials should not be used as filling under floor slabs. Well-heated buildings are unlikely to be affected because of heat loss to the ground, but not so the outer walls of poorly heated houses, nor houses under construction—the concrete ground floor slabs of the latter are particularly vulnerable until glazing is complete. The slabs tend to heave more than the external walls, and service connections and wall to floor connections can be damaged. Unheated garages and outhouses, often built on concrete slabs and footings shallower than those of adjoining houses, are likely to be similarly affected.

Organic soils and made-up ground

Peats and other soils containing a lot of organic matter in the form of decaying vegetation vary greatly in volume as their water content changes. They are also very compressible and settle readily even under their own weight. Made-up ground behaves in much the same way and settles for many years unless it is good material, carefully placed, and compacted in thin layers. The bearing capacities of sites filled by end-tipping are often poor and variable; these sites should never be built upon unless deep foundations passing through the fill can be provided. Poorly compacted fill is unsuitable for foundations no matter how long it consolidates under its own weight.

Large-scale movements

Some foundation movements result from deep-seated instability in what would otherwise be good foundation soil. They can be caused by natural or geological phenomena, artificial agencies or by a combination of both.

(i) *Slopes and landslips.* Clay soils on sloping sites are likely to move downhill, albeit slowly, if the angle of slope exceeds about 1 in 10. Larger landslips occur intermittently above cliffs, river valleys and deep cuttings.

(ii) *Swallow holes.* In chalk and limestone areas cavities in the bedrock can form by the action of underground streams or water courses dissolving the rock away. If the overburden collapses into a cavity, as is common where the overburden is sandy, a 'swallow hole' is formed at the surface with consequential damage to buildings above or nearby. Because water movement is essential to swallow hole formation, soakaways should always be placed at a safe distance from buildings in these areas.

(iii) *Mining subsidence.* Large settlements must be expected in mining areas as the ground subsides over workings. Usually the ground surface 'stretches' as the front of a subsidence approaches and buildings start to tilt towards it. The tilt eventually decreases but the settlement increases as the ground beneath is affected. If structural damage is of a minor character buildings slowly return to a more or less plumb position at a lower level. To withstand the horizontal and vertical forces acting on a building during subsidence the construction should be either rigid, using a heavily reinforced concrete raft, or extremely flexible; movements can then be accommodated without serious damage to structure or finishes. Usually, small brick houses on quite thin rafts reinforced to resist the horizontal forces can survive moderate movements without undue damage.

Movement and damage

The effect of a differential vertical movement h between any two points A and B on a building depends on their distance apart. It is therefore more useful to measure differential movement as 'angular distortion', defined as h/AB (Fig. 6). Brickwork and plaster quickly show the effects of differential movement and from the limited data available the onset of cracking can be associated with an angular distortion of about 1/300. This is equivalent to a differential movement of 10 mm over a span of 3 m. Warehouse and factory buildings of framed construction can usually tolerate larger angular distortions but at about 1/150 structural damage may be expected unless the joints have been specially designed to tolerate movements.

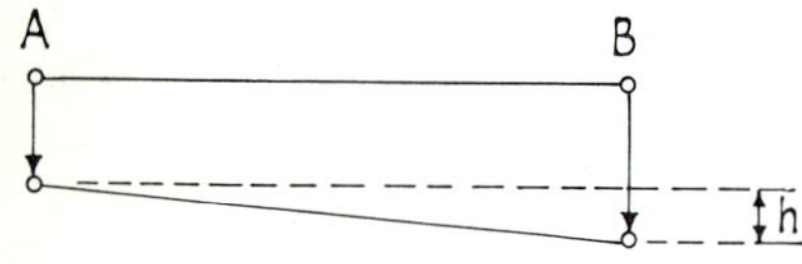

Fig. 6 Angular distortion

Soils and foundations: 2

This Digest outlines the soil and other investigations necessary in assessing the suitability of sites for new buildings. Digest 63 deals with the movements resulting from interactions between soils and foundations; foundation design is discussed in Digest 67.

The need for site investigation

An investigation of the site is needed when any building work is envisaged that is likely to be affected by conditions below ground level. This applies equally to alterations and extensions of existing work and to new schemes. It is of particular importance at the present time, when the possible use of sites that have been avoided as building land in the past is being considered. This may seem to be stating the obvious, but in fact it is known that schemes have been prepared, and in some cases work actually started, before any consideration has been given to below-ground conditions at the site and whether they are suitable for the project in hand. The extent of the investigation will depend largely on the situation, size and type of the proposed work.

This Digest is intended to emphasise the need for an early appraisal of the site so that any special measures that may be required to deal with difficult conditions can be planned at the outset. Many other aspects which bear on the intended use of the site, and which will need to be considered, such as leases, wayleaves, limitations on use and so on, are outside the scope of the Digest.

Outlined below are the main stages that can be undertaken in an investigation. A comprehensive study of the procedure is contained in the British Standard Code of Practice CP 2001 : 1957, *Site investigations*.

Stage 1. Information available off site

The local authority should be regarded as a main source of information at this stage. Its intimate knowledge of the soil and general conditions in the area, particular difficulties that can arise and local building practice can be invaluable.

Older editions of the Ordnance Survey maps, ancient maps and records from other local sources can also provide useful clues to the positions of features that might cause difficulty, such as infilled ponds, ditches and streams, disused pipes, sites of old buildings and services. Current maps, plans and records that might also be studied to advantage are set out in the Appendix.

Stage 2. Site reconnaissance

Useful information can be obtained from a surface exploration of the site and its surroundings. Nearby buildings should be looked at to see if they show any signs of distress or damage, and if possible the depth and types of foundations used should be established. The information gathered in this stage can be grouped under the following headings:

Topography

In walking over the site, nearby quarries, cuttings, ditches and slopes should be examined. These should show whether the soil conditions vary. Topographical features such as a step in the line of a valley may indicate a geological fault or a zone where shattered rock has been more rapidly eroded than the surrounding rock. Broken and terraced ground on hill slopes, especially when steep, may be caused by landslips. While gentle slopes usually make good building sites because of the drainage they promote, slopes greater than about 1 in 10 should be examined for soil creep. Slowly moving clay slopes are often responsible for the tilt of walls, fences and trees. If the ground is heavily grazed, cattle terraces along the contours are indicative of downhill creep.

Low-lying flat areas in hilly country may have been the sites of lakes and suggest the possible presence of soft silty soils and peats. On the other hand, mounds and hummocks in more or less flat regions frequently indicate former glacial conditions and the possibility of boulder clay and glacial gravel overlying solid strata.

In limestone or chalk country, craters or gentle depressions usually indicate swallow holes formed by the collapse of sandy or loamy soils into the fissured rock below. Examples may be seen in the Mendips, in parts of Pembrokeshire and Monmouthshire, on the edge of

Fig 1 Damage caused by trees on clay

the London basin, and wherever chalk or limestone occur. Local farmers will be familiar with them because they are potential cattle traps—to be fenced off or kept filled in.

Unstable conditions can also occur where springs issue from fine sand and silt strata on grassy slopes. The sand frequently erodes when turf is removed for building purposes, and ground of this nature is best avoided.

Where a polygonal pattern of cracks about 25 mm wide is seen in the ground surface during a dry summer, the surface soil is shrinkable. Shallow depressions round mature trees in open ground, broken kerb-stones and frequent repairs to paved surfaces close to trees in developed areas are further indications of shrinkage due to drying (Fig. 1). Larger cracks, more or less parallel to each other, are frequently indicative of deeper-seated movements such as are caused by mining, brine pumping, or landslips.

Flooding

Sites along river valleys and coasts may be liable to flooding from time to time, and the highest recorded flood levels should be established. Dry inland valleys may also be flooded as a result of heavy rain or melting snow, particularly when the soil beneath the snow is frozen.

Groundwater

The water levels in nearby ponds, streams and rivers noted in a site reconnaissance may not be a good guide to the level of the water in the adjoining country; near coasts and estuaries the groundwater level may fluctuate with the tide. The groundwater conditions in each of the underlying strata should be determined, as excavation below water level, particularly in fine sands, is difficult and costly. A layer of clay at foundation level may also confine a lower bed of sand with a high water pressure and cause uplift during excavation.

Because of constructional difficulties, foundations on sands and gravel should, where possible, be kept above the groundwater level. Where basements are to be constructed, precautions should be taken to ensure that they do not 'float' if the water level rises during construction. For purposes of basement design it should be assumed that the water level in clay soils extends to the ground surface.

Some soils contain sulphates; under certain conditions these may cause corrosion of buried concrete, iron and steel. If the presence of sulphates is suspected, the groundwater should be analysed, care being taken to avoid contamination of the samples by surface water.

Vegetation

The kinds of vegetation growing on a site can provide further useful pointers to the level of the water table in the ground for at least part of the year. For example, naturally occurring reeds, rushes, cotton-grass, poplars and willows indicate a high water table, whereas bracken and gorse usually indicate a well drained soil. The presence of stunted specimens of heather and gorse often means that the water table is occasionally too high for the good health of these plants. However, many plants tolerate a wide range of soil types and climatic conditions, so that the conclusions drawn from such inspections should preferably be checked with other evidence.

A change in the vegetation over quite a small area may indicate an important change in the subsoil or rock formation, and show where more detailed soil investigations should be made. Unless it cannot be avoided, individual buildings should not be founded in varying soils because of the probability of differential settlements.

The existence of peats and swamps will usually be obvious. Such sites should be avoided unless deep and extensive foundations are justified. The shrinkage of clays caused by plant roots is discussed in Digest 63. Most of the damage is caused by large trees and, in the London area, poplars have been particularly troublesome because they grow rapidly and are often planted closely spaced to form screens. If trouble from this source is anticipated, buildings close to the site and particularly near trees should be examined carefully.

The absence of vegetation typical in surrounding areas may suggest made-up ground, and a fairly superficial examination may reveal brick, wood, rubble and other kinds of mineral or chemical matter.

Stage 3.   Site history

Previous uses of a site may suggest acceptable types of foundation for the proposed new works or in extreme cases make it clear that the site is unsuitable for new building. In areas where there have been underground workings, such as worked-out ballast pits, quarries, old brickworks, coal mines, mineral workings and brine pumping, enquiries should be made to establish their exact location. Reports of damage to structures, services and drainage systems should be investigated.

Sites previously cleared of orchards or woodland may be of dried clay and this must be allowed to re-absorb moisture and swell before construction is started—unless specially designed deep foundations are to be provided.

A good deal of site history will of course be apparent from site reconnaissance and the study of maps and local records. The latter should be searched for evidence of landslips, floods, bomb craters, disused drainage systems and so on. In urban areas, the positions of sewers, drains and services should be verified to avoid risk of damage by new foundations later.

Stage 4.   Soil investigations

The previous Digest showed that foundation movements are of two kinds: those caused by the foundation load acting on the soil, and those caused by soil movements generated by other factors. The objects of soil investigations are to determine the strength and deformation characteristics of the soil under load and to identify the conditions in which soils are otherwise susceptible to movement. Some guidance on both will have been obtained by visual examination of the exposed surface of the site, while other simple tests, not involving detailed measurement, can yield useful results.

Scope of the investigations

In plan
For buildings of up to four storeys, sufficient inspection pits should be dug near the boundaries of the site to establish the soil profile. On a sloping site, or one with rock near the surface, other pits may have to be dug nearer the centre so that any variations in profile may be noted.

In depth
In the last Digest the depths were given to which soils are stressed by different types of foundation. Soil investigations should therefore be made to these depths at least.

If the inspection pits or the information obtained from other sources mentioned above suggest that piled foundations will be necessary then the investigation must be made to greater depths and a specialist firm should be consulted. In investigations for raft and piled founda-

Table 1 Soil identification

Soil type	Field identification	Field assessment of structure and strength	Possible foundation difficulties
Gravels	Retained on No. 7 BS sieve and up to 76·2 mm Some dry strength indicates presence of clay	Loose—easily removed by shovel 50 mm stakes can be driven well in	Loss of fine particles in water-bearing ground
Sands	Pass No. 7 and retained on No. 200 BS sieve Clean sands break down completely when dry. Individual particles visible to the naked eye and gritty to fingers	Compact—requires pick for excavation. Stakes will penetrate only a little way	Frost heave, especially on fine sands Excavation below water table causes runs and local collapse, especially in fine sands
Silts	Pass No. 200 BS sieve. Particles not normally distinguishable with naked eye Slightly gritty; moist lumps can be moulded with the fingers but not rolled into threads Shaking a small moist lump in the hand brings water to the surface Silts dry rapidly; fairly easily powdered	Soft—easily moulded with the fingers Firm—can be moulded with strong finger pressure	As for fine sands
Clays	Smooth, plastic to the touch. Sticky when moist. Hold together when dry. Wet lumps immersed in water soften without disintegrating Soft clays either uniform or show horizontal laminations Harder clays frequently fissured, the fissures opening slightly when the overburden is removed or a vertical surface is revealed by a trial pit	Very soft—exudes between fingers when squeezed Soft—easily moulded with the fingers Firm—can be moulded with strong finger pressure Stiff—cannot be moulded with fingers Hard—brittle or tough	Shrinkage and swelling caused by vegetation Long-term settlement by consolidation Sulphate-bearing clays may attack concrete and corrode pipes Poor drainage Movement down slopes; most soft clays lose strength when disturbed
Peat	Fibrous, black or brown Often smelly Very compressible and water retentive	Soft—very compressible and spongy Firm—compact	Very low bearing capacity; large settlement caused by high compressibility Shrinkage and swelling— foundations should be on firm strata below
Chalk	White—readily identified	Plastic—shattered, damp and slightly compressible or crumbly Solid—needing a pick for removal	Frost heave Floor slabs on chalk fill particularly vulnerable during construction in cold weather Swallow holes
Fill	Miscellaneous material, e.g. rubble, mineral, waste, decaying wood		To be avoided unless carefully compacted in thin layers and well consolidated May ignite or contain injurious chemicals

tions, samples for soil testing should normally be obtained from boreholes; these are a more economic and practical proposition than deep inspection pits. Shrinkage of clays is to be expected to depths of about 2 m and they should invariably be examined to these depths. Where trees are growing on clays, boreholes to about 5 m are needed.

Information required

Information is required on the following soil characteristics and the depths at which they are observed:

> soil type
> soil uniformity
> whether deformable or breakable in the fingers
> whether gritty, smooth, plastic, sticky
> whether homogeneous, fissured or shattered
> presence of organic or foreign matter
> colour, smell
> depth of water table.

Soil identification

In this Digest it is possible only to discuss the qualitative aspects of soil identification. Table 1 classifies soil types as they are identifiable in the field and indicates their suitability for various foundations.
Quantitative tests in the field and the laboratory will, however, be necessary where a comprehensive account of soil conditions is required, and reference should be made to CP 2001 : 1957 and BS 1377 : 1967, *Methods of testing soils for civil engineering purposes.*

Sampling

Equipment for sampling depends on whether inspection pits or boreholes are taken out and this, as suggested above, depends upon the depth of investigation necessary. For boreholes the simplest tool is the post-hole auger; with this two men can easily reach 7–8 m in soft or firm clays. It is important that the samples are extracted in as undisturbed a form as possible and that they are immediately protected from drying, because otherwise wrong conclusions may be drawn about the soil conditions. Inspection pits are of course more accessible than boreholes and allow removal of clay from the walls with a spade or knife. In sands and gravels, casing tubes prevent collapse of the borehole and percussive tools are necessary, and the services of specialist firms should be used.

Appendix—Sources of information
British Ordnance Survey Maps

The most useful are likely to be the large and medium scale maps and plans, ranging from 1/1250 to 1/25 000. Six-inch (1/10 000) maps cover the whole of Great Britain and 25-in. plans (1/2500) cover all areas except waste and mountainous districts. These maps and plans may be purchased from the Main Agents:

Cook, Hammond & Kell Ltd
22-24 Caxton Street } for England and Wales
London SW1

Thomas Nelson & Sons Ltd
18 Dalkeith Road } for Scotland
Edinburgh EH16 5BS

Further information is available from The Director General, Ordnance Survey, PO Box 32, Romsey Road, Maybush, Southampton SO9 7BR

Air photographs

Air photographs covering England and Wales at scales of 1/10000 and 1/20000, and enlargements of these, can be obtained from the Air Photographs Officer, Department of the Environment, 2 Marsham Street, London SW1. In Scotland enquiries should be addressed to the Department of the Environment, Directorate of Scottish Services, Argylle House, 3 Lady Lawson Street, Edinburgh EH3 9SD. Details of the area and grid references should be quoted.

Geological Survey Maps

The geology of much of Great Britain has been recorded on 1 in. (1/50000) scale maps and that of many coal-mining areas and the County of London on 6 in. (1/10000) maps. Manuscript 6 in. (1/10000) maps of other areas may be inspected at the Institute of Geological Sciences Library in London, and at the appropriate Geological Survey Offices in Leeds, Edinburgh and Belfast. Much additional information is available as memoirs, reports and handbooks. Requests for lists of maps, currently available, and other enquiries should be addressed to The Director, Institute of Geological Sciences, Exhibition Road, South Kensington, London, S.W.7.

Meteorological Office publications

These are available as daily, monthly and annual records of the climate in Britain and are generally obtainable from H M Stationery Office. Enquiries should be addressed to the appropriate office of the Meteorological Office.[*]

Admiralty charts and publications

The charts show high and low water marks for Home waters and the levels of the sea and river beds with reference to a low-water datum. The Authority for chart information is the Hydrographer of the Navy, Ministry of Defence, Old War Office Building, Whitehall, London S.W.1.

Mining records

In coal-mining areas the Divisional Offices of the National Coal Board should be consulted for details of active and prospective workings and for records of abandoned workings. Records of abandoned mines of other than coal and oil-shale are held by the Department of Trade and Industry.

***Advisory offices of the Meteorological Office:**

Meteorological Office	Meteorological Office	Meteorological Office
Met 0.3	Tyrone House	26 Palmerston Place
London Road	Ormeau Avenue	Edinburgh
Bracknell	Belfast	EH12 5AN
RG12 2SZ	BT2 8HH	

References

British Standard Code of Practice, CP 2001 (1957), 'Site investigations'.

British Standard BS 1377: 1967, 'Methods of testing soils for civil engineering purposes'.

Digest 63   Soils and foundations: 1

Digest 67   Soils and foundations: 3

Digest 95   Choosing a type of pile.

Digest 174   Concrete in sulphate-bearing soils and groundwaters.

Soils and foundations: 3

This Digest considers the factors governing the choice of foundations for buildings and shows in what circumstances the choice is dependent upon soil conditions and other site characteristics such as vegetation and nearby buildings. The Digest should be read in conjunction with Digests 63 and 64, which deal with soil and foundation behaviour and site investigations.

Construction types

Both the type of construction and the subsequent use of the building are important determinants of foundation design. Strip footings are usually chosen for buildings in which the load is carried mainly on walls. Pad footings, piles or pile groups are more appropriate when the structural loads are carried by columns. If differential settlements must be controlled to within fine limits, shallow strip or pad footings (except on rock or dense sand) will probably be inadequate and surface rafts may have to be considered as an alternative.

Foundation loading

The forces carried at ground level by each load-bearing wall or column of a building should be calculated, taking account of how roofs and floors are supported. Superimposed loadings are specified in the Building Regulations 1972, and BS CP 3, Chapter V; Part 1: 1967, *Dead and imposed loads*, and BS 648: 1964, *Schedule of weights of building materials,* enables calculation of dead load forces.

A typical two-storey semi-detached house of about 85 m² area, in cavity brickwork, with lightweight concrete or clay block partitions, timber floors and a tiled roof, has a mass of 100 tonnes (approx. 1000 kN) excluding the weight of the foundations. The loads at ground level are, in kN/m, approximately: party wall 50, gable end wall 40, front and back walls 25 and internal partitions less than 15. While the use of modern materials tends to reduce the *total* loads, some forms of construction, e.g. the cross-walled type, may exert higher foundation loads on *individual* walls than is the case with more traditional forms.

When the foundations have been designed, their weight must be added to the loadings already calculated so as to obtain the total bearing pressures on the soil beneath.

Soil data

Inspection pits, samples from boreholes and simple field tests suggested in Table 1 of *Digest 64* should provide data enabling near-surface soils to be identified and their strengths estimated. Geological maps and local records will often indicate whether the soils beneath pits and boreholes are likely to influence foundation design.

Adjacent buildings

New construction must not put the stability of existing properties at risk. If new foundations are placed close to those of an existing building, the envelopes of stressed soil will overlap and the loads on the soil will increase. As a result, clays may consolidate further and cause cracking of both buildings by differential settlement.

When an excavation is made, the stability of adjacent buildings may be threatened unless the excavation is adequately supported. This is particularly important with sands and gravels which derive their support from lateral restraint. Water draining towards excavations below the water table can also disturb the ground beneath adjacent buildings by increasing the effective weight of the soil and by carrying fine particles away.

When piles are to be driven the effects of vibration on surrounding soils must be considered.

Vegetation

Clays which have supported mature trees for many years will be in a dried state and likely to swell when the trees are felled (*see Digest 63*). Buildings erected before swelling is completed will therefore be subject to uplift. Unless the resulting differential movements can be accommodated by the structure or eliminated by the foundations, it will be necessary to defer

Table 1. Choice of foundation

Soil type and site condition	Foundation	Details	Remarks
Rock, solid chalk, sands and gravels or sands and gravels with only small proportions of clay, dense silty sands	Shallow strip or pad footings as appropriate to the load-bearing members of the building	Breadth of strip footings to be related to soil density and loading (*see* Table 2). Pad footings should be designed for bearing pressures tabled in CP 101 : 1972. For higher pressures the depth should be increased and CP 2004 : 1972 'Foundations' consulted	Keep above water wherever possible. Slopes on sand liable to erosion. Foundations 0·5 m deep should be adequate on ground susceptible to frost heave although in cold areas or in unheated buildings the depth may have to be increased. Beware of swallow holes in chalk
Uniform, firm and stiff clays: (1) Where vegetation is insignificant	Bored piles and ground beams, or strip foundations at least 1 m deep	Deep strip footings of the narrow widths shown in Table 2 can conveniently be formed of concrete up to the ground surface	
(2) Where trees and shrubs are growing or to be planted close to the site	Bored piles and ground beams	Bored piles dimensions as in Table 3	Downhill creep may occur on slopes greater than 1 in 10. Unreinforced piles have been broken by slowly moving slopes
(3) Where trees are felled to clear the site and construction is due to start soon afterward	Reinforced bored piles of sufficient length with the top 3 m sleeved from the surrounding ground and with suspended floors, or thin reinforced rafts supporting flexible buildings, or basement rafts		
Soft clays, soft silty clays	Strip footings up to 1 m wide if bearing capacity is sufficient, or rafts	*See* Table 2 and CP 101 : 1972	Settlement of strips or rafts must be expected. Services entering building must be sufficiently flexible. In soft soils of variable thickness it is better to pile to firmer strata (*See* Peat and Fill below)
Peat, fill	Bored piles with temporary steel lining or precast or *in situ* piles driven to firm strata below	Design with large safety factor on end resistance of piles only as peat or fill consolidating may cause a downward load on pile (*see Digest 63*) Field tests for bearing capacity of deep strata or pile loading tests will be required	If fill is sound, carefully placed and compacted in thin layers, strip footings are adequate. Fills containing combustible or chemical wastes should be avoided
Mining and other subsidence areas	Thin reinforced rafts for individual houses with load-bearing walls and for flexible buildings	Rafts must be designed to resist tensile forces as the ground surface stretches in front of a subsidence. A layer of granular material should be placed between the ground surface and the raft to permit relative horizontal movement	Building dimensions at right angles to the front of long-wall mining should be as small as possible

construction until swelling is complete, which can mean a delay of many years.

To avoid this delay the following alternative methods should be considered :

(i) Anchoring the building by reinforced bored piles sleeved from the ground over their top 3 m, and providing suspended floors. Beams spanning between the piles must be well clear of the ground surface.

(ii) Using flexible framed construction without brickwork or plastering.

(iii) Making the building rigid by either constructing a basement or reinforcing the foundations and brickwork.

Buildings erected close to standing trees should be supported on piles of sufficient depth, and for small structures bored piles are likely to be the most economical. Where nearby vegetation is insignificant, strip footings 1 m deep are usually adequate.

Choice of foundation type

Subject to the factors already considered, Table 1 gives the most appropriate foundations for buildings of not more than four storeys for a variety of typical soil conditions.

Economic and constructional factors

The availability of builders' plant can influence the relative costs of foundation types, particularly for small works. Large works may require specialist equipment; if so, the foundation design chosen should make the fullest use of it.

Experience has shown that, in shrinkable clay areas, short piles bored with mechanical augers are competitive in cost with traditional strip footings of the required depth, even for quite small contracts. For single house contracts bored piles are usually slightly more costly than strip footings, but the extra cost is justified by the additional safety obtained.

In any case, simple costing on a labour-plus-materials basis may be misleading. Construction of bored piles can often continue through the winter months when the trenches for strip foundations would be waterlogged or damaged by exposure to frost. On large sites, once all vegetation has been cleared, piles can sometimes be placed in one operation in summer, leaving above-ground construction to continue through the following winter. In addition to use on clay sites, bored-pile foundations with pre-cast ground beams should be useful for low-rise industrialized systems; in cross-wall house construction more than 10 piles are rarely necessary.

Constructional problems must also be considered. Elimination of water from excavations other than by pumping from drainage sumps can be expensive. Foundations on sands and gravels should be kept above the water-table whenever possible; if not, water should be drained or pumped away from the excavation rather than towards it so as to reduce the risk of erosion at its face. If the sides of excavations in coarse-grained soils are not sufficiently shallow to prevent collapse, support by timbers or sheet piles is necessary. Cuts in firm clay and chalk normally stand unsupported for short periods at steeper slopes than in granular soils, although some very soft clays lose much of their strength when disturbed. Clay and chalk surfaces deteriorate when exposed to water and frost and should be protected by concreting. Ground which has softened must be removed. Filling often has to be placed within the foundations of a building to bring the ground level up to that necessary for pouring concrete floor slabs. This cannot be compacted with mechanical plant, particularly if the fill level is higher than that of the ground outside, without risk of damaging the brickwork. The fill should therefore be chosen for its ability to bed well under light compaction. Alternatively, the fill should be placed and compacted before the foundations are laid.

Table 2. Minimum width for strip foundations

Soil type	Field assessment of structure and strength	Minimum width (mm) for total load (kN/m) of not more than:						
		16	24	32	40	48	56	64
Gravels and sands	Loose	300	450	600				
	Compact	225	225	300	375	450	525	600
Silts	Soft	450	675	900	Silts are frequently combined with sands or clays. Values for composite types are given in the Building Regulations			
	Firm	300	450.	600				
Clays	Very soft	450	675	900				
	Soft	360	525	675				
	Firm	260	325	375	450	560	675	750
	Stiff or hard	225	225	300	375	450	525	600
Peat	Soft	Footings should be in firm ground beneath the peat. Rarely does investigation show that the peat itself will support foundations						
	Firm							
Chalk	Plastic	Assess as clay above						
	Solid	Equal to the width of the wall						
Fill		To be determined after investigation						

Table 3. Load-carrying capacity of bored piles (in kN)

Strength classification	Diameter of pile (mm)	Length of pile (m)				
		2·5	3	3·5	4	4·5
Stiff	250	40	48	56	64	72
(Unconfined shear strength	300	50	60	70	80	90
more than 70 kN/m²)	350	65	77	90	102	115
Hard	250	55	65	75	85	95
(Unconfined shear strength	300	70	82	94	106	118
more than 140 kN/m²)	350	95	108	120	132	145

Note: The figures are for clay which increases in strength with depth to the 'stiff' and 'hard' classifications near the bottom of the piles. The figures should not be applied to piles in other situations.

Design

Strip footings

The recommendations for strip footings in Table 2 are in SI units which correspond approximately to the provisions of the Building Regulations (1972) and BS CP 101 : 1972, *Foundations and substructures for non-industrial buildings of not more than four storeys*. The table gives minimum values for the width of strip foundations in soil types identified in Table 1 of *Digest 64*.

Pad footing and rafts

Pad footings and surface rafts should be designed using the bearing capacities tabled in CP 101 : 1972. They should be proportioned wherever possible so that the centre of area of the foundation is vertically beneath the centre of pressure of the imposed loads, and the thickness and reinforcement should be calculated to resist all shear and bending forces. An angle of spread of load of 45° to the vertical may be assumed.

The bearing pressure on soil beneath deep rafts or basements is reduced because of the amount of soil excavated. Further, because of the associated walls, deep rafts possess considerable rigidity. These two advantages may be somewhat offset, however, by the greater expense of deep rafts compared with piles and ground beams.

Note: The design of raft foundations requires considerable knowledge and care, and expert advice should be taken.

Short bored piles

Table 3 gives load-carrying capacities for bored piles in shrinkable clay soils. A simple method of design not requiring laboratory measurements of clay strength is referenced (Ward and Green) below. It is most important that an adequate thickness of compressible material such as loose ash is placed between the ground beams and the clay surface to allow for relative movement. Concrete floor slabs, if used, should be isolated from the beams and main walls. Suspended floors are preferable when swelling is expected.

Other foundations

Except for major works, when expert advice must be taken, the foundations of all extensions, garages and bay windows should be of the same form and at the same depth as the foundations of the main building.

Further reading

Digests

63 Soils and foundations : 1
64 Soils and foundations : 2
142 Fill and hardcore

Papers

Green H Building on shrinkable clays—some applications of short concrete piles *JRIBA* 1952, **59** (April) pp. 212–213

Green H Long-term loading of short-bored piles. *Geotech.*, 1961, **11** (1) (March) pp. 47–53.

Lacey W D and Swain H T Construction theory : design for mining subsidence *Archit. J.*, 1957, **126** (October 10), pp. 557–570.

Ward W H House foundations, *JRIBA*. 1947, **54** (February), pp. 226–235

Ward W H and Green H House foundations : the short-bored pile. Public Works and Municipal Services Congress 1952

Fill and hardcore

This digest discusses: materials for fill and methods of placing, the suitability of existing filled sites for building, the design of foundations for use on filled ground; materials for hardcore.

New fills

Materials The best filling materials are those with particles that are hard and durable, not subject to decay or breakdown by the weather or chemical attack, and that can readily be placed in a compact and dense condition, so that consolidation after placing is small and takes place quickly. Materials with fairly large particles, such as hard rock wastes, gravels, and coarse sands, drain readily and consolidate quickly. If the material is well graded (ie contains a proportion of finer grains with the coarser grains) it can be placed in a very dense condition.

The smaller the predominant particle size, the longer is the time required for natural consolidation to take place under the weight of the fill. Fills made up of very fine sand, silty soils, inert industrial wastes and the like may require a period of years in which to settle if they have not been properly compacted during placing and the bearing capacity of the surface layers of the fill is likely to be smaller than for a fill comprising coarse-grained materials. Fills containing a large proportion of very fine-grained material (eg clay) take even longer to consolidate sufficiently. The bearing capacity of the surface layers may also be low. Chalk that has been excavated for house foundations is sometimes used as a fill beneath solid floors. If it is taken from near the surface it will frequently be soft and wet, liable to frost heave and, like clay, may take a long time to consolidate.

Household and industrial wastes may contain materials liable to decay or to decompose. Particularly undesirable are metal containers which rust and leave large voids, and plastics containers which compress but do not decompose. A large proportion of combustible material in a fill involves a risk of underground fires of which there may be no sign on the surface. The support afforded in such cases is likely to be very varied, especially where cavities have been left during placing or have formed as the result of decay, decomposition, combustion, or erosion by water. Sites of this nature are best avoided for building purposes, but if they must be used the foundations must be piled down to natural ground.

Site preparation If possible, surface water should be removed from ponds or depressions (by pumping if necessary) and arrangements made for the permanent drainage of the whole area. If the natural soil strata include compressible layers (eg beds of soft clay or peat) the weight of the fill may consolidate these layers and distort the surface of the fill however well it is placed. This settlement will be additional to that resulting from consolidation of the fill itself. Time will be required for this movement to take place; alternatively, if it is practicable, the compressible strata may be removed.

The consolidation of fill on impervious ground will be much slower than if the underlying ground is free-draining; an adequate drainage layer at the bottom of the fill will speed up the process.

On sloping sites the original ground should be well terraced to ensure general stability, and drainage should be provided throughout the site. The stability of the existing slope should be checked, and the possibility of a landslip being caused by the filling must be considered. The first layer of fill should preferably be of a granular nature to provide drainage. On very soft natural ground such a layer might also be necessary to provide a running surface for vehicles used in placing the fill.

Placing The common method of placing layers up to 1 m thick and rolling is better than loose tipping but fills containing clay or silt are only partly compacted by this method and appreciable consolidation may continue for some time. If domestic refuse is used, settlement can be restricted to about 10 per cent of its original height by controlled tipping in layers; the settlement of a 3 m fill can be expected to be complete in 5 to 10 years.

End tipping, in which thick layers are deposited in a loose condition, gives satisfactory fills only with well-graded sands and gravels and hard, slow-weathering rock waste. Occasional rolling of the top surface compacts only a thin upper layer, leaving the main body of the fill loose.

If the fill consists wholly of hard granular material, satisfactory support can be readily obtained. Loose layers not more than 400 mm thick systematically rolled with a heavy smooth roller will give a sufficient degree of compaction. The water content of the material is not important.

If the fill consists mainly of granular materials but contains a small proportion of clay, more care is necessary to keep the water content at, or a little above, the optimum (a condition which may be roughly defined as one in which the material will just bind together when squeezed hard in the hand). The layers for rolling should be not more than 200 mm thick.

If the fill contains a high proportion of clayey material, the layers should be spread at about 150 mm thickness. Large lumps will need to be broken down and special compaction equipment may be required. The water content should be controlled at, or a little above, the optimum, and construction of the fill during wet spells should be avoided.

Fills of mixed composition should, where possible, be disposed in horizontal layers over the area of the site. If there is only a limited amount of good granular material, it should be placed between layers of poorer fill; this will reduce the time required for the poor materials to consolidate naturally. With clay fills, the top layer also should consist of good granular material, since this will cut down shrinkage troubles which would arise if the clay were exposed to drying during spells of dry weather.

Settlement New fills will usually be made of whatever materials are available locally. Information available on the settlement of various types of fill shows (Fig 1) that the rate of movement decreases rapidly with time and that after about two years is relatively small. Two years can therefore be taken as the minimum period before building on fills consolidating under their own weight, although if substantial movements are likely to continue for a very long time, special foundations will probably be necessary. When the fill is compacted in layers, however, the settlement is considerably reduced and it may be possible to start building immediately after completion of the fill.

For a given material, the total settlement under its own weight increases with the depth of the fill, whilst the period during which movements take place increases in greater proportion. Where the depth of fill varies over the site, differential settlements are likely to occur even after the shallow portions of the fill have reached virtual equilibrium.

Existing fills

If the detailed history of a fill is known, a general assessment of its strength can be made by considering the nature of the material, the depth of fill and method of placing, the nature of the underlying ground and the drainage conditions. Otherwise, a careful examination of the site is needed.

Site investigation A careful examination should be made for indications of the presence of large cavities, such as old pit-shafts, sewers, etc. A broad assessment of the bearing capacity of an existing fill can be made by inspecting the sides of trial pits dug to the full depth of the fill to identify the materials and, by driving in a peg or stake, to obtain a field assessment of the strength or density.[1] At the same time the underlying ground can be examined.

If the fill is composed of hard granular materials it is likely to give good support. If it contains clay and compressible materials the support given will depend largely on the degree of consolidation: if large voids are found, the support is likely to be poor. If the filling is in a soft condition[1] the bearing capacity will be low[2] and further settlement may occur, but if it has been in place for some time and is found to be solid and hard the support is likely to be satisfactory.

A closer assessment of the properties of the fill may be obtained from field tests and, where possible, from tests on undisturbed samples. Small-scale loading tests may give a useful lead if their limitations are borne in mind; such tests reflect the properties of the layer of fill to a depth below the test plate of only $1\frac{1}{2}$ to 2 times the diameter of the plate and give no idea of the condition of deeper layers. A useful indication of the conditions throughout the depth of some types of fill can be given by the resistance to penetration of a probing tool, eg a split-barrel sampler.[3]

Quantitive measurements of the in situ density of the fill combined with laboratory compaction tests,

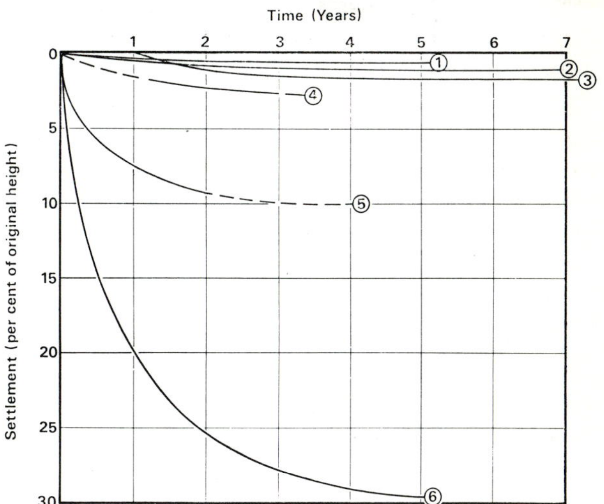

Description of curves

No.	State of compaction	Material
1	Well-compacted	Well-graded soil
2	Medium	Rock fill
3	Lightly-compacted	Clay and chalk
4	Uncompacted	Sand
5	Uncompacted	Clay
6	Well-compacted	Mixed refuse

Fig 1 Observed settlement of fills

consolidation tests and shear tests,[3] have proved useful, but small-scale laboratory tests are of little use because of the variability of most fills. Fills formed from ashes, industrial wastes and old plastered brickwork may contain soluble salts such as sulphates, which, dissolving in the ground water, may attack concrete floor slabs, buried pipelines etc. Chemical analysis is advised to check the need for special precautions.[4]

Design of foundations

Foundation design should aim at keeping differences in settlement within limits that the superstructure can stand without harmful distortion. At the same time either the total settlements should be restricted or special arrangements made so as to ensure that connections to the building, such as drains, etc, are not damaged. The type of foundation chosen[1, 5, 6] should be one that will keep the movements as small as possible. If a certain amount of differential movement of the ground is expected the foundations taken together with the structure should be made sufficiently rigid to redistribute the load and so keep the relative settlements of the structure small. Alternatively, the structure should be flexible enough to deform without damage.

Special conditions apply to buildings on fill because settlements may arise from two causes. Firstly, there is the settlement due to the weight of the fill which may result from consolidation of soft underlying strata as well as from consolidation of the body of the fill. Secondly, there is the further settlement which may result from the additional load of the structure. With deep fills the weight of the fill itself may be the main cause of the consolidation, the building load being quite small in comparison. Support might then be removed from beneath the foundations at places where the fill settles most.

In view of the comparatively large and variable compressibility of many existing fills, the foundations for structures on them require particular attention and some allowance must be made for this in designing the structures themselves. For example, long structures should be avoided or divided into compact parts to ensure flexibility of the whole. This is specially necessary when parts of the site carry different loadings, as for instance with buildings of a varying height or occupancy.

A raft or piled foundation[7] is generally required where the fill is unsound. Where some uneven bearing cannot be avoided, even though the fill has undergone some consolidation, a concrete raft reinforced top and bottom is particularly suitable for light structures such as houses.

If the bearing capacity is found to be good enough to take wide strip footings, these should be reinforced in both directions and with ties to prevent displacement and increase stability.

If the filling, although affording moderately good support, is thin and overlies firm natural strata, it may be preferable to use a beam and pier or beam and pile foundation with suspended floors.

Where the fill is loose and of poor supporting value, the foundations, particularly those of fairly heavy buildings, must be taken through to firm natural ground by piers or piles with the structure carried on beams tied together and bonded to the piers or piles. In deep fills, which can be readily penetrated, piles are often preferable. In assessing the design loads for piles, allowance should be made for 'negative skin friction', ie the downward drag on the piles which will result from consolidation of the fill after the structure is completed.

Special care should be taken with buildings sited on filled ground near the interface with the natural soil because the consolidation of the fill decreases towards the interface. Placing buildings partly on natural ground and partly on fill must be avoided unless the loads are carried down to the natural ground by piers or piles.

The relative movements between the building and services entering it and between various sections of piping contained in the fill merit careful consideration. If the movements are not likely to be large, the use of strong rigid pipes (eg cast-iron) with flexible connections[8] may be sufficient. In more severe cases it may be necessary to use flexible pipes[9] or to carry the services on piles bearing on a firm stratum below the fill.

Hardcore

A bed of hardcore may be needed on a site to fill depressions and to raise the finished level of an oversite concrete slab after removal of turf and other vegetation; on soft or wet sites it may be necessary to provide a firm working surface and to prevent contamination of the lower part of the wet concrete during placing and compaction. It can reduce the rise of moisture from the ground, though not to such an extent as to eliminate the need for a reliable moisture barrier.[10]

Ideally, material used as hardcore should be chemically inert and not affected by water but few of the materials available at reasonable cost satisfy these requirements completely. The soluble sulphate content in representative samples of hardcore for use on a wet site and which will be in contact with normal Portland cement concrete should not exceed 0·5 per cent. Some materials swell when wetted and expand when exposed to frost during construction; this combined with expansion of the concrete by sulphate attack has been known to move enclosing walls and to break and lift floor slabs.

Chemical action and swelling can occur only if water is available to the material and the suitability of a material prone to either depends on the probability of a high water table or surface water flooding and the efficiency of the site drainage. Material that swells on wetting should not be used as hardcore under these conditions. Concrete can be protected against excess sulphates by a layer of bitumen felt or plastics sheeting placed on the hardcore; alternatively, or in addition, more resistant types of cement can be used.[4]

If a hardcore bed is intended to reduce the rise of ground moisture, the proportion of fine material, at least in the lower layers, must be as low as possible. The layers should not be rolled to such an extent that the subsoil is forced up into the voids; it may be necessary to put down a first layer of clinker to serve as a base for a larger aggregate which, in turn, must be blinded with fine material.

Materials suitable as hardcore
Blastfurnace slag Slag produced during the manufacture of pig-iron is widely used. It contains sulphides, which may cause unpleasant smells on a wet site and will eventually oxidise to sulphates. Weathered slag from an old bank may already contain sufficient sulphates to cause trouble.

Brick or tile rubble Rubble from brick or tile works is clean, hard and chemically inert but it may contain little or no fine material and may not consolidate readily.

Old brick rubble varies greatly in quality. If the bricks are soft and crumble easily or if the rubble contains a high proportion of mortar, it will consolidate easily to give a good working surface; the layer will not, however, be an efficient barrier to the rise of ground-water. When required for use on a wet site, it should be examined for the presence of sulphates in the form of gypsum plaster.

Clinker Hard-burnt clinker in which the material has been fused and sintered into hard lumps is good but furnace residues that are not hard-burnt, 'breeze' or 'ashes', may contain too high a proportion of un-burnt coal or sulphates for safety on a wet site. The amount of unburnt material should not exceed about 20 per cent.

Colliery shale Only well-burnt colliery shale is suitable, recognised by its deep brick-red colour; dark, sometimes black, material is unburnt and may swell when it gets wet. Some shales, even though well-burnt, contain sulphates.

Concrete rubble Although suitable for use in all conditions, this material may be obtainable only in very large lumps and will have to be supplemented by other fine material.

Gravel Well-graded gravel meets most of the requirements but it is expensive. Coarse screened gravel, say all over 40 mm, may be obtainable more cheaply and could be used for the bottom layer, covered by a layer of well-graded material and rolled to a firm surface.

Oil shale Spent waste shale from oil refining is a by-product of a controlled chemical process. It is tough and resists disintegration by weather but its soluble sulphate content is high. If precautions appropriate to the wetness of the site are taken against sulphate attack of concrete, this material is suitable for use as hardcore.

Quarry waste The material is clean, hard and safe to use; it may be unevenly graded and therefore difficult to consolidate to give a good working surface. Many quarry owners can often offer crusher-run stone and similar products which should be economical near to the quarry.

Pulverised-fuel ash PFA is recovered from the flue gases produced by boilers fired with pulverised coal. Some 75 per cent of the ash is carried in the flue gases, the remainder, a coarser material known as furnace bottom ash, drops to the bottom of the combustion chamber and is collected from there. If not required in the conditioned state, the PFA is transported hydraulically with furnace bottom ash to lagoons where they are left to drain. They may then be transferred to stockpiles. The material is light-weight, and when conditioned with the correct amount of water and compacted, has self-hardening properties.

References

1 BRS Digest 64 Soils and foundations: 2

2 Civil Engineering Code of Practice No 4, Foundations. Published by Institution of Civil Engineers.

3 BS 1377:1967 Methods of testing soils for civil engineering purposes (with amendment PD 6412, May 1968).

4 BRS Digest **174** Concrete in sulphate-bearing soils and groundwaters.

5 BRS Digest 63 Soils and foundations: 1.

6 BRS Digest 67 Soils and foundations: 3.

7 BRS Digest 95 Choosing a type of pile.

8 BRS Digest 130 Drainage pipelines: 1.

9 BRS Digest 131 Drainage pipelines: 2.

10 BRS Digest 54 Damp-proofing solid floors.

Choosing a type of pile

This Digest discusses the problem of which piling system to choose from the many that are available when a job requires piled foundations. It emphasises the need for a full investigation of ground conditions before coming to a decision whether piles should be used or not, and stresses the importance of consulting a foundation engineer at an early stage. Types of piles and methods of installation are considered, and a series of questions and answers is included, designed to lead to the correct choice of pile.

There are many different piling systems available, offering piles of various materials and methods of construction; the architect with no particular knowledge of the subject may find difficulty in recognising what would be a suitable type of pile to use.

It should first be decided, however, whether piling or some other type of foundation would be better in the given circumstances. This demands adequate information about the ground conditions, the loading to be imposed and the settlement that can be tolerated. For any structure of large size, the problems involved in making this decision require specialist knowledge of soil mechanics and foundation engineering; even with small structures, it is advisable to take the decision on technical grounds rather than by rule of thumb. Piling does not offer an automatic solution to foundation problems and each case must be judged on its merits. In many cases piles are unnecessary, in some cases they cause problems, but by and large a properly designed piled foundation does give a satisfactory performance. Reference should be made to Civil Engineering Code of Practice No. 4 (1954), *Foundations* (now being revised for publication as British Standard CP 2004).

The foundation system needs to be taken into account when designing the superstructure if an economical form of building is to result. Therefore the ground conditions and the type of structure envisaged should be discussed with the foundation engineer at an early stage, so that if piles are preferred the superstructure will be designed accordingly.

Many factors enter when choosing the type of pile to be used, but since, broadly speaking, the different forms of piling have been devised to suit different ground conditions, it is the nature of the ground which largely determines the type.

Site investigation

Unless the ground conditions are known accurately, it is impossible to reach a worth-while assessment of the situation and so make a reasoned decision about the type of foundation to be used. This means that information about the identities, strengths and compressibilities of all the strata likely to affect the foundations should be available (see Digest 64, *Soils and foundations: 2* and CP 2001 : 1957, *Site investigations*).

The sequence of strata determines how the piles will transmit load to the ground, whether by end-bearing or by skin friction. With an end-bearing pile, the shaft usually goes through soft deposits and the base or point rests on bedrock or penetrates some distance into a stratum of dense sand or gravel so that the pile acts as a column (see Figs. 1 and 2). A friction pile is embedded entirely in cohesive soil and obtains its support mainly by the adhesion or 'skin friction' of the soil on the surface of the shaft (Fig. 3). With piles of both types, since the strata below the level of the pile points must carry the load, the site investigations should extend to a depth of at least one-and-a-half times the width of the structure below the anticipated pile point level. The presence of water-bearing sands or gravels should be determined, especially if the water is moving or is under artesian pressure. The choice of the pile material will also be influenced by the presence of chemically aggressive groundwater or soil which may cause deterioration of the pile material and, in the case of harbour work, by the likelihood of attack by marine borers.

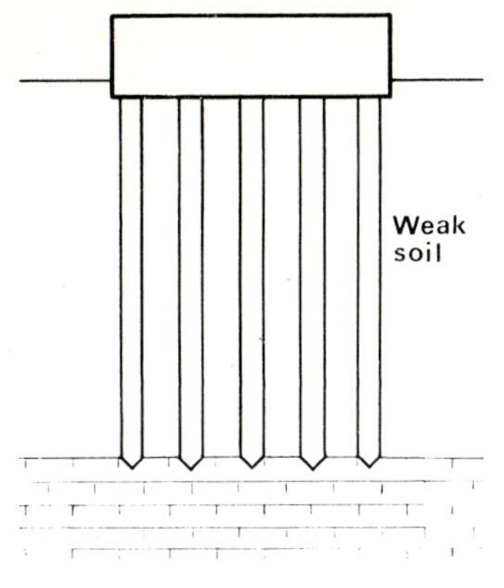

Fig 1

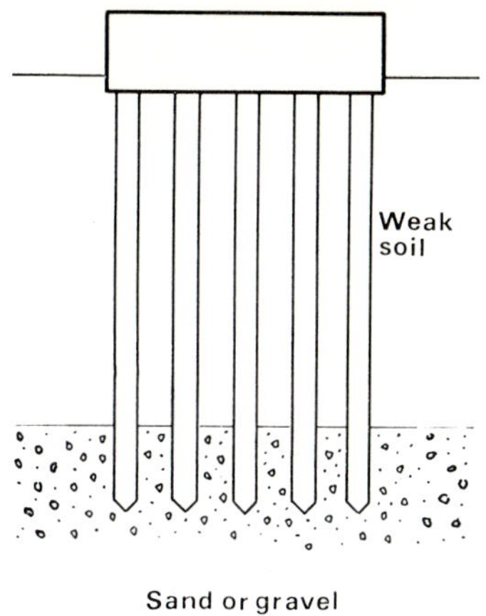

Fig 2

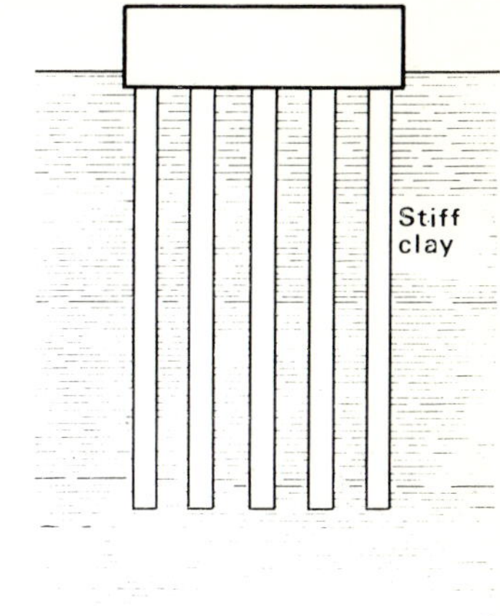

Fig 3

Types of piles

Apart from the division between end-bearing and friction piles, there is a clear distinction between piles which are installed by forcing the soil to move out of the way, such as happens when a stake is driven into the ground (these are known as 'displacement' piles) and those which cause no displacement because a hole is bored or excavated in the soil and the pile is formed by casting concrete in the hole (these may be termed 'non-displacement' or 'replacement' piles). **H**-section steel piles are usually thought of as 'small-displacement' piles.

Preformed solid piles of timber or reinforced concrete, and concrete or steel tubes or 'shells' with the lower end closed are clearly displacement piles. A displacement pile can also be formed by driving a steel tube with a detachable shoe or plug. As the tube is withdrawn, the shoe is left behind, while concrete poured into the tube runs out at the now open bottom to fill the void. The concrete is thus cast directly against the soil and is compressed or tamped in some way, depending on the particular proprietary system of piling. The hardened concrete forms the pile.

Replacement piles can be formed by driving an open tube, the bottom edge of which may be formed to a cutting edge, and extracting the soil core as the tube is sunk. When a suitable bearing stratum is reached, the tube is cleaned out and filled with concrete. A very common method of making a replacement pile is to form a borehole by auger or percussion methods, whichever is suitable, using temporary casing or lining tubes if needed to keep the hole open, and then to fill the borehole with concrete, withdrawing the casing tubes as the concrete is placed.

Preformed or cast-in-place piles may therefore be of either displacement or replacement type. Almost all cast-in-place displacement piles, bored piles and shell piles driven with the aid of a mandrel are installed by specialist piling contractors, but it is the usual practice for a general civil engineering contractor to install preformed timber, reinforced concrete, steel **H** piles and shell piles not requiring a mandrel.

Timber and precast reinforced concrete piles are usually made long enough to ensure that they will not need to be lengthened to accommodate normal ground variation or errors, so that waste can result. Lengthening, on the other hand, may involve expensive delays. Steel piles and tubes are easily shortened or lengthened and shell piles can be adjusted to length by the addition of further lengths of shell. Cast-in-place piles of either displacement or replacement type are formed to the length required.

Precast concrete piles require either a casting yard on the site, or to be transported to the site. Neither of these alternatives is convenient in urban areas. Ready-mixed concrete is generally used for cast-in-place piles on an urban site. Pile shells usually present no transport problem since they may be brought to the site in convenient lengths.

Methods of installation

The traditional method of installing a pile is by means of a drop hammer, raised by a winch and allowed to fall on the pile head. Drop hammers are still in common use, together with compressed air, steam and diesel hammers. Loud noise and some vibration often accompany hammer-driven piling operations, but driving under carefully controlled conditions does not necessarily give rise to harmful vibrations, and noise can be reduced if the pile is driven by an internal drop hammer.

When the soil conditions are suitable, piles and casing tubes can be driven by vibration. The vibrator-driver brings the pile into vibration in the direction of its longitudinal axis, at a frequency and amplitude dependent on the particular system in use. This causes the resistance of the soil to be broken down, permitting the pile to sink into the soil under the combined weight of the driver and the pile. The chief source of noise in vibration driving is from the motors.

Boreholes up to 600 mm in diameter can be sunk using percussion tools on tripod rigs with compressed air or diesel winch motors. If the pile is short and of small diameter, a tractor-mounted auger might be used, but for bored piles over 600 mm in diameter, large truck- or crane-mounted augers or excavating rigs, using percussion tools and grabs within heavy casing tubes, would be used. The boring operation itself is relatively quiet but there will be noise from the motors. Piles used for underpinning may be installed by progressively adding short lengths to a pile, which is jacked into place. Small tripod rigs, needing no more than 2·5 m headroom, can form bored piles for underpinning.

Large piling rigs impose heavy loads on the soil, and the temporary work needed to get the rig to its worksite over weak ground can become a major item. Similarly, if a large rig is needed, the on-costs per pile may become unacceptably high if only a few piles are to be installed.

Influence of the ground

When choosing a type of pile, the foundation engineer will ask a series of questions and will look for the answers in the information provided by the site investigation. Some of these questions will already have been asked when consideration was being given to whether a piled or some other type of foundation should be used; knowledge of the feasibility of using a particular piling system will influence that decision.

In very many circumstances, the choice of piles for buildings and bridges could be made by satisfying the following questions:

1 Is the soil sensitive to an important degree?

Many alluvial soils of recent origin are sensitive, i.e. they are reduced in strength when remoulded. Highly sensitive soils need careful handling, and if piling must be used both the foundation engineer and the contractor should be guided by local experience.

2 Can heaving of the ground be tolerated?

When displacement piles are driven, the soil in the vicinity is displaced and heaving may occur. Generally this does not matter on an open site, except that redriving each pile after adjacent piles have been driven will be advisable if the piles are end-bearing. Heaving caused by displacement piles may result in damage to adjacent existing structures and the associated lateral thrust may cause bursting if displacement piles are driven within a cofferdam. In such cases, replacement piles or 'small-displacement' piles are required.

3 Does the soil contain boulders or beds which would impede piling operations?

With ground of this sort, it is best to look at the design again to see if piling can be avoided. If piling is still considered to be the best solution, then only trial will show if the boulders can be pushed aside by driven preformed piles or shell piles. If beds of resistant material must be penetrated, this may be done either by preboring or by 'spudding', which is the process of breaking through the obstruction by dropping or driving a strong steel probe, the pile then being inserted in the resulting hole and driven on. Boulders or resistant beds can be tackled by percussion tools and grabs working inside a tube which, if left in place, would follow normal caisson methods of foundation construction. If the tube is withdrawn as concrete is placed, the resulting foundation would be regarded as a pile.

4 Can the noise and vibration of hammer driving be tolerated?

If the answer is 'no', then clearly some form of pile not requiring driving must be used. The choice of an alternative will be limited to some form of bored or excavated pile.

5 If the pile is to be end-bearing, will the base or point be in water-bearing granular soil?

When a replacement pile, such as a bored pile or a tube from which the soil core is removed, is sunk to water-bearing sand or gravel, it is very difficult to prevent the ground around the base becoming loosened by inflow of water into the borehole. A preformed driven pile would be the better choice if it satisfies other requirements.

6 How strong is the soil through which the pile shaft will pass?

Clearly, in the case of a cast-in-place displacement or replacement pile, in which the concrete is cast directly against the soil, the soil itself has to be strong enough to support the column of wet concrete until it sets. Ground which is so weak or loose that a casing is required to prevent inflow of soil into a borehole needs special care to avoid the formation of 'waists' or gaps in the piles when the casing is withdrawn. Normally the use of a sufficient head of concrete of adequate workability, combined with continuous checks of the concrete level during extraction of the casing, is sufficient to ensure sound workmanship. In particularly adverse conditions, as, for example, where there is a strong groundwater flow, the use of a permanent casing may be advisable. A preformed pile can be used, either a displacement pile (r.c., closed tube, or shell pile) or a replacement pile, such as a permanent tube, open at the bottom and driven with simultaneous extraction of the soil core.

If the soil is strong enough, cast-in-place displacement or replacement piles are possible, depending on the decisions on the previous questions.

If the soil is a firm clay, a borehole may be formed without requiring a casing, and the base may be enlarged if required by undercutting without the need of roof support. This is the typical condition in London Clay and enables economical bored pile methods to be used.

7 *Is the soil chemically aggressive?*

Certainly the material of which the pile is constructed must be chosen with regard to durability in its given situation, and if either the groundwater or the soil is likely to attack pile materials this must be considered. The need for a particular material may determine the choice of the pile type; for example, where very highest quality concrete must be used, this demands a factory-made precast unit. The advice of chemists should be sought, but in the long run practical experience of similar conditions forms the best guide.

Economic considerations

It has been mentioned earlier that the type of structure should be suited to the type of foundation. Therefore, when comparing the cost of piling with another form of foundation, the overall cost of the structure should be considered.

Where two or more types of pile would serve equally well, then invitations for tenders should be spread among competing specialist piling systems. Prospective tenderers should in all cases be given the fullest possible information about the site (its location and the details of the site investigation) and the settlement and loading demands of the structure.

Meeting structural requirements

The ability of any pile to carry load and the amount it settles under load are functions of the ground and the actual shape and method of construction of the pile. For an end-bearing pile, the cross-sectional area at the point or base is important. For a friction pile, the area of shaft surface and the adhesion or 'skin friction' of the soil on the surface determine the load carried. Various specialist piling systems make provision for increasing the base area, for compressing cast-in-place concrete against the soil, for corrugating the shaft surface and so on, with the intention of improving the bearing capacity of the pile. Loading tests are usually necessary to check the claims made for any proprietary system.

The type of structure may have an overriding influence on the choice of pile type. Where part of the length of the pile will be left standing above the ground, as in harbour or bridge work, or where the pile has to be placed through water, a preformed pile is needed and the choice between a displacement or a small-displacement pile (in this case probably an open-ended tube or box pile) often depends on the problems of installation. Bracing is not easily fixed to reinforced concrete piles, but can be bolted or welded to steel piles.

The ease with which the head of the pile can be bonded into a pile-cap varies with the type of pile. Precast reinforced concrete and prestressed piles have to be stripped down to expose bars for bonding. On the other hand, cast-in-place concrete piles and shell piles filled with concrete are formed to the correct elevation and bond-bars inserted. Steel **H**-section piles may need shear connectors to transfer the load to the pile.

Conclusion

Whether the structure is a building or a civil engineering construction, the approach to the design of the foundation is substantially the same. Discussions with a foundation engineer should take place as early as possible in the design stage so that the form of superstructure and foundation can be correctly chosen to suit each other. The following factors must be considered at this point:

1   The expected loading and tolerable settlement of the structure.

2   The necessity of protecting adjacent property from settlement, heaving and noise.

3   The information resulting from an adequate site investigation.

If a piled foundation is chosen as a result of design studies, the choice of the type of pile will follow the process outlined in this Digest and will no doubt be made simultaneously with the decision to use piling.

2 Floors

Design of timber floors to prevent decay

Most species of timber, especially softwoods, are liable under certain conditions to attack by the 'dry rot' fungus, *Merulius lacrymans,* which feeds on the timber, destroying it in the process. In persistently wet conditions, wood may be attacked by the 'cellar fungus', *Coniophora cerebella*; the effect is often called 'wet rot'.

An outbreak of dry rot can spread quickly, cause serious damage, and prove troublesome and costly to eradicate; prevention, therefore, is all-important. There are two possible courses—use timber that is naturally resistant to dry rot (eg Western red cedar, Californian redwood or yellow cedar) or take precautions in design and construction with less resistant species. To be immune from attack, the moisture content of the timber must be kept below 20 per cent (based on its oven-dry weight). Accordingly, the timber must be nearly in this condition at the time of fixing and everything possible must be done to keep the moisture content below 20 per cent thereafter. For the most part, this can be achieved by careful attention to the principles of good building construction—full provision against the entry of rain and ground moisture, thorough drying of the structure on completion, and efficient heating and ventilation to minimise condensation. Nevertheless, in the more inaccessible parts of a building—under floors, behind skirting and panelling, etc.—unsuspected damp conditions may persist unless precautions are taken.

Selection and storage of timber

Only sound, well-seasoned timber should be used; any that shows signs of incipient decay should be rejected. If possible, the conditions under which it has been stored should be ascertained and nothing allowed on the site which comes from a yard in which careless stacking or unclean conditions are tolerated.

Care taken in storage at the timber yard will be largely wasted unless it is matched by similar care on site. If the timber cannot be stored inside a building, it should be stacked on bearers so as to avoid contact with the ground. Dry, well-seasoned timber should be closely stacked, to minimise absorption of atmospheric moisture, and wholly covered with tarpaulins or polythene sheeting around and under the stack as well as over it.

If the timber is moist when it reaches the site, it should be open-piled so that air can circulate round each piece to promote further drying; the stacks should have top cover but the sides should be left open so as not to hinder air flow through the stack. If the stacks are within a building, windows should be left open.

Timber that has been treated with water-borne preservatives will not have been re-dried in a kiln unless this was specially ordered; normally, therefore, it will need to be open-piled and air-dried.

The moisture content of floor joists and battens should not exceed 22 per cent at the time of fixing; under normal drying conditions it will then quickly fall to 20 per cent and lower.

To prevent undue drying shrinkage, floor boarding and blocks, at the time of laying, should have a moisture content not more than about 2 per cent higher than they are expected to attain in use in a dried-out building; this final value is about 14 per cent where heating is intermittent, or 12 per cent where heating is continuous; softwood boarding may be fixed at a moisture content of up to 18 per cent if some gapping of the joints is acceptable.

Types of floor
Unventilated ground floors

Wood block floors. Wood blocks laid in hot-applied bitumen or pitch adhesive require no additional protection against rising moisture; the adhesive, in which the blocks are dipped, should form a continuous layer and keep them out of contact with the concrete. Cold bitumen emulsions, however, cannot be relied on to give adequate protection, and blocks laid with this type of adhesive should be protected by a sandwich type damp-proof membrane (*see* Digest 54) or by an underlay of mastic asphalt, using an adhesive which is compatible with the asphalt.

Board or strip flooring. Board and strip flooring may be nailed to battens fixed to the concrete sub-floor by one of the methods illustrated in Fig 1. Protection by an impervious membrane is essential for these floors.

A surface membrane will protect the floor against rising ground moisture and also against the effects of residual construction moisture in the screed, but it will be pierced by the legs of the flooring clips or by the flooring nails. A sandwich membrane need not be pierced but sufficient time must be allowed for the concrete above the damp-proof layer to dry before the boards are fixed. The time necessary for this will depend on the density and thickness of the concrete and on the drying conditions, but since in the absence of heating facilities it may be a very long period, the sandwich method is perhaps less useful in practice. Failure to dry out the floor screed thoroughly may lead to decay in these floors and the risk will be greatly increased if impervious floor coverings, such as rubber, linoleum, plastics, etc., are laid prematurely.

The damp-proof membrane should make a watertight joint with the damp-proof course in the walls, etc., the membrane being extended vertically if necessary. Materials suitable for use as damp-proof membranes are listed in Digest 54.

All battens should be impregnated with preservative. Brush application of a preservative to the underside of the boarding is also recommended; for this purpose it will normally be advisable to avoid creosote or any preservative which might creep or bleed through and disfigure the exposed surface of the flooring.

Tongued and grooved boarding should be used, so that water used for washing down, or spilled accidentally, will not readily pass through the joints. If any water reaches the underside of the boarding, it will be slow to evaporate and dangerous conditions may persist for a considerable time.

Ventilated ground floors

Ground floors of boarding on joists have a relatively large air space beneath them. To minimise the evaporation of water into this space, the sub-floor should be covered with a bed of good-quality dense concrete or a hot applied damp-resisting coating (solum) to BS 2832; the latter should be laid on a well-compacted base of hardcore, blinded with ashes to form a level surface free from fissures. The surface of the concrete bed or solum should not be lower than the level of the surrounding ground, or alternatively, the site should be drained so that inundation of the area cannot occur. The air space should be thoroughly ventilated and, if this is efficient, no preservative treatment of the timber is necessary.

For efficient cross-ventilation, there should be a clear depth of 150 mm between the underside of the floor joists and the site covering and sufficient air-bricks to provide at least 3000 mm^2 *open area* per metre run of external wall. The airbricks should be placed as high as possible on opposite walls and sleeper walls should be built honeycombed. Care should be taken to avoid unventilated air pockets such as may occur near a bay window; ducts should be formed under hearths and under solid floors wherever these might interrupt cross-ventilation.

Airbricks should not be built in to the north walls of buildings which are situated in a frost hollow, particularly if the underbuilding or crawl-space is a deep one; otherwise, there is a risk that in some weather conditions condensation may occur on the underside of the floor near the airbricks. The risk will be greatest below rooms that are unheated or intermittently heated, such as bedrooms. On sloping sites where a deep underbuilding cannot easily be avoided, the need for a second damp-proof course near ground level should be considered. If a terrace around the building is cut into the hillside, it may be desirable to bring the level of the concrete bed or solum at least up to that of the terrace, for in these circumstances drainage of the terrace may not always be able to cope with the volume of rain-water run-off from the hillside. The concrete bed or solum should be stepped so as to coincide approximately with the finished level of the surrounding ground.

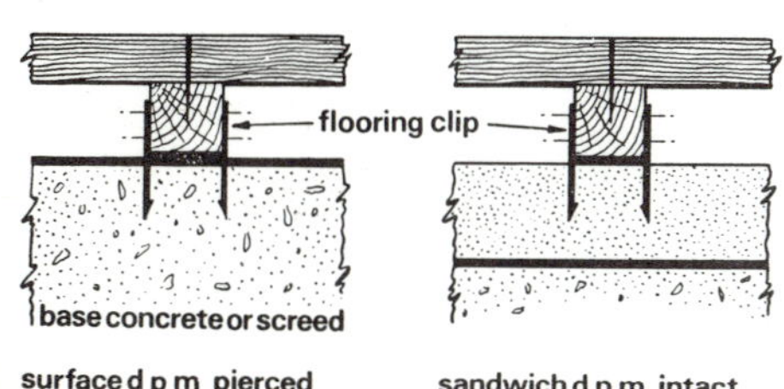

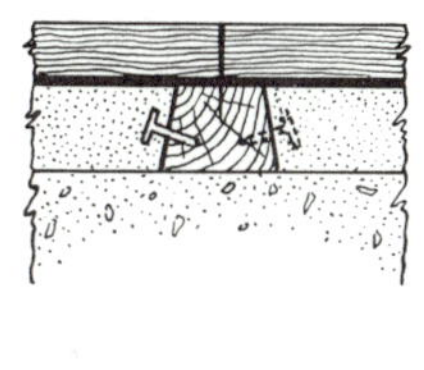

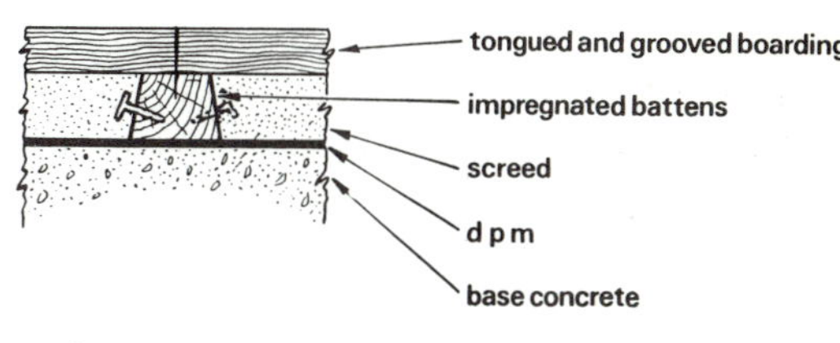

Fig 1 Board and batten fixing

Tongued and grooved boarding should be used because, if under-floor ventilation is as efficient as it should be, objectionable draughts would pass through the open joints of plain-edged boarding.

The thermal insulation of floors of this type can be improved by providing a layer of insulating material below the boarding. In a properly constructed and ventilated floor, this should not increase the risk of dry rot.

A damp-proof course should be provided in all sleeper walls and other supports that are in contact with the ground. The ends of joists should be cut back clear of the external walls (*see* Fig 2).

Upper floors

Dry rot in upper floors is rare because they are not exposed to moisture rising from the ground and the air above and below them is usually dry. No special precautions are required except that timbers built into walls that might become damp should be treated with preservative.

Board and batten floors (Fig 1) might be used on upper floors of *in situ* reinforced concrete, hollow beam or similar construction, but the damp-proof membrane would normally be omitted. However, structural floors of these types, possibly with the addition of a concrete screed, may contain a considerable quantity of residual construction water, and if a lightweight concrete is used the drying-out time will be substantially prolonged. Flooring should not be laid until drying-out is well advanced. Brush application of a preservative would then give temporary protection to the battens until drying-out is complete; the treatment could usefully be extended to the underside of the floorboards. Pressure impregnation, although not essential, would not only give protection during drying-out but would also continue to protect the battens from damage by accidental spillage, etc.

Wood block and parquet flooring laid on concrete, hollow tile or similar structural upper floors do not require special protection against fungal attack, but the concrete and screeds should be dry before the flooring is laid.

Floor coverings

Impervious floor coverings such as linoleum, rubber and plastics can safely be laid over floors that have been properly constructed in accordance with the foregoing recommendations, provided that residual construction moisture is not trapped below them. There may be some risk, however, in covering old timber floors, even when these have behaved satisfactorily for many years; the timber may have been kept at a safe moisture content mainly by evaporation from the upper surface, in which case the laying of an impervious covering might allow the moisture content to rise above the danger limit. Laying such coverings on a new boarded floor above a screed or slab which has not properly dried out could have a similar effect.

Skirtings

The absence of any protection on the back surface of a skirting and early painting of the exposed surface have the effect of raising the moisture content in skirting boards fitted to newly plastered walls that have not dried out. The risk of fungal attack is greatest when the wall is plastered right down to the floor and when skirtings are fixed above solid floors that are still wet.

All timber to be fixed in contact with new walls should, therefore, be given two brush-applied coats of preservative on the contact face. Impregnation would give still better protection; the use of preservatives that are soluble in water or certain organic solvents would allow for the normal painting of exposed surfaces.

Preservative treatment

Even with sound design and construction, it is necessary in some situations to treat the timber with a preservative. The effectiveness of this in eliminating or reducing the risk of attack will depend upon the thoroughness of the treatment. Any such treatment should be regarded as providing an extra margin of safety, the small cost of which, in comparison with the high cost and inconvenience entailed by fungal attack, is worth while in any situation where dampness, however slight, can occur. The treatment is further justified by the protection it affords against insect attack.

Timber to be treated with preservative should be clean and dry. The preservative may be applied either by brushing or by impregnation. Where permanent dampness cannot be avoided, timber

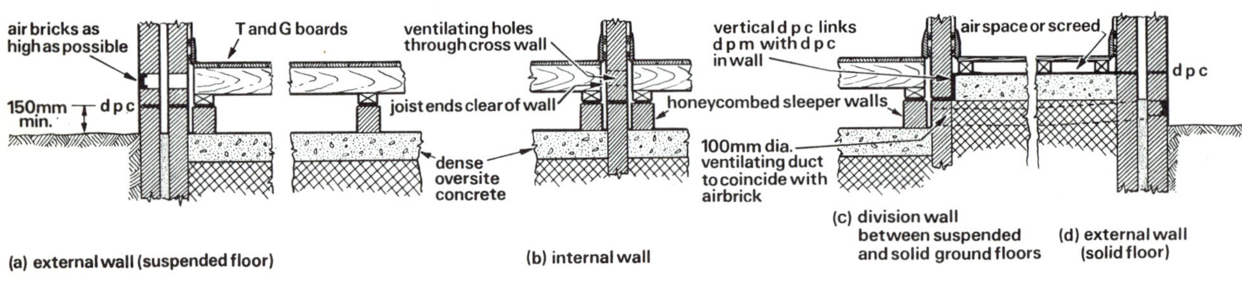

Fig 2 Ventilated ground floors

should be impregnated. The maximum protection will be obtained if the timber is impregnated after all cutting and boring has been done; if this is not possible, preservative should be brushed liberally on to any cut surface.

Brush application gives less protection and should be used only where dampness is likely to be slight and temporary, as, for example, in those parts of a new building that can be expected to dry out fairly quickly.

Some care is needed in choosing the preservative to be used. In some situations the persistent smell of creosote and other preservatives of the tar oil type may be objectionable; also they tend to creep into and stain adjacent materials such as plaster. Alternatives are available in the organic solvent and water-borne types of preservative; see references below.

Method of determining the dryness of the sub-floor

A simple form of instrument measures the relative humidity of a pocket of air in equilibrium with the sub-floor (Fig 3). It consists of a paper hygrometer in a vapour-tight mounting housed in a well insulated box. Provision is made for sealing the edges of the instrument against the surface to be tested—Plasticine is a convenient material for this purpose—and for reading the hygrometer scale whilst the instrument is in position on the floor.

The instrument should be sealed firmly to the floor surface for not less than four hours to allow the entrapped air to reach moisture equilibrium with the concrete base before a reading is taken. A longer period is preferable and can sometimes be obtained conveniently by placing the instrument in position overnight and taking the reading in the morning. If readings are to be taken at several points with only one instrument available, subsequent positions may be covered by impervious mats (bitumen felt, polythene sheeting, etc.) about one metre square, laid down when the instrument is placed in its first position, to speed up the later readings.

When the hygrometer readings fall to 75–80 per cent, the sub-floor can be assumed to be sufficiently dry for flooring to be laid.

The use, instead of a hygrometer, of anhydrous copper sulphate (which turns blue in the presence of moisture) or of anhydrous calcium chloride (which liquefies), is not recommended.

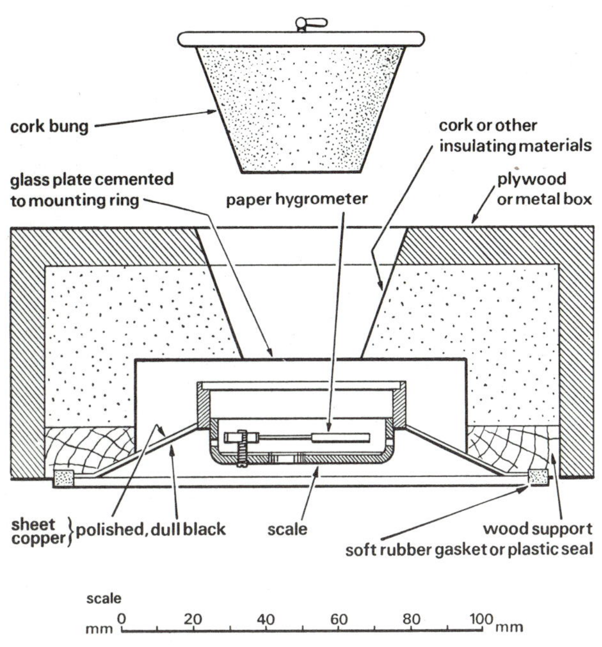

Fig 3 Apparatus for measuring dampness

Further reading

Decay in buildings—recognition, prevention and cure Tech Note 44 :1969 Princes Risborough Laboratory (free on application to Distribution Unit, BRE)

Decay of timber and its prevention K St G Cartwright and W P K Findlay; HMSO (London) 1967 : £2·47

BS 144 :1973 *Coal tar creosote for the preservation of timber*

BS 913 :1973 *Wood preservation by means of pressure creosoting*

BS 1282 :1959 *Classification of wood preservatives and their methods of application*

BS 3051 :1972 *Coal tar creosotes for wood preservation (other than creosotes to BS 144)*

BS 3452 :1962 *Copper/chrome water-borne wood preservatives and their application*

BS 3453 :1962 *Fluoride/arsenate/chromate/dinitrophenol water-borne wood preservatives and their application*

BS 4072 :1966 *Wood preservation by means of water-borne copper/chrome/arsenic compositions*

BS CP 98 :1964 *Preservative treatments for constructional timber*

BS CP 102 :1963 *Protection of buildings against water from the ground*

Damp-proofing solid floors

This digest discusses briefly site drainage and the principles of damp-proofing solid floors. Tables show the effects of ground moisture on floor finishes and the protection afforded by various damp-proof membranes.

Principles of damp-proofing

Oversite concrete and floor screeds do not keep back all ground moisture. This rises towards the surface, where it either evaporates or accumulates under a less pervious material.

The tolerance of some common floor finishes to moisture is classified in Groups **A**–**D** (*see* Table 1), which range from finishes which combine the functions of flooring and damp-proofing to those for which reliable protection against damp is always needed.

For the intermediate groups, in particular Group **C**, some judgement must be exercised in deciding whether or not to provide a membrane. Similarly, a choice must be made between hot-applied and cold-applied membranes for use under the more moisture-sensitive floorings. Consideration of the following factors will help to assess the degree of risk.

Figs. 1-3. Some typical effects of rising dampness and adhesion failure.

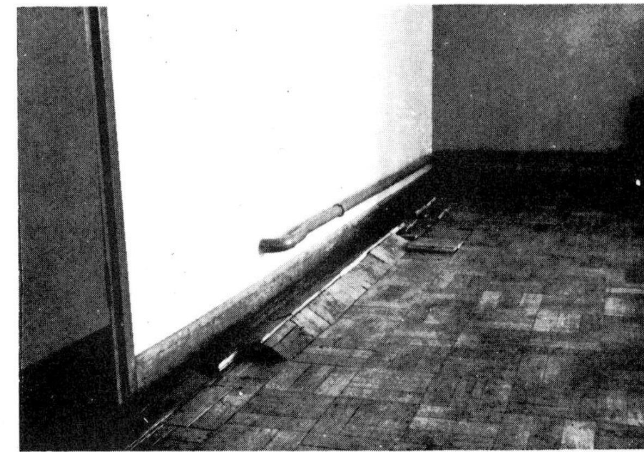

Table 1   Effects of rising moisture on floor finishes

Group	Material	Properties
A Finish and damp-proof membrane combined	Pitch mastic flooring Mastic asphalt flooring	Resist rising damp without dimensional or material failure
B Finishes that can be used without extra protection against damp	Concrete Terrazzo Clay tiles	Transmit rising damp without dimensional, material or adhesion failure
	Cement/rubber latex Cement/bitumen Composition blocks (laid in cement mortar)	Transmit rising damp slowly without dimensional or material failure and usually without adhesion failure
	Wood blocks (dipped and laid in hot pitch or bitumen)	Transmit rising damp slowly without material failure and usually without dimensional or adhesion failure. Only in exceptional conditions of site dampness is there risk of dimensional instability
C Finishes that are not necessarily trouble-free but are often laid without protection against damp	Thermoplastic flooring tiles PVC (vinyl) asbestos tiles Acrylic resin emulsion/cement Epoxy resin flooring	Under severe conditions dimensional and adhesion failure may occur. Thermoplastic flooring tiles may be attacked by dissolved salts
D Reliable protection against damp needed	Magnesium oxychloride	Softens and disintegrates in wet conditions
	PVA emulsion/cement	Dimensionally sensitive to moisture; softens in wet conditions
	Polyester resin flooring Polyurethane resin flooring Rubber Flexible PVC flooring Linoleum Cork carpet Cork tile	Lose adhesion and may expand under damp conditions
	Textile flooring	Dimensional and material failure, and usually adhesion failure occur in moist conditions
	Wood block laid in cold adhesives Wood strip and board flooring Chipboard	Acutely sensitive to moisture with dimensional or material failure

Wetness of site

This is influenced by the nature of the soil, the presence or absence of hardcore beneath the site concrete, and the slope of the ground. On some sloping sites, foundations can interfere with natural drainage, with consequent build-up of water pressure on the underside of the floor. In these cases, a ditch dug along the upper sides of the building, draining to a lower level, will help. The ditch may be filled, if it has a land drain in the bottom, with coarse material up to top-soil level.

Solid floors should not be subjected to water pressure unless specifically designed to withstand it; the problem then becomes one of tanking, and reference should be made to British Standard Code of Practice CP 102.

Temperature to be maintained in the building

The effect of a temperature gradient is to concentrate moisture in the cooler part and to lower the moisture content of the warmer part. Thus, if the floor surface is colder than the ground below it, moisture will tend to rise and collect near the surface;

protection against dampness is then advisable. In a permanently heated building, the risk is less, though with floor warming, a sandwich damp-proof membrane is necessary whatever the finish.

Sea-salt contamination of concrete aggregate and fill

Sea salt has two undesirable effects: it tends to keep the floor damp and it reacts with lime from the cement, liberating caustic alkali, which can contribute to an attack on floorings and their adhesives. Where sea salt is present, a damp-proof membrane is necessary.

Position of membrane

Surface membranes permit floor laying to follow completion (including cooling or hardening) of the membrane with no delay for drying time. They do not interfere with the bond between screed and base concrete; a screed can, therefore, be laid monolithically to a thickness of only 12–25 mm, or in separate construction to a minimum thickness of 40 mm (*see* Digest 104). The thicker surface membranes may also serve as a levelling compound on a well-finished concrete slab without screed or on a slightly irregular screed. Surface membranes are, however, susceptible to damage by subsequent trades; with mastic asphalt or pitchmastic membranes there is a risk of indentation under heavy loads. They should therefore be laid only just in advance of the floor finish.

If sulphates are present in hardcore under the floor or in the ground water, a surface membrane will give no protection against sulphate attack of the base concrete or screed and the risk of sulpho-aluminate expansion will be high.

Sandwich membranes located between concrete base and screed should be protected against damage by subsequent trades by laying the screed as soon as possible after completion. Hot-applied membranes can be covered by a screed as soon as they have cooled; cold-applied solutions or emulsions are likely to require several days to apply and harden. The membrane effectively eliminates any bond between screed and base; a screed laid on a sandwich membrane should therefore be at least 50 mm thick.

Any of the sheet materials recommended for use as sandwich membranes can be used below the site concrete but the base on which they are to be laid must be structurally sound, to avoid fracture of the membrane, and the surface should be sufficiently smooth to avoid puncturing the membrane. A blinding layer of weak concrete or sand, on a consolidated bed of hardcore, provides a suitable base. A membrane in this position protects the site concrete and screed against attack by any sulphates present in hardcore or ground water and permits bonding of screed to concrete base. It has, however, the disadvantage that both the concrete base and screed must be allowed time to dry before a moisture-sensitive finish can be laid.

Table 2 Materials for membranes

Material	Standard or grade	Comment
Hot-applied		
Mastic asphalt	BS 1410: Mastic asphalt for flooring (natural rock asphalt aggregate) BS 1076: Mastic asphalt for flooring (limestone aggregate) BS 1451: Coloured mastic asphalt for flooring (limestone aggregate)	May be used as a floor finish (*see* BS CP 204); if used as an underlay to a floor finish, the thickness should be not less than 12 mm. A compressible underlay is not recommended but glass-fibre may be used. All are suitable for surface membranes.
	BS 1097 (limestone aggregate) BS 1418 (natural rock asphalt aggregate)	When loaded, can withstand hydrostatic pressure; suitable only for sandwich membranes.
Pitch mastic	BS 1450: Black pitch mastic flooring BS 3672: Coloured pitch mastic flooring	Normally used as a floor finish but may form a surface membrane to protect other finishes. Its indentation characteristics make it less suitable than mastic asphalt.
Pitch	BS 1310: Coal tar pitches for building purposes (Grade R & B 40)	Should be laid on a primed surface to give an average thickness of 3 mm (3 kg/m^2); suitable only for sandwich membranes.
Bitumen	Should have a softening point of 50–55°C; this corresponds to a penetration number of 40–50 at 25°C	
Cold-applied		
Bitumen solutions, coal tar pitch/rubber emulsion, or bitumen/rubber emulsion	Not defined by any BS Specification or Code	BS CP 102 recommends 0·6 mm minimum thickness but this is for broad guidance only. The solids content will usually have been adjusted to give adequate coverage by two or three coats; the material should not be thinned by dilution or spread in thinner coats than recommended by the manufacturer. Suitable only for sandwich membranes.
Pitch/epoxy resin		Although applied in thin layers, the material is strong enough and its adhesion to concrete is usually sufficient to make it a satisfactory base for a variety of floor finishes. It cannot tolerate any cracking in the surface to which it is applied. Generally used as a surface membrane.
Sheet material		
Polythene film	0·12 mm thick (500 gauge)	Suitable for sandwich membranes. Joints must be properly sealed; the welting method normally used appears satisfactory. Where there is risk of damage by subsequent screed-laying operations, material about twice as thick (1000 gauge) may be used with confidence.
Composite	Polythene and bitumen, self-adhesive; thickness is in excess of 1·5 mm	Adhesion simplifies joint treatment and reduces risk of tearing.
Bitumen sheet	Bitumen sheet to BS 743	Suitable for sandwich membranes. Joints must be properly sealed.

Properties required of damp-proof layers

A damp-proof membrane must be impervious to *liquid* water; it must also have at least as high a resistance to the passage of water *vapour* as the flooring it is designed to protect.

Hot-applied membranes are considered to have the higher vapour resistance. Cold membranes of adequate thickness also have a high vapour resistance, but in practice they may have thin spots.

An integral waterproofer incorporated in the concrete or screed is not a satisfactory alternative to a damp-proof layer.

Suitable damp-proofing materials are listed in Table 2. Site experience shows that not all are equally effective but if the materials are laid to the recommended minimum thicknesses, dampness troubles are more likely to arise from poor workmanship and bad detailing than from any deficiencies in the membrane materials. If flooring laid over a membrane shows early trouble, the cause is unlikely to be a defective membrane; the latter will usually require several months to show in the form of a flooring failure.

Drying

A long delay before laying the finish is unavoidable for all sandwich membranes, because the screed has to be left to dry out. Even under favourable conditions this may take as long as one month per 25 mm thickness. Thus, a 50 mm thick screed would take at least two months to dry. Where a membrane is introduced below the site concrete, a 150 mm thick slab and screed would take at least six months to dry sufficiently for a moisture-sensitive finish to be laid. A device for testing the dryness of a screed is described in Digest 18.

Continuity

All damp-proof layers must be continuous with the damp-proof courses in adjoining walls and with damp-proof membranes in adjacent floors. Where necessary, a vertical damp-proof course must be provided to link membranes at different levels, as shown in Fig 4.

A floor that is finished in part with material that needs no protection against ground moisture and elsewhere with material that does require protection, must be provided with continuous damp-proofing of the whole area, or each area of moisture-sensitive flooring must be protected by tanking—*see* Fig 5.

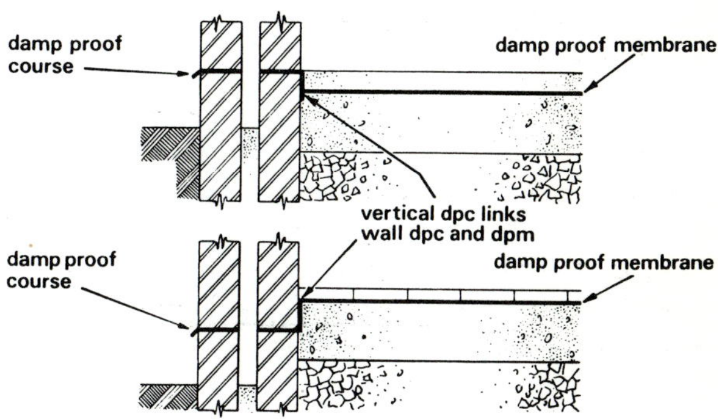

Fig 4

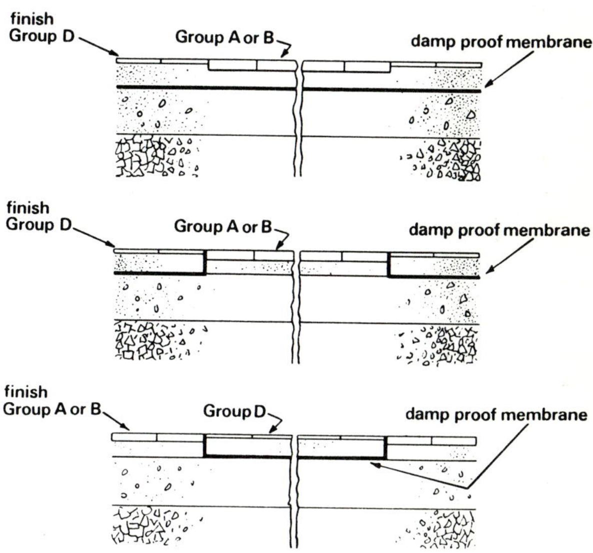

Fig 5

Floor screeds

Considerations in establishing the necessity for a screed: smooth surface—falls—thermal and sound insulation—accommodation for services.
Constructions: monolithic—separate—unbonded—floating—heated screeds.
Types: dense concrete—synthetic anhydrite—cement and sand modified with admixtures—lightweight aggregate concrete. Base preparation—thickness—mix design—division into bays—workmanship—curing and drying. Surface texture and finish.

Rational design of a floor screed begins with consideration of the floor as a whole element—structural base, screed and finish together. If the floor has to incorporate a barrier against rising dampness—ground-water under a solid floor or condensation on the underside of a suspended floor—if it has to accommodate any special services or a floor-warming system, if it must meet a standard of thermal or sound insulation, the whole element must be designed with these requirements in mind. All too often it appears that a structural slab is designed, sometimes even laid, before such special needs are considered, leaving them to be dealt with by means of various sandwich layers and screed. It should be emphasised that a screed does not provide the final wearing surface of the floor.

The need for a screed

Smooth surface—It is often possible to make the surface of an *in situ* concrete base smooth enough to suit the proposed floor finish. *In situ* bases and some that are of precast units may be designed so that an adequate surface can be produced simply by means of a levelling compound. In any other circumstances, a screed will be necessary.

Falls—The most efficient and economical way of providing surface falls is to form them in the structural base. A screed laid on a level slab and finished to falls must not be thinner in any part than the appropriate minimum thickness recommended below; over most of its area it will be wastefully thick, merely to provide falls.

Sandwich constructions—Damp-proof membranes and insulating materials are often incorporated in sandwich constructions between structural base and screed. If the screed is properly designed and laid, a satisfactory job results, but alternative design solutions should be examined before making a final selection. For example, in the range of floor constructions for sound insulation shown in Digest 103, the cost of a resilient layer, a floating screed and a 'hard' floor finish may be higher than that of a 'soft' floor finish on a solid base—and the risk of failure may be greater.

In general, the fewer the layers that comprise a floor, the less will be the risk of failure.

Thermal insulation—A floor screed of lightweight concrete provides better insulation than dense concrete. But a comparison of efficiency and cost with other means of providing the same standard of insulation should form part of the process of selection, bearing in mind that under thin materials, for example, flexible pvc flooring, a dense topping to the screed might be needed.

Services—The thickness of a floor screed can conveniently accommodate services—pipes and cables. In first cost, it probably provides the cheapest solution. Should any fault develop in a pipe or cable, however, or if access is required for any other reason, the floor finish must be disturbed and the screed broken up with risk of damage to other services. Location of a fault is likely to prove difficult. Because screed thickness is reduced above an embedded pipe, cracking along this line is more likely.

Underfloor warming—Heating cables or pipes can be embedded in a floor screed to provide a space-heating system. Some of the problems associated with heated screeds are discussed on page 41.

Dense concrete floor screeds

Bond and thickness—The thickness of a dense concrete screed should be related to the state of the base on which it is to be laid, i.e. its age when the screed is to be placed and the strength of the bond likely to be attained.

Monolithic construction—When a screed is laid on an *in situ* concrete base before it has set (within three hours of placing), complete bonding is obtained. Maximum restraint is thus provided on the shrinkage of the screed and, because screed and base can shrink together, there is a reduction in the potential differential shrinkage between them. A thickness of only 12 mm is all that is necessary; layers thicker than 25 mm should be avoided if the effects of the shrinkage forces arising from the screed are to be restricted. Undoubtedly monolithic construction offers the best chance of eliminating problems of cracking and curling but the decision to lay the screed monolithically with the base must be made at the design stage so that the work may be planned to make this type of construction possible. Protection of the completed screed during subsequent building operations needs also to be assured.

Separate construction—Once the base has set and hardened, monolithic construction can no longer be used. The strength of the bond between screed and base will then depend on the thoroughness with

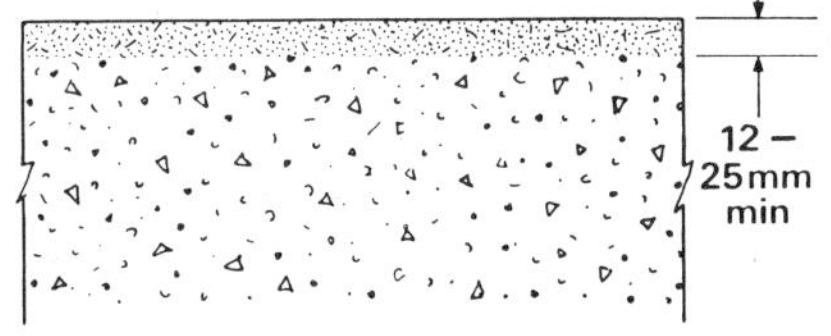

Monolithic—screed placed within three hours of base

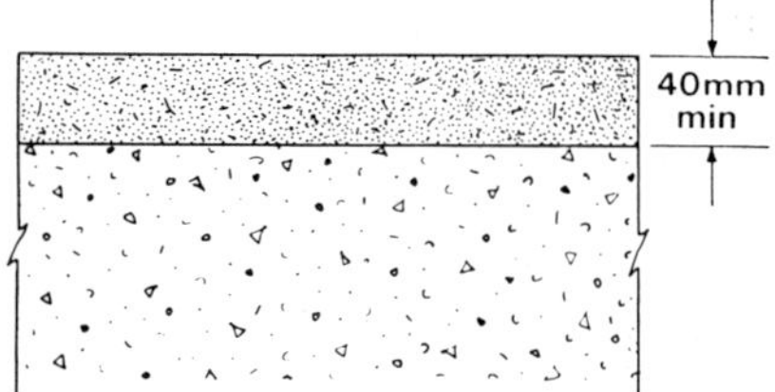

Separate—more than three hours' interval

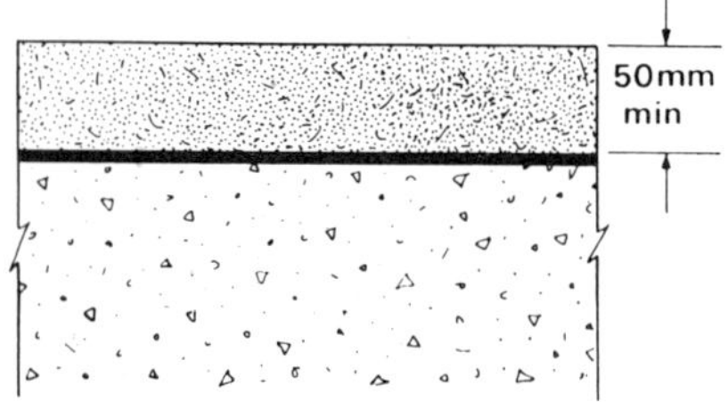

Unbonded

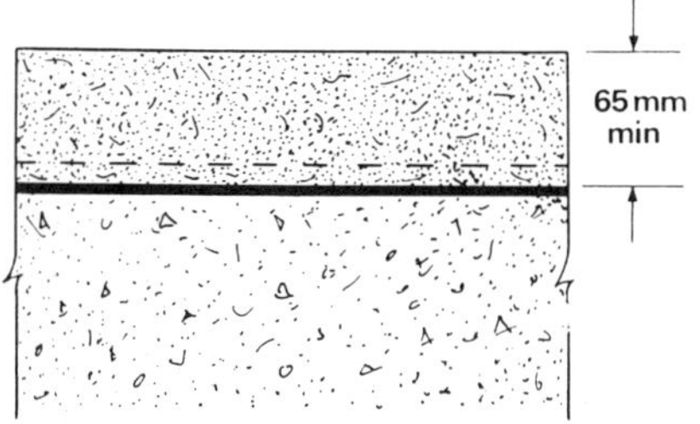

Unbonded and with heating cables

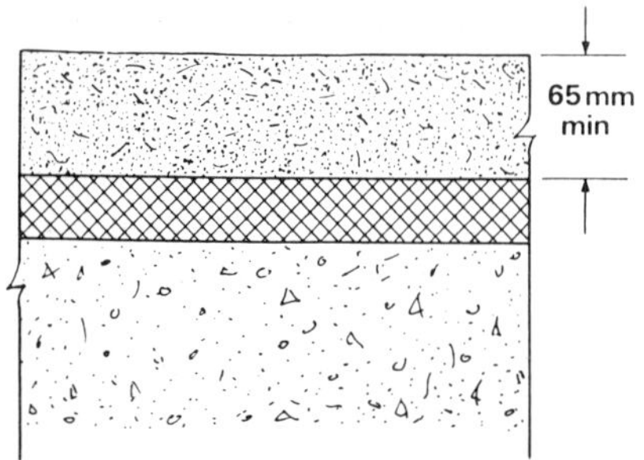

Floating—on compressible material

which the base has been prepared. For maximum bond, the base should be thoroughly hacked by mechanical means, cleaned and damped to reduce suction, then grouted immediately before the screed is placed. A bonding agent may be used as an alternative to grouting. Obviously the amount of preparation considered to be adequate will vary enormously, but assuming a good average standard has been achieved, a minimum thickness of 40 mm should suffice.

Unbonded construction—There are many bases with which it is impossible to achieve a bond, for example damp-proof membranes and concrete that has been impregnated with oil and grease. A similar situation exists where a screed is to be laid on a base of concrete containing a water-repellent admixture or on a base of weak concrete. In the latter case, the bond achieved may be strong but the restraint imposed by a weak material can be only slight. Because there is no useful bond it is necessary to have a screed at least 50 mm thick, or, if it contains heating cables, at least 65 mm thick.

Floating—A screed laid on a compressible layer of thermal or sound-insulating material should be at least 65 mm thick, or, if it contains heating cables, at least 75 mm thick.

Division into bays—To avoid random cracking and to reduce curling, it used to be considered necessary to lay all dense screeds in bays.

It is, however, difficult to ensure that screeds laid in bays will not show slight curling or unevenness at the junctions of bays. These show as undulations when thin floorings are used, particularly in large areas and if the surfaces are highly polished. Undulations are very difficult and expensive to eliminate but random cracks are generally considered to be easily repaired.

It is, therefore, no longer recommended that screeds should be laid in bays unless they contain underfloor warming cables (see page 41) or are intended to receive an *in situ* floor finish.

When a screed is to be laid in bays, the size of each bay should be related to the thickness of the screed which in turn has been shown to be dependent upon the state of the base on which the screed is laid. In monolithic construction, screed bays should coincide with the bays in the structural base. For other constructions the bay size should not exceed 15 m².

The ratio between the sides of the bays should be as near 1 : 1½ as possible; long narrow bays should be avoided. Edges should be vertical and should abut closely at all joints. When the bays shrink the gaps left between them will, although small, be sufficient to cater for most likely thermal movements. Expansion joints are required only where similar joints are provided in the main structure and these are unlikely to be needed at intervals of less than 30 m.

Mix design—Mixes made with Portland cement will start to shrink soon after placing and will continue to do so for a long time. The effects of shrinkage can be minimised by good mix design, by restraint (i.e. bond strength), by limiting differences in shrinkage that can occur between layers of concrete of different mix proportions or states of dryness and by careful curing (see page 42).

Mix proportions should be specified by weight of the dry materials. For screeds up to 40 mm thick, cement and sand 1 : 3–4½ (by weight) is suitable. The mixes less rich in cement are likely to have lower shrinkage. For thicker screeds, fine concrete of 1 : 1½ : 3 (cement : fine aggregate : coarse aggregate) using 10 mm maximum coarse aggregate is suitable. It is sometimes argued that a good steel-trowelled finish cannot be achieved on a fine concrete screed. Where there is likely to be difficulty, the fine concrete can be surfaced with a thin layer of a cement and sand mix. This layer should be about 5–10 mm thick and it should be placed before the fine concrete has set. Alternatively, the amount of sand in the mix can be increased to a small extent provided that the overall cement/aggregate ratio is unchanged, or a levelling compound may be used at a late stage in the work.

Materials

Cement—Portland cement to BS 12 is commonly used and is satisfactory for most conditions.

High-alumina cement gives higher early strength but it must not be mixed with any other cement nor should calcium chloride or any other admixture be used with it.

Aggregates—Good grading helps to improve workability and permits low water/cement ratios. The most suitable aggregates are those given in BS 882. Material conforming to the grading of Table 1, BS 1199, is also suitable, but the finer material for finishing coats and plastering is unsuitable for floor screeds.

Water—Only sufficient water to produce a mix that permits thorough compaction with the means available should be used. A simple test is to squeeze a handful of the mix—it should ball together but it should not be possible to squeeze water out of it. A lower water content is possible in a mix to be compacted by machine than in one that is to be hand-compacted. Although the mix should be as dry as possible, failures have arisen because mixes have been too dry to permit adequate compaction. This usually results in a generally weak screed or at least a weak layer in contact with the base. This weak layer leads to poor adhesion and also increases the differential shrinkage stresses that can be set up between the top and bottom of the screed. Shrinkage differences of this kind coupled with poor adhesion may cause curling at the corners and edges of the screed. Lower water/cement ratios can be achieved with workability aids.

Heated concrete screeds—Careful attention should be paid to the recommendations for reducing drying shrinkage and minimising its effects. These relate particularly to mix design, a low water/cement ratio, thorough compaction and curing. If monolithic laying can be organised, this will involve the least risk of failure, but if a separate construction is adopted, preparation of the base should be thorough (*see* page 41) so as to obtain the benefit of the maximum bond that can be achieved. However, in many installations no bond is possible between screed and base. Although the thickness of a heated screed commonly lies within the range of 65–75 mm, which is regarded as adequate for a floating screed, the thermal gradient across the thickness of the screed increases the risk of curling. For this reason, the higher figure (75 mm) should be adopted for a floating screed that is also to be heated, and the recommendations given above must be carefully followed.

When an electrically heated screed is to be placed on thermal insulating material or on a damp-proof membrane, it is convenient first to place a thin bed of cement and sand about 10–15 mm thick, on which the heating cables can be placed without risk of contact between cables and insulation or membrane. The screed must be carefully and thoroughly packed around the cables. During the screed-laying operations the cables should be monitored so that faults can be rectified before the screed has set and hardened.

Heated concrete screeds should be divided into bays not greater than 15 m². Where cables pass from one bay to the next, they should be looped and wrapped as the cable supplier recommends to avoid fracture by shrinkage of the screed; pipes may be either wrapped or coated with bitumen.

Floating screeds—The use of floating screeds to reduce the transmission of impact noise through upper floors was discussed in Digest 103. The screed should be not less than 65 mm thick, or if it incorporates heating elements, 75 mm thick. The screed material must be prevented from seeping through the sound-insulating layer or penetrating at the joints to form solid bridges between the screed and the base slab; a continuous layer of impervious material—polythene sheeting, building paper, etc.—well lapped at all joints is effective for this purpose.

It is usual to provide 20–50 mm mesh wire-netting in the floating screed. This is normally laid directly on the impervious sheeting to protect the sheeting and the quilt from mechanical damage during the operation of placing the screed.

Because the curling of separate bays in a floating screed can lead to difficulty with the floor finish, room-size panels are often adopted, but if these exceed 15 m² by a substantial amount it must be accepted that the risk of curling and cracking is increased.

It has been suggested, page 41, that a thin layer of cement and sand can be used to bring a fine concrete screed to a smooth surface. If this is too strong or too thick it could encourage a floating screed to curl.

Curing—For at least the first week after the screed has been laid it should be covered with, for example, polythene sheets or tarpaulins to prevent drying as much as possible. During this period of curing, the screed gains strength and the onset of drying shrinkage is delayed. When drying is allowed to start, the material is then better able to resist the shrinkage stresses.

Drying—Because the amount of water used in making the screed is necessarily greater than is required to hydrate the cement, there is always surplus water to be removed before moisture-sensitive floorings can be laid successfully. Apart from the adverse effects of moisture on these floorings, the water in the screed will contain alkalies, derived from the cement, which can cause chemical attack on some flooring materials and adhesives. The only way to eliminate excess water in the screed is to allow it to evaporate. The slower the rate of drying the lower is the risk of cracking and curling. No particular drying period can be specified because the time required will depend on many factors, such as the quality and thickness of the screed and the heating and ventilation within the building. A rule-of-thumb method often quoted is that one month should be allowed for every inch (25 mm) of screed thickness. This figure is useful for forward planning of building operations but no finish should be laid until the screed is shown to be sufficiently dry by a reliable moisture test. Various test methods are discussed in BRS Digest 18 and CP 203.

The temptation to use hot-air blowers, underfloor heating, or other means of accelerating drying should be resisted at least for the first four weeks or so.

Other screed materials

Synthetic anhydrite—Synthetic anhydrite (anhydrous calcium sulphate) is a by-product of hydrofluoric acid manufacture and has different properties from natural anhydrite. It is used with specially graded aggregate—1 : 2½ by volume—to form screeds having the following characteristics:

(a) Almost all the water used is combined as water of crystallisation. The amount left to evaporate is thus very small and under normal drying conditions the laying of moisture-sensitive floorings can be considered from about ten days after placing the screed.

(b) The material loses strength when it becomes wet. It must therefore be laid only on bases that are dry and likely to remain so.

(c) Drying shrinkage is very low. Large areas can be laid without joints and with no risk of curling.

(d) Minor damage can be repaired quickly and easily.

(e) The material is not bonded to the base and so for level bases preparation is unnecessary.

(f) Drying out during the first two days must be avoided or there will be insufficient water to combine with the anhydrous calcium sulphate. Dustiness and failure to gain sufficient strength may result.

(g) The minimum thickness is 25 mm.

(h) Over compressible material, e.g. a glass silk quilt, in floating floor construction, the minimum thickness should be 40 mm.

(i) A screed that contains underfloor warming cables should be not less than 30 mm thick; a greater thickness will be needed to provide thermal storage capacity in an off-peak system.

Modified cement and sand—Because the basic characteristics of dense screeds impose limitations on design, there have been many attempts to increase the versatility of the screed by the use of admixtures. The best known of these have water-repellent properties (for example, metallic soaps) and are often assumed to provide adequate substitutes for damp-proof membranes. Not only is this assumption false but they introduce drying problems because the movement of water by capillary action is inhibited.

There are many proprietary screeds which are basically dense screeds modified by the inclusion of materials in emulsion form. Materials that have been used alone or in combination are bitumen, polyvinyl acetate, acrylic resins and some synthetic rubbers. The inclusion of these is aimed at improving adhesion to the base, thereby permitting thinner screeds to be laid. These emulsions tend also to improve the resistance to cracking, despite the fact that emulsion/cement/aggregate mixtures tend to have higher drying shrinkage than similar mixes containing water instead of emulsion. It is often claimed that these screeds dry out faster than unmodified ones. As they are usually thinner, less drying time may be required but no general guidance on this issue can be given.

Lightweight aggregate concrete—As a floor screed, its most important property is its low density which enables it to be used in the thickness required to provide falls or to accommodate services with only the minimum increase in the load imposed on the structural floor. Advantage can also be taken of its thermal properties. It is not suitable for use as an electrically heated screed. Lightweight aggregates and some of the properties of fully compacted concrete made with them are covered in Digest 123. Screeds of no-fines lightweight aggregate concrete are not considered.

This type of screed should be not less than 40 mm thick. If the proposed floor finish is incapable of spreading point loads (for example, thermoplastic or pvc (vinyl) asbestos tiles, flexible pvc, rubber or linoleum), a topping of cement and sand at least 10 mm thick will be required. To reduce the risk of cracking and to minimise the effects of shrinkage, the topping should be not richer than 1 : 4 and it should be laid monolithically with the lightweight screed.

Generally

Surface texture—The surface finish should be related to the kind of flooring that is to be laid on the screed. The slightly rough surface left by a screeding board drawn across screeding battens will be suitable for many floorings, for example, magnesium oxychloride. A wood float finish may be necessary for others, for example, wood block, but for most sheet and tile floorings the smooth dense surface produced by steel trowelling is necessary. The standard of the finish will depend ultimately on the skill of the operative.

Level—Some useful information on tolerances in terms of what can reasonably be expected of the floor surface is given in CP 204 : 1965. It is important to bear in mind that insistence on very close limits may result in higher costs. The limits for floor surfaces given in the code are generally applicable to screeds. Briefly, the permissible deviation from datum depends on the area involved ; for large open areas, a deviation of 10 mm can be tolerated. Localised deviations from datum level of 3 mm in any 3 m would normally be acceptable in a nominally flat floor.

Hollowness—If the adhesion of a screed to its base is examined by tapping the surface with a rod or hammer, a hollow sound indicating poor adhesion might be detected in some areas, most commonly at the edges and corners of bays. Only if the hollow areas have actually lifted by a visible or measurable extent, so that there is a risk of fracture under superimposed loads, should the screed by considered unsatisfactory on this account.

Clay tile flooring

This digest describes methods of laying clay floor tiles to avoid risk of failure by arching or ridging. Cement mortar bedding on a separating layer of sheet material gives satisfactory service. This method, and the newer ones of laying by the 'thick bed' (also sometimes referred to as 'dry bed' or 'semi-dry bed') technique or by using thinly-applied adhesive compounds, are described below. More detailed guidance can be found in BS CP 202: 1972, Tile flooring and slab flooring.

Classes of tile available

For general use

Two classes are available, floor tiles and quarry tiles the difference being as stated in BS 1286:1974: 'floor tiles are manufactured to narrower limits than quarries and are made by the compaction of powder method, usually from refined and blended ceramic powders but some are made from neat clay powders; quarries are made from stiff plastic clays by extrusion or other plastic process'. Both types are divided into two classes, as shown in the following table:

	Maximum water absorption (%)
Ceramic floor tiles	
fully vitrified	0.3
vitrified	4.0
Quarry tiles	
class 1	6.0
class 2	10.0

For special uses

Slip resistance. Improvement in slip resistance can be obtained with ribbed or studded surfaces, particled surfaces (shot-faced or pin-head finish) or with non-slip aggregates incorporated in the wearing surface. 'Slip-resistant' tiles are more difficult to keep clean. In some areas, industrial kitchens, for example, there is always the possibility of greasy films being formed on the surface, and unless extra care is taken to eliminate these films, the benefit of the special surfaces will probably be nullified.

Chemical resistance. Clay tiles in general have good resistance to chemical attack but for use in severe conditions tiles should meet the requirements of BS 3679: 1963.

Frost resistance. Most floors will not be subjected to frost attack in normal service but if this is likely, e g if used externally, the frost resistance of the tiles should be verified by reference to the manufacturer. There is no standard method of test.

Bases for tiling

Rigidity

The base on which the tiling is to be laid must be rigid and stable, e g concrete slabs, precast concrete or hollow clay units; most wood joist-and-board floors are not suitable.

Falls

Any falls required in the finished floor should be formed other than in the bedding of the tiles although the thick semi-dry method of bedding offers some opportunity to form slight falls in the thickness of the bedding.

Prepared at Building Research Station, Garston, Watford WD2 7JR
Technical enquiries arising from this Digest should be directed to Building Research Advisory Service at the above address.

46

Smoothness

Thick semi-dry bedding can be applied directly to a good 'spade-finished' surface of concrete. The thickness of bed can accommodate some irregularity in the surface of the base. For other methods of laying, the base must have a smooth surface; a structural concrete slab can be finished to a satisfactory surface by mechanical means but it is more usual to apply a screed or levelling compound.

Methods of laying

Good laying techniques are designed to prevent bonding between the bedding and the base so that relative movement between them is not restrained.

Bedding on a separating layer (see Fig 4).

A separating layer of sheet material such as polythene or building paper is spread over the base—whether concrete slab or screed. Adjacent sheets should be lapped 100 mm to prevent any bond between bedding and base. The tiles are then laid on the sheet in as thin a bedding of cement and sand mortar as will produce a level floor. A thickness of 13 mm is common but a thinner bed, say 10 mm, may be possible on a smooth base. Where heavy traffic or sharp impacts are likely, the thickness of bedding may be increased to a maximum of 20 mm but its thickness should be less than that of the tiles. The only purpose of a separating layer is to avoid forming a bond between bedding and base. Tiles may therefore be bedded directly on to a damp-proof membrane or on to any surface treated so as to eliminate bonding, e g surfaces coated with curing compounds. An exception is mastic asphalt, where a separating layer lapped 50 mm is required wherever cement/sand bedding layers are used.

'Thick bed' method

No special preparation of the base is required but it may be necessary to dampen the surface to reduce suction. A semi-dry mix of cement and sand not richer than 1 : 4 should be packed on to the base to a thickness which should not be less than 20 mm but may be as thick as 75 mm; a thickness of 40 mm produces good results. The area of bedding placed in one operation should allow grouting and tiling to be completed whilst the bedding is still plastic. The surface of the compacted bedding should be spread with a 1 : 1 cement and sand slurry about 3 mm thick. The floor tiles are normally laid dry and are tapped into the slurry. A separating layer is not required with this method.

'Thin bed' (adhesive) fixing

Proprietary cement-based and other types of adhesive are available. They should conform to the requirements of Appendix B of Part I of BS CP 212. They should be used according to the instructions of the manufacturer, at a thickness not exceeding

Fig 1   Arching of floor tiles

Fig 2   Ridging of floor tiles

5 mm. Adhesives bond the tiling to the base but the bedding material is sufficiently resilient to tolerate some slight movement between tiling and base without ill effect.

Arching of tiles

The above methods of laying avoid risk of failure of floor tiling by arching or ridging (see Figs 1 and 2). A feature of this type of failure is the way in which the tiles separate cleanly from the bedding; there should be no risk of confusion, therefore, with failures caused by sulphate attack on the concrete base, where the whole floor may bulge upwards (see Digest 75).

Shrinkage of the screed

A newly laid screed shrinks during drying, the tiles do not; indeed some tiles may undergo a small but non-reversible expansion early in their life. If the tiles are firmly bonded by their bedding to the screed surface, considerable stresses can be set up which in the most severe cases are eventually relieved by areas or rows of tiles lifting. In typical instances, failure would occur during the first year after laying. Thin tiles rise more readily than thick ones.

Thermal movements

Failures similar in appearance can be caused by thermal movement in conditions that exclude screed drying shrinkage as a cause of failure, that is in new tiling laid on an old concrete base, or in tiling which has behaved satisfactorily for many years. It may occur after a prolonged cold spell because concrete contracts two or three times as much as clay tiles and a sufficiently low temperature may produce sufficient stress in the tiling to cause lifting. Drying shrinkage and differential thermal movement often act together in causing failure; the stresses set up by one may be insufficient to produce lifting until augmented by the stresses of the other.

Expansion of tiles

In common with most ceramic materials clay tiles may expand slightly due to gradual uptake of moisture, although with highly vitreous low porosity tiles the expansion is insignificant. The mechanism of expansion is not fully understood but it probably involves both physical adsorption of water vapour and chemical hydration. Expansion is most rapid in the first few days after manufacture but can continue at a reduced rate for many years. Long-term expansion is the most likely cause of failure of tiled flooring when this occurs after some years of satisfactory service (see Fig 3).

Special requirements for bedding

Bitumen bedding

If the base is to be subjected to high temperatures, e g around boilers and heating installations, bituminous bedding is recommended. This should consist of 1 part of aqueous bitumen emulsion and $2\frac{1}{2}$ parts of soft dry sand by volume, The bedding should be laid only on a dry surface, primed by brushing a coating of bitumen emulsion over it. The minimum of water should be added to the composition to improve the workability of the mix.

Fig 3   Failure caused by long-term expansion of tiling

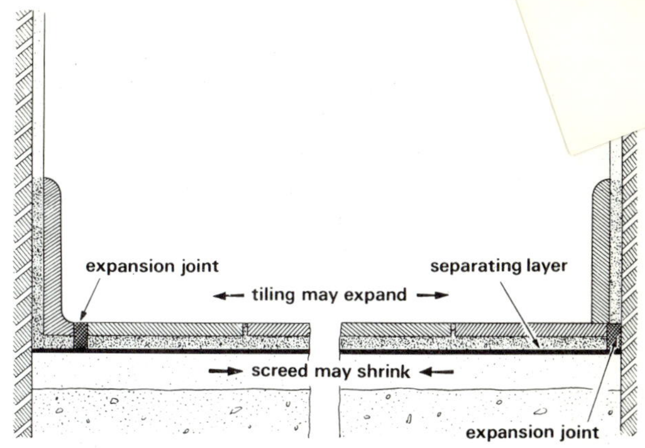

Fig 4   Use of expansion joints (left) between tiling and skirting and (right) between tiling and wall

Corrosion-resistant bedding

Where there is risk of chemical attack, special corrosion-resistant jointing compounds should be used and in the severest cases the compound should be used for bedding as well as jointing. More detailed recommendations are given in Digest 120. It should be borne in mind that chemical attack can derive from seemingly harmless liquids such as milk and fruit juices.

Expansion joints

Expansion joints with suitable compressible material should be provided around the perimeter of every clay tiled floor and at abutments against piers, machine bases, etc. They may be formed either between tiling and coved skirtings, or against walls, where they are lapped and concealed by a skirting (see Fig 4). If the area of flooring requires it, additional expansion joints should be provided at intervals of about 4.5 m across both the length and width of the floor.

Heated floor

Clay tiling is suitable for use on floors that incorporate a heating system. If the bedding is to be other than cement and sand, its suitability for use on a warmed floor should be verified with the manufacturers. The thermal resistance of the tiles is not significantly different from that of concrete but the heating engineer should be informed of the proposed thickness of tiles and bedding. The provision of expansion joints as outlined in the preceding paragraph is important.

Sheet and tile flooring made from thermoplastic binders

Sheet and tile floor finishes for non-industrial premises are nowadays frequently made from mixtures consisting essentially of fillers, pigments and thermoplastic binders. This Digest discusses materials other than linoleum and natural rubber together with some of the problems that have arisen in their use, and the sub-floor requirements. A section at the end describes the constitution of the various types of material, and their manufacture and availability, and will provide a background for an understanding of the problems discussed in the earlier sections.

The information given on the composition, manufacture and uses of flooring made from thermoplastic binders should be read in conjunction with the relevant British Standard Specifications and Codes of Practice. These are as follows:

BS 2592 : 1955   Thermoplastic flooring tiles
BS 3260 : 1969   PVC (vinyl) asbestos floor tiles
BS 3261 : 1960   Flexible PVC flooring
BS CP 203 : 1969   Sheet and tile flooring

In addition to the materials listed above, brief reference is made in this Digest to PVC floorings not covered by BS 3261, and to synthetic rubbers. The latter are introduced for the sake of completeness, though the Station's experience of their behaviour in practice is not yet sufficient to provide detailed information.

Sub-floors

All the types of flooring described will conform to the contours of the sub-floor which should, therefore, be even, smooth and rigid. Otherwise the appearance of the finished floor will be affected and any raised portions will show signs of wear more quickly than the rest of the floor. Any sub-floor can be levelled and smoothed, and suspended timber floors can be made more rigid by the use of suitable underlays. The choice of these depends on the situation in which they are to be used and detailed consideration of underlays such as mastic asphalt, emulsion/cement mixtures, plywood and hardboard is given in BS CP 203 : 1969.

Whilst these sub-floor requirements are comparatively easily met, the most difficult problem associated with sub-floors is the exclusion of moisture. Dampness in sub-floors is the commonest cause of premature failure of floorings fixed with adhesives. The problem arises because the finishes, the adhesives and in some cases the sub-floor (e.g. magnesium oxychloride cement or timber) can be affected by moisture. The commonest sub-floor for any of the materials under discussion is, however, concrete used in the form of a screed (see Digest 104). It is with moisture in this screed that the following remarks are concerned.

Moisture in the screed can derive from water used to mix the screed or base concrete on which it is laid, and from water that can on wet sites rise through concrete in contact with the ground. In each case the water contains alkalis derived from the Portland cement, and can soften all adhesives normally used and give rise to considerable moisture expansion of some of the finishes. With thermoplastic and vinyl asbestos tiles, alkaline moisture produces a curious effect which will be discussed later.

Water rising from the ground or from the base concrete can be prevented from reaching the screed by taking the precautions discussed in Digest 54 and detailed in BS CP 102, Protection of buildings against water from the ground.

In making the cement/sand screed the amount of water used is always much greater than that required to hydrate the cement. Nearly all the excess water must be removed before the floor finish is laid, irrespective of whether the finish, e.g. thermoplastic and vinyl asbestos tiles, has good tolerance to moisture. The only way to eliminate excess water from the screed is to allow it to evaporate. No particular drying period can be specified because the time required will depend on many factors, such as the quality and thickness of the screed, and the heating and ventilation within the building. A rule-of-thumb method often quoted is that one month should be allowed for every inch of the screed thickness. This figure is useful for forward planning of building operations but no finish should be laid until the screed is shown to be sufficiently dry by a reliable moisture test. Various test methods have been devised and of these a method based on the use of a paper hygrometer fitted into an insulated box has been found by the Station to be both simple and reliable. Details of the apparatus and method of use are given in Digest 18 and BS CP 203 : 1969.

Tests that are of little value are those relying on the use of anhydrous calcium chloride, anhydrous copper sulphate or phenolphthalein. In the presence of moisture, calcium chloride becomes visibly damp and copper sulphate turns blue, but the amount of moisture necessary to produce these changes is so small that they can occur even when the screed is sufficiently dry to install the floor finishes. It is generally believed that drops of an alcoholic solution of phenolphthalein placed on concrete will always turn purple when the concrete is too wet to lay the finish. Unfortunately, the colour change can occur even when the concrete is sufficiently dry, and in certain circumstances it will even remain colourless on wet concrete.

The effect of alkaline moisture is the same for all flexible floorings. The edges of the sheet or tile lift and bubbles appear in the surface. Usually the adhesive remains firmly attached to the underside of the finish but leaves the concrete surface completely. Occasionally the adhesive is attacked so severely that it is converted to a brown or black liquid that oozes up between the joints.

Thermoplastic and vinyl asbestos tiles are much more resistant to alkaline moisture but sometimes the adhesive is attacked in the manner just described. Alkalis can, however, produce failure peculiar to these tiles. Water rising through the base concrete penetrates the joints and on evaporation deposits salts, mainly sodium carbonate, from the concrete on the edges of the tiles; the salts tend to creep inwards from the joints over the surface of the tiles, and when dry, characteristic white bands about 25mm wide are formed around the edges, which have been described as 'window framing'. This condition, though unsightly, is not necessarily harmful, and the salt deposits can be removed by careful cleaning. In severely damp conditions, however, thermoplastic tiles can absorb these salts, and the pressures set up within the tiles as the salts crystallise are sometimes sufficient to cause delamination and ultimately powdering of the edges of the tiles.

Uses

Thermoplastic tiles may be used in most non-industrial situations. The tiles are not resistant to grease and oil but are satisfactorily used in domestic kitchens.

Vinyl asbestos tiles tend to have better abrasion resistance than thermoplastic tiles. They have good grease and oil resistance and are more suitable for domestic kitchens or even industrial canteens: they are not, however, suitable for industrial kitchens.

Flexible PVC flooring is produced in such a wide variety of forms that a suitable type can be found for most non-industrial situations and in some instances for industrial situations also. Some are tough enough to withstand truck traffic but it is difficult to fix them so as to prevent the stretching and splitting this produces. In places where a jointless floor is desirable, such as in industrial chemical laboratories or hospitals, it is possible to make a continuous floor finish impervious to liquids by hot air welding of adjacent sheets. Sometimes sheets are butt jointed and welded; in other cases a rod of material of similar composition to the sheets is welded into a gap left between the sheets or one formed by edge trimming.

A disadvantage of PVC flooring is that it is more easily damaged by cigarette burns than some other materials, but it is doubtful whether any floor finish should be expected to withstand abuse of this kind.

The characteristics of synthetic rubber flooring are much the same as those of natural rubber flooring but they have the advantage of higher resistance to solvents, grease and oils, and freedom from the surface crazing often associated with natural rubber flooring.

Adhesives

There are few difficulties associated with adhesives for flooring so long as it is realised that adhesives cannot act as damp-proof membranes. Even on a dry base there are some problems involved in laying flexible PVC flooring on adhesives. Tar from an adhesive can diffuse slowly through the thickness of the flooring to produce indelible brown stains on the surface. Adhesives containing tar are not therefore recommended.

Some adhesives cause flexible PVC flooring to shrink and produce unsightly gaps around the tiles. This shrinkage is quite different from that shown by dimensionally unstable material (BS 3261 : 1960 includes a test for this) and is produced by movement of plasticiser from the tile into the adhesive. The shrinkage is high (0·7 per cent has been observed) and because the plasticiser softens the adhesive there is a tendency for the tiles to move under traffic and for very wide gaps to appear.

The problem can be solved by using an adhesive that is resistant to plasticisers. Water emulsions of synthetic resins or synthetic rubbers form the basis of most suitable adhesives but the choice of adhesive must normally be left to the manufacturer of the flooring, since he is likely to use a range of plasticisers any of which might bleed into particular adhesives.

Maintenance

To retain the initial appearance of the floor finishes discussed in this Digest it is necessary to maintain them by polishing. The polishes available may be divided into two categories: (i) emulsions in water of resins, waxes or blends of these; and (ii) pastes or liquids in which wax is dissolved in white spirit, turpentine or other solvent. Emulsion polishes can be used on all the materials without risk of damaging the finish, but solvent-based polishes should never be used on thermoplastic tiles since the solvents soften the binders in these. Excessive use of polish should always be avoided, since it leads to slipperiness and high dirt retention. Worn or dirty coats of polish can be removed by washing with a solution of a neutral detergent and subsequently rinsing with clean water.

Two common kinds of markings which may present problems of flooring maintenance are those produced by black rubber in footwear, castor tyres and protective thimbles on the legs of metal furniture, and those produced on the floorings of motor showrooms by rubber tyres.

The scuff markings first referred to can be removed by rubbing with scouring powder and fine steel wool and, on materials other than thermoplastic tiles, by wiping with a cloth moistened with turpentine or white spirit. The markings in motor showrooms, however, are indelible as they are caused by interaction between the antioxidants in the rubber tyres and the floor finish. Here, prevention is the only real answer, although careful choice of pattern in the flooring can help to mask the effect. Mats, metal strips, or thin sheets of polyethylene are effective safeguards in preventing contact between tyres and floors.

Composition

Thermoplastic binders

These are termed thermoplastic because they soften when heated and harden again on cooling. At normal temperatures they are usually brittle unless blended with plasticisers. Large quantities of flooring employ the following binders:

Pitch and petroleum bitumen—dark brown to black substances obtained from the distillation of coal and petroleum respectively.

Gisonite—a dark brown naturally occurring asphaltic material.

Coumarone resins—pale yellow substances found in coal tar naphtha. Extensive processing, including polymerisation, is necessary before they can be used for flooring.

Vinyl resins—white resins synthesised from acetylene or ethylene. Important members of this group are polyvinyl chloride (PVC), copolymers of vinyl chloride and vinyl acetate containing up to 15 per cent of the acetate, and copolymers of vinyl chloride and vinylidene chloride.

Plasticisers

Plasticisers are usually oily liquids having high boiling points and low volatility. They are used to increase the pliability of the binder during the manufacturing process and to provide some measure of control over the flexibility of the finished product.

Fillers

These are used mainly to reduce material costs but they have subsidiary uses, e.g. carbon black may sometimes act as a reinforcing agent or even as a means of providing an electrically conducting material, and asbestos fibre helps to improve the handling properties of hot material during the manufacturing process. Most of the powdered fibres used are of mineral origin, e.g. calcium carbonate and china clay. Cellulosic fillers such as sawdust are seldom used since products containing them are more sensitive to moisture: further, they sometimes lead to high dirt retention by the finish.

Stabilisers

Stabilisers are used mainly to prevent degradation of some of the thermoplastic binders during manufacture, but they also protect the finished flooring subsequently when this is exposed to strong sunlight.

Manufacture and Availability

Thermoplastic flooring tiles

Thermoplastic tiles are made by heating and masticating binders, plasticisers, short-fibre asbestos, powdered mineral fillers and pigments. The resulting hot mass is passed between rollers, which combine heat and pressure, to produce a sheet of appropriate thickness from which the tiles are produced, usually by die-cutting. At ordinary temperatures the material is brittle so that it cannot be marketed in sheet form. The binders range from petroleum bitumen and pitch in the very dark coloured tiles through gilsonite to coumarone-indene and coumarone-styrene resins in the lightest colours. Many of these tiles are now produced with a small proportion of plasticised PVC; they are more flexible and more resistant to grease. The dark binders and their associated fillers and pigments are cheapest and this is reflected in the lower cost of the dark tiles. They are produced in a variety of plain colours and marbled patterns; other patterns, e.g. one simulating the grain and colour of timber, have been made.

PVC (vinyl) asbestos tiles

PVC (vinyl) asbestos tiles formed a natural development from thermoplastic tiles. They are manufactured in the same way with approximately the same ratio of binder to filler: the difference lies in the binder, which consists mainly of vinyl chloride polymer or of copolymers of vinyl chloride with vinyl acetate. The binder allows the production of tiles having clearer, brighter colours and greater flexibility than thermoplastic tiles, though the material is too inflexible to allow of its marketing in sheet form.

Flexible PVC flooring

Flexible PVC flooring is made in both sheet and tile form and in the same thicknesses as vinyl asbestos tiles. The binders used are polyvinyl chloride, copolymers of vinyl chloride or blends of these. It should be noted that although the binding medium may not be entirely polyvinyl chloride it is common practice to refer to the binders as PVC: to avoid confusion this description will be used here. Powdered minerals are the most commonly used fillers but some flooring contains a small proportion of asbestos fibre. The ratio of binder (plasticised resin) to filler can vary very widely; some materials contain as much as 85 per cent of binder, whereas others may contain only 25 per cent or even less. The prices of the finished products when compared on the basis of equivalent thicknesses give some indication of the proportion of binder, as this is the most expensive ingredient.

The most important processes in the manufacture of PVC flooring include the following:

(1) The binder and its compounding ingredients are heated and masticated to form a hot mass from which sheets are produced by calendering. Sheets up to 3 mm thick can be satisfactorily produced in this way, but it is often easier to manufacture thinner sheets which are then pressed together in layers or laminated with other materials such as bitumen-saturated paper felt, cellular rubber and cellular PVC. When similar sheets of thin material are laminated together it is often difficult and sometimes impossible to distinguish the finished product from one made to the full thickness in one operation. Frequently the backing has the same composition, apart from pigmentation, as the face, so as to allow trimmings or offcuts, which would otherwise be scrapped, to be re-used.

(2) A cold paste of the plasticised binder is spread on to a supporting layer such as needleloom felt or hessian. The paste is then heated and a tough film is formed when it gels. This process lends itself to the production of embossed patterns.

(3) Pelletised material is formed into blocks and these blocks are sliced into tiles. Particular patterns, e.g. simulated terrazzo, can be produced by this method.

Methods (1) and (2) are used for the production of PVC flooring having a jute hessian backing. Hessian may also be used to support material during production, in which case it is stripped from the product before it is distributed.

Synthetic rubbers

There is a small but increasing use for flooring of a group of materials generally referred to as synthetic rubbers. These materials resemble natural rubber flooring in appearance and can be produced on conventional rubber manufacturing equipment. Tiles may be cut from calendered sheet or produced individually in moulds to give the same wide range of thicknesses and sizes that are available for natural rubber flooring. The binders used include chlorosulphonated polyethylene, styrene-butadiene copolymers, and butyl and nitrile rubbers.

Corrosion-resistant floors in industrial buildings

This Digest replaces Nos. 73 and 74 from the First Series.

The floor finish in industrial buildings is provided primarily to protect the structural floor from the effects of traffic, abrasion, impact loads and corrosive liquids. Mostly factories are dry and a good quality concrete as described in Digest 47 is sufficient protection. With wet processes, though, problems arise because the liquids may attack concrete.

Attention in this digest is directed at suspended, reinforced concrete floors which can become dangerous when the reinforcement is attacked, but much of what is discussed is equally applicable to solid ground floors where the problems of structural protection are less acute. Between the structural base and the flooring there may be layers of material to provide levelling, liquid-proofing, thermal insulation, etc, but they are regarded as part of the base. Timber floor systems are not discussed. The term concrete, when used without qualification, refers to concrete made from Portland cement to BS 12.

Classifying broadly, concrete is resistant to alkalis, mineral oils and many salts, so a concrete floor finish is satisfactory protection when these substances are used. On the other hand, concrete is attacked by acids, vegetable and animal oils, sugar solutions and by some salt solutions. When these liquids are to be used they must be prevented from reaching the concrete base.

The risk of corrosion of the steel reinforcement also arises. If acids or salts gain access to the steel, corrosion is inevitable. Good quality concrete and adequate cover, which ensure protection in normal circumstances, cannot be relied upon when liquids severely corrosive to concrete are concerned and it becomes necessary to adopt other means to ensure that the liquids do not reach the structural base. Some solutions, such as common salt, which do not attack concrete, are corrosive to steel; where saline wash waters are frequently used, therefore, a liquid-tight layer must be provided below the flooring.

Where conditions are very corrosive, full protection is needed; where they are not corrosive a good concrete finish is sufficient. Difficulty in making a choice arises in factories handling liquids that attack concrete only slowly. High quality concrete does not readily absorb liquids, so if a high standard of cleanliness is maintained, the liquids will not be in contact with the concrete for sufficient time for attack to occur.

The variety of conditions in industry is so large that general recommendations can only have limited applicability; experience in similar factories must often be sought for guidance.

Floor design

Layout

The first consideration in the design of a floor system is the layout of the plant and the manufacturing processes which are to take place on the floor. The best way to prevent corrosion of the floor is to prevent the spilling of corrosive liquids, but large quantities of mildly corrosive liquids, such as the broth from cooked meat, are frequently and deliberately discharged on to floors. Even in the best designed plants spillage will occur if only by accident and it is wise to group the likely spillage points so that maximum protection need be provided in limited areas only.

If the continuity of the floor finish is to be broken by service pipes passing through the floor, the pipes should be grouped because a good seal can be made more effectively and economically around one large hole than around several smaller ones. Plant layout will often be determined by factors other than flooring requirements but the flooring system should, as far as possible, be considered as part of the plant, even though the design may have to cater for several types of corrosive conditions.

Drainage

The most valuable precaution that can be taken in designing a floor is to provide adequate drainage. All flooring, no matter how well laid, has small depressions which will collect liquors and become centres of corrosion. To prevent these pools forming the floor surface should be sloped by an amount depending on the kind of process, on the amount of pedestrian and truck traffic and on the flooring material. The slope selected will always be a compromise; the steeper the slope the more readily will liquors drain off, but safety and convenience must be considered. A steep slope increases the risk of slipping, more effort is required to move trucks and drums, and, most important of all, trucks may run down it out of control.

If the surface is smooth and there is little spillage, a slope of 1 in 80 may be used, but this must be regarded as the minimum, and slopes up to 1 in 60 should be the aim; where there is much spillage or the surface is rough a slope of 1 in 40 will be necessary; anything steeper is inconvenient and dangerous. The direction of falls must be planned with the traffic flow in mind, so that the traffic will move across, rather than up and down, the slope.

Provision of adequate falls is controlled largely by the position of the drains which, in turn, is controlled by the following considerations. Drains should be close to the main sources of spillage but they should be as far as possible from any source of vibration and from columns and beams. They should be constructed so as to permit examination and repair without having to cut the floor; they should not pass below any major item of equipment. Many of these requirements may be conflicting but a good design will balance the various features.

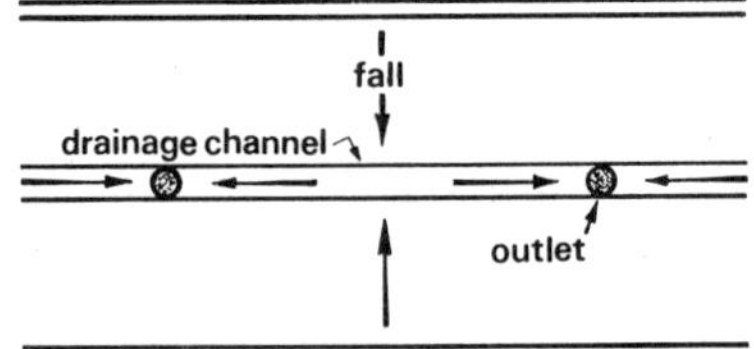

Fig 1 Transverse slope

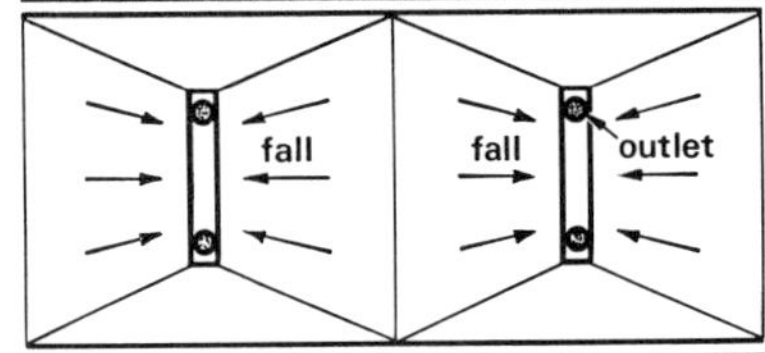

Fig 2 Longitudinal slope

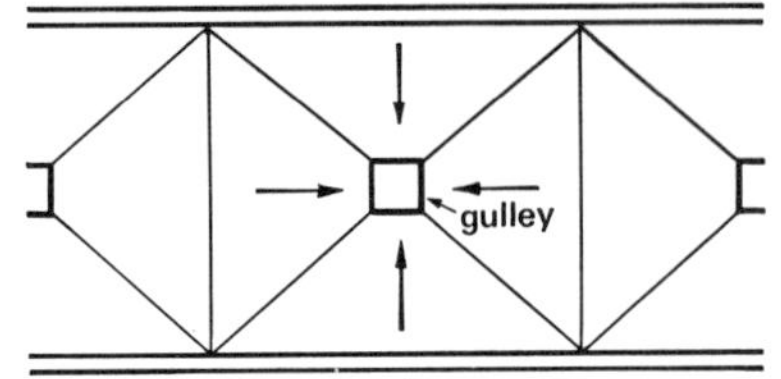

Fig 3 Saucer slope

The floor slopes to the drains may be:

(1) Transverse, in which the floor slopes to a channel running along the length of the building (Fig 1).

(2) Longitudinal, in which the floor slopes to centre channels across the width of the building (Fig 2).

(3) Saucer, in which the floor is divided into a series of rectangular dishes with a central pipe drain (Fig 3).

A mixture of all these types of slope may sometimes be needed.

The fall must not be formed within the thickness of the flooring, the surface of the structural base should be constructed to the necessary slope.

A floor constructed with the correct falls should not have undrained pockets where parts of the plant rest on or pass through the floor. All openings should preferably have a 150 mm high kerb around them (Fig 4), but where this is not possible, for example, at a stair, the edge with no kerb should be at the lower side of the slope. Projections occurring in a slope should be shaped and placed so as to give minimum resistance to flow (Fig 5).

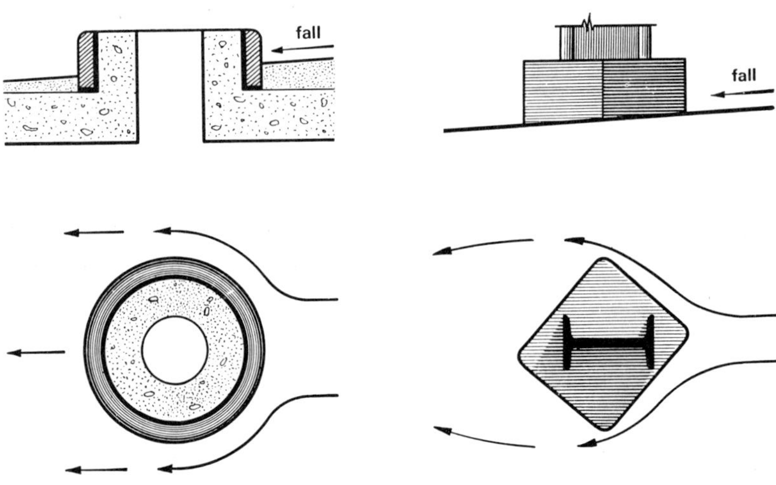

Fig 4 Opening with kerb

Fig 5 Shape and place projections to give least resistance to flow

Membranes

The ideal in corrosion-resistant construction to protect the structural base from liquid penetration would be to use a completely impermeable flooring material. It is usually impossible, however, to eliminate all leaks: minor defects in workmanship and movements in the structure can cause fissures through which liquids pass quite readily. If a leak occurs in an apparently undamaged floor its location cannot usually be traced without severe disruption of the floor system and a consequent risk of further damage.

The most satisfactory seal is a membrane laid between the base and the flooring, the base having been constructed with a fall so that any liquids reaching the membrane flow down to the drain. The membrane must be impervious and resistant to the liquids that will come into contact with it, and be sufficiently flexible and strong to resist damage from movements caused by loading. The membrane must be continuous around drains and at the points where service ducts pass through the floor. The siting of services must form part of the original floor design so that the membrane can be sealed around holes when it is laid; sealing is much more difficult later. If there is a possibility of changes in the siting of services at a later date, spare service ducts should be provided in the original design. Figs 6 to 9 show examples of details. The following membrane materials are those most commonly used—the final choice will depend on the conditions in each particular application.

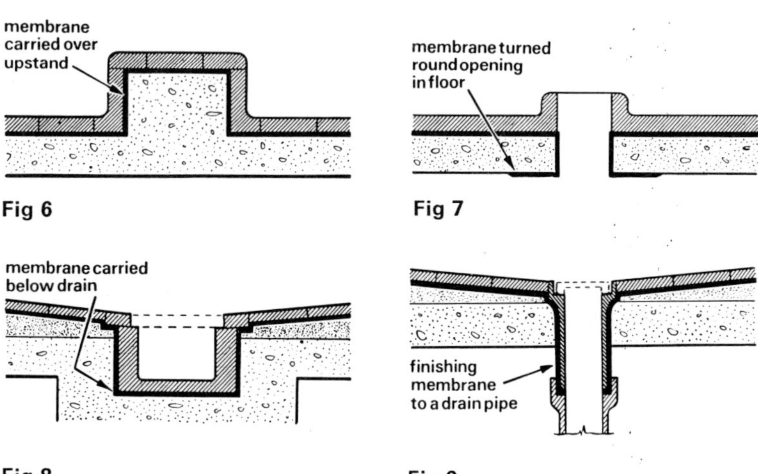

Asphalt The most common membrane is acid-resistant asphalt, often laid on a layer of saturated felt. The finished thickness of the asphalt should not be less than 10 mm laid in two coats with all construction joints broken. Reliance must not be placed on the joint between asphalt and steelwork because a liquid-tight seal cannot be guaranteed.

Bituminous felt Bituminous felt can be used as a membrane; joints should be lapped 75 mm and sealed with hot bitumen. At least two layers should be laid, with maximum staggering of joints. Internal angles, into which it is difficult to work the material, should be built up with fillets before the felt is laid. External corners should be rounded or splayed.

Polythene, pvc or synthetic rubber Membranes of polythene, polyvinyl chloride or synthetic rubber are available in thicknesses from 0.2 mm upwards. The material should be as thick as possible to lessen the risk of puncturing during subsequent floor-laying operations.

The Flooring

The most suitable flooring is one comprising ceramic materials bedded and jointed with a material resistant to the corrosive agents to which it is likely to be exposed. The ceramic materials may be low porosity tiles or engineering quality bricks, the latter being more suitable for areas in which traffic is heavy or where impacts may occur. Tiles and bricks for use in severely corrosive conditions should meet the requirements of BS 3679 Acid Resisting Bricks and Tiles.

Bedding and jointing Materials

The choice of bedding and jointing materials is wide. The following notes describe briefly the more common types of cement used for bedding and jointing ceramic products. Their properties and resistance to corrosion are indicated in Table 1.

Portland cement A mortar of Portland cement mixed with sand (1 to 3 by weight) and gauged with a small quantity of water can be used for bedding and jointing; adhesion to the bricks and tiles is moderately good but the joint is not liquid-tight.

Supersulphated cement Supersulphated cement is a mixture of granulated blastfurnace slag and calcium sulphate with a small proportion of slaked lime or Portland cement. It is used the same way as Portland cement but requires rather richer mortar mixtures and the mortar must be kept damp for several days after laying. Additional lime, calcium chloride or Portland cement must not be mixed with it. Joints are not liquid-tight.

High alumina cement Mixed into a mortar with sand, high alumina cement can be used for bedding and jointing but joints are not liquid-tight. Whilst setting, the heat evolution is high and the mortar is likely to dry and crack unless kept wet for about 24 hours. The material loses strength and resistance to abrasion in wet conditions at temperatures above 30° C. High alumina cements must not be mixed with Portland cement or lime.

Silicate cement Silicate cements are the oldest of the 'chemical' (as distinct from building) cements. The cementing material is silica gel, which is prepared by mixing sodium silicate and powdered quartz or brick dust with sodium silicofluoride. The cement sets quickly and must therefore be made in small batches so that it can be used before setting; no addition of fillers, water or Portland cement should be made. After setting, the material should be brushed with hydrochloric acid to neutralise the alkali liberated in the setting process. Silicate cements resist all acids except hydrofluoric but are not resistant to alkali; they are eroded slowly by cold water and rapidly by hot water, but become porous when thoroughly dry. They behave best when always wet with acid. Liquids can penetrate the joint between brick or tile and jointing. Silicate cement is suitable for bedding and jointing.

Sulphur cement. Sulphur cement is a mixture of sulphur and sand, sometimes with the addition of hard pitch, tar, natural resin or gums. The material is prepared by thoroughly mixing sand in molten sulphur and immediately pouring the mix into wide joints between bricks or tiles. The cement gives a good but brittle joint. It has good resistance to water, alkali and acids, except concentrated nitric and sulphuric. Resistance to vegetable oils, some solvents and sulphur-reducing bacteria is low. It is not suitable for bedding. Because the properties of sulphur change at 96° C it should not be used where this temperature will be reached.

Cashew nut resin cement A resin is produced from the brown liquid which occurs between the shell and the kernel of the cashew nut. This resin, when mixed with a filler and a catalyst, sets to form a flexible cement which resists all acids except concentrated nitric and sulphuric and fairly concentrated alkali. Unlike most of the other cements mentioned its resistance to organic solvents is low. The joints formed between bricks and tiles are impermeable to liquid.

Phenol formaldehyde resin cement Phenol formaldehyde resin cement is supplied as a syrup with an inert filler to which a catalyst is added just before use. The cement sets hard in the cold but its resistance to corrosion is greatly improved if the joint is heated gently. The cement adheres well to brickwork and is reasonably water-tight; it will resist most acids except concentrated nitric and sulphuric. It resists dilute but not concentrated alkali. The jointing should not be heated above 180° C.

Table 1 Properties of jointing materials

Material	Resistance to wear	Hardness	Adhesion to smooth clean surfaces	Liquid-tight	Resistance to deterioration through the action of						
					Acids & alkalis (satisfactory pH range)	Organic acids	Organic solvents	Mineral oil	Animal and vegetable oil	Sugar	Water
Portland cement BS 12	G	Hard	F	Np	7–12*	P	VG	VG	P	P	VG
Supersulphated BS 4248 : 1968	G	Hard	F	No	3·5–12	G	VG	VG	G	G	VG
High alumina cement BS 4248 : 1968	G	Hard	F	No	4·5–9	F	VG	VG	F	F	G
Silicate cement	F	Hard and brittle	G	No	1–8	G	VG	G	G	G	P
Sulphur Cement†	P	Hard and brittle	P	No	1–10	G	G	G	P	G	VG
Cashew nut resin cement	G	Hard and tough	G	Yes	1–12	G	P	P	P	G	VG
Phenol formaldehyde resin cement	G	Hard and tough	G	Yes	1–10	G	G–P	G	G	G	VG
Rubber latex cement	G	Resilient	VG	Yes	Depends on cement	G–F	P	G–F	F	G–F	VG
Bitumen cement†	G–F	Plastic	G	Yes	1–11	G	P	P	P	G	VG
Furane resin cement	G	Hard and tough	G	Yes	1–12	G	G	G	G	G	VG
Epoxy resin cement	G	Hard	VG	Yes	Depends on hardener	G–P	G–P	G	G	G	VG–G

†Not for bedding

*Somewhat better acid-resistance is obtainable with the use of sulphur-resisting Portland cement to BS 4027 : 1966 (see Digest 90)

Rubber latex cement. This cement is prepared from a mixture of inert aggregate, rubber latex (natural or synthetic) and either Portland or high alumina cement. It adheres well to bricks and tiles and the joint is liquid-tight. Resistance to corrosion is generally good, but depends on the type of cement used. When the cement has been etched from the top of the joint the rubber swells and gives a liquid-tight joint which protects the cement lower in the joint. Rubber latex cement can be used for bedding and jointing.

Bitumen cement. Bitumen cement is a mixture of bitumen and an inert filler and must be poured into joints immediately after mixing. The material is resistant to dilute acids and alkalis and is attacked by oils and solvents; it is not suitable for bedding. The material forms a good, plastic joint but the bitumen softens above 50° C.

Furane resin cement A resin is produced from the husks of maize which, when mixed with an accelerator and fillers, sets to give a hard solid mass. Furane resin cement resists alkalis, organic solvents and all acids except nitric and sulphuric. Its properties are broadly those of phenol formaldehyde resin cement but it is easier to handle and requires no heat treatment.

Because of their high cost, phenol formaldehyde resin cement, cashew nut resin cement and furane resin cement are normally used only for jointing.

Epoxy resins. Epoxy resins set to a hard mass when mixed with a hardener. The chemical resistance and the jointing formed by the material depends on the hardener used, the amount and type of filler used and the degree of compaction within the joint. Epoxy resins adhere well to tiles and bricks and liquid-tight joints can be formed. Tiles and bricks can be bedded in epoxy resin mortars.

Laying tiles and bricks

The laying of tiles and bricks in any of these cements is a highly specialised operation. The cost of most of the cements is high so it is worth while selecting the ceramic units closely to size from the batch to make the joints as narrow as possible and so save cement. (The actual joint size will depend on the cement: to ensure that the gap is completely filled, the denser cements require a wider joint.) To reduce costs it is common practice to bed the tiles in the cheaper, less-resistant cements and to point only the top of the joint in the more expensive materials, but this can be regarded only as a 'second-class' technique.

The usual method of laying is with a simple broken joint and with the continuous joint across the slope so that there is no continuous runnel of jointing down which liquors can drain. This method has the disadvantage that if the traffic is moving across the slope, the truck wheel will move parallel to the tile joint and may cause crumbling along the edge of the tile. It is better for the traffic to cross the tiles at an angle by using diagonal, or if the tiles are rectangular, herringbone bond. Herringbone bond, however, needs very accurate tiles to obtain a thin joint. For good floor construction it is better to use special tiles than to cut the tiles on the job, but if cutting is necessary the cut tiles should be at the top rather than at the bottom of the slope because there will be less concentration of liquor at the top to attack the wider joints.

In-situ flooring

When conditions do not justify the use of a fully resistant construction an in-situ flooring material may be adequate. Details of concrete, mastic asphalt, pitch mastic and rubber-latex cement are given in BS CP 204: 'In-situ floor finishes.' Resin materials are also available; usually spread in layers up to 6 mm thick, their resistance to corrosion depends on the resin system and on the amount of filler used, but they are not usually resistant to the constant discharge of hot liquids. All in-situ floorings have wider use on ground floors than on suspended floors. When considering their use it must always be borne in mind that jointless flooring is normally more difficult to repair than flooring of tiles and bricks, and some jointless flooring cannot be repaired at all if corrosive liquids or water are present.

Further reading

British Standard Code of Practice CP 204 : 1965 In-situ Floor Finishes.

BRS Digest 79 Clay Tile Flooring, HMSO, Feb. 1967.

Dairy Floors, HMSO 1967 (Ministry of Agriculture, Fisheries and Food).

Heat losses through ground floors

Digest 108 explained the basis of calculating 'standardised' U-values and dealt in some detail with wall and roof constructions. For ground floors, either solid or suspended, it is not possible to calculate U-values from first principles but the value of a basic construction can be adjusted according to the nature of the floor finish and any insulation.

Solid ground floors

A solid floor, laid in contact with the ground, with or without a bed of hardcore, is exposed to the air on only one face. The heat flow from inside the building to the outside air is indicated by the flow lines in Fig 1. The greater the distance it has to travel the less is the quantity of heat lost. A U-value for a solid ground floor must, therefore, take account of the size and edge conditions of the slab.

The results of an examination of this problem, published over twenty years ago,[1] still form the basis of the conventional method of dealing with heat loss calculations through ground floors as set out in the IHVE 'Guide'.[2] Table 1 shows the basic U-values for a range of sizes and shapes of solid floor in contact with the ground; the values given are

applicable to dense concrete floors, with or without a bed of hardcore. Because the thermal conductivities of ground and slab are similar, the values may be used for slabs of any thickness. They will not be affected by a hard dense floor finish such as granolithic concrete, terrazzo, clay tiling etc or by a thin finish of little insulation value such as thermoplastic tiles.

In applying these values, the full temperature difference between inside and outside should be used. During the early life of a building, heat flows into the ground to raise it to its final equilibrium temperature; because of the high thermal capacity of the ground, this may take 6–12 months, with an increased demand on the heating during the early period. Steady-state conditions are applicable only to a narrow band round the edge of the floor slab but, nevertheless, the convention is adopted of basing heat loss calculations on a U-value for the whole floor and this will not lead to great error after the ground has been warmed to its equilibrium temperature.

The effect of moisture content on the thermal conductivity of masonry materials is discussed in Digest 108,[3] which describes the standard assumptions to be made for moisture contents of walls in protected and exposed situations. No such assumptions are required for solid ground floors.

Table 1 U-values for solid floors in contact with the earth

Dimensions of floor	Four exposed edges		Two exposed edges at right-angles	
metres	*W/m² °C*		*W/m² °C*	
Very long × 30	0·16*	*6·25*	0·09	*11·11*
× 15	0·28*	*3·57*	0·16	*6·25*
× 7·5	0·48*	*2·08*	0·28	*3·57*
150 × 60	0·11	*9·09*	0·06	*16·67*
× 30	0·18	*5·55*	0·10	*10·0*
60 × 60	0·15	*6·66*	0·08	*12·5*
× 30	0·21	*4·76*	0·12	*8·33*
× 15	0·32	*3·12*	0·18	*5·55*
30 × 30	0·26	*3·84*	0·15	*6·66*
× 15	0·36	*2·77*	0·21	*4·76*
× 7·5	0·55	*1·82*	0·32	*3·12*
15 × 15	0·45	*2·22*	0·26	*3·84*
× 7·5	0·62	*1·61*	0·36	*2·77*
7·5 × 7·5	0·76	*1·32*	0·45	*2·22*
3 × 3	1·47	*0·68*	1·07	*0·93*

* Applies also to any floor of this breadth and losing heat from two parallel edges (breadth then being the distance between the exposed edges)

Figures in italics are reciprocals of the U-values ie the air-to-air resistance

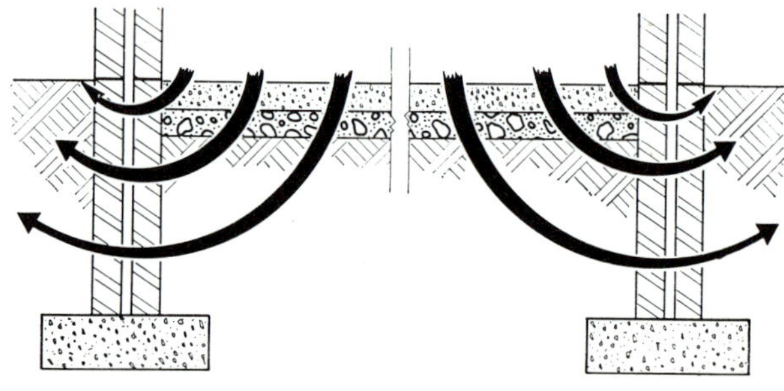

Fig 1 Heat flow through ground slab

Insulation of a solid ground floor

If a floor finish or screed affording some useful degree of thermal insulation is to be used, the U-value of the floor can be calculated accurately enough for most design purposes by taking the reciprocal of the basic U-value, ie the total air-to-air resistance (shown in italics in Table 1), adding the resistance of the additional material and then calculating the reciprocal of the combined resistances to obtain the U-value of the floor.

Assuming, for example, that a concrete floor size 15 m × 7·5 m exposed on all four edges is to be covered with 20 mm softwood flooring:

From Table 1, the basic U-value is 0·62 W/m² °C and its total resistance, the reciprocal, is   1·61

From Table 5, the thermal resistivity of softwood is 7·7 m °C/W; the thermal resistance of 20 mm

material is therefore $\dfrac{7\cdot7 \times 20}{1000}$ = 0·15

total air-to-air resistance of insulated slab = 1·76
U-value = 1/1·76 = 0·57

The effect of an overall layer of insulation (Fig 2) can be calculated in the same manner, by adding the thermal resistance of the insulation to that of the basic floor. The efficiency of the insulation is not, however, constant over the whole area of the floor because the greatest loss through an uninsulated floor is from the edges and the cost of overall insulation is seldom justifiable. An alternative that will give nearly as good results is to treat only the edges

Table 2 Corrections to Table 1 for edge-insulated floors

Dimensions of floor	Percentage reduction in U for edge insulation extending to a depth of:		
metres	**0·25 m**	**0·5 m**	**1·0 m**
Very long × 30	3	7	11
× 15	3	8	13
× 7·5	4	9	15
60 × 60	4	11	17
30 × 30	4	12	18
15 × 15	5	12	20
7·5 × 7·5	6	15	25
3 × 3	10	20	35

of the slab. This can be done in various ways (Fig 3). A vertical layer of insulating material (a) can be used; this should extend from finished floor level down to a depth of not less than 250 mm, but can with advantage be taken down to the top of a strip foundation. Alternatively (b) a horizontal strip about one metre wide can be laid in conjunction with a vertical strip through the full thickness of the floor around all exposed edges. Insulating material used in any of these positions should be of a type that is unaffected by moisture either in its performance or durability, or it should be protected from ground moisture. Data sheets which set out the properties of many insulating materials are included in 'Thermal insulation of buildings'.[4]

Corrections to Table 1 to allow for the effects of edge insulation as in Fig 3 (a) are given in Table 2. The detail shown in Fig 3 (b) will have a performance at least as good as with the same amount of insulation used as in 3 (a).

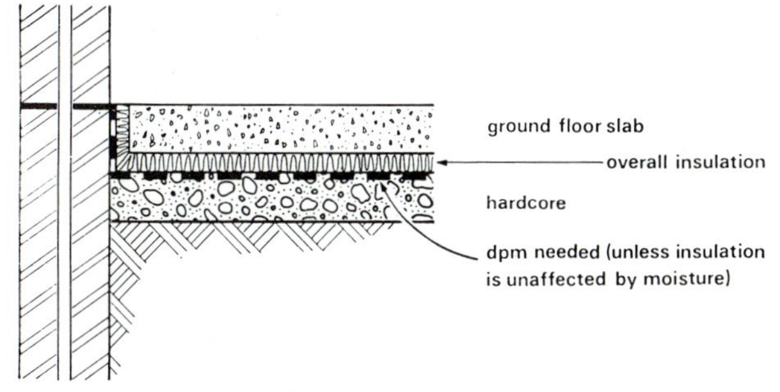

Fig 2 Overall floor insulation

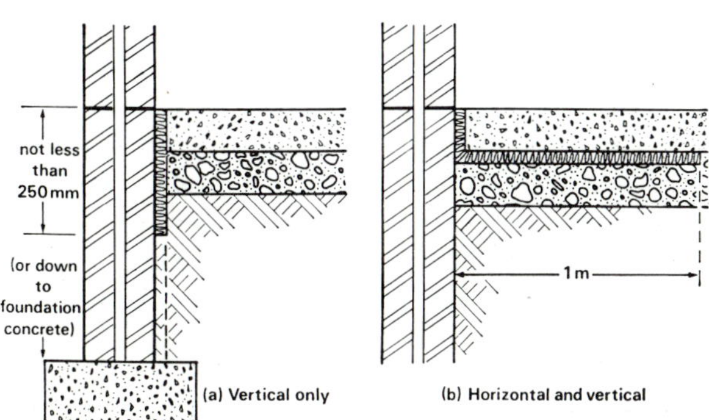

Fig 3 Edge insulation

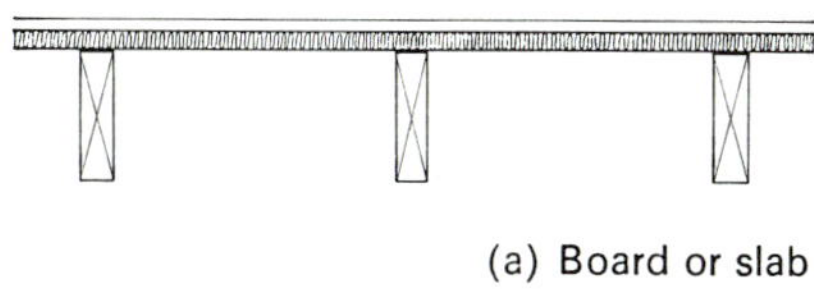

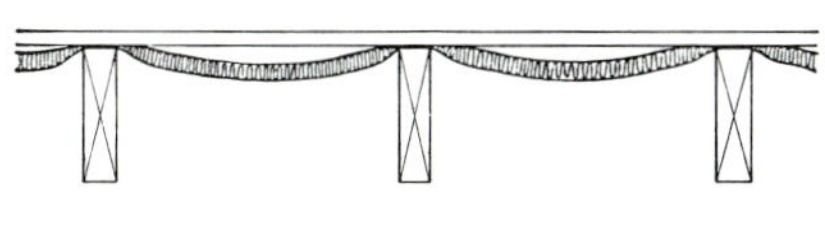

Fig 4 Insulation above joists

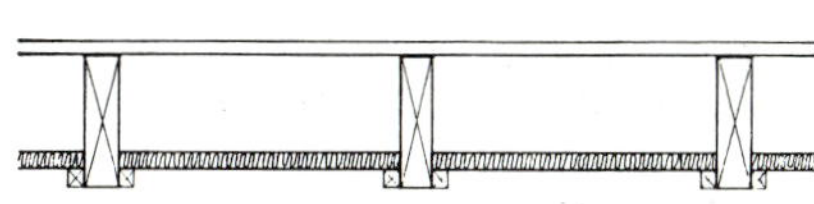

Fig 5 Insulation between joists

Table 3 Suspended floors directly above ground

Dimensions of floor	Basic thermal resistance (any floor structure) $R_{si}+R_a+R_e$	U-values: timber floors		
		Bare or with linoleum, plastics or rubber[1]	With carpet or cork[2]	With any surface finish and 25 mm quilt[3] (as Fig 4b)
metres	$m^2 °C/W$	$W/m^2 °C$	$W/m^2 °C$	$W/m^2 °C$
Very long × 30	5·35	0·18	0·18	0·16
× 15	2·82	0·33	0·33	0·26
× 7·5	1·67	0·53	0·52	0·37
150 × 60	7·13	0·14	0·14	0·12
× 30	4·58	0·21	0·21	0·18
60 × 60	5·90	0·16	0·16	0·14
× 30	4·02	0·24	0·23	0·20
× 15	2·54	0·37	0·36	0·28
30 × 30	3·42	0·28	0·27	0·22
× 15	2·34	0·39	0·38	0·30
× 7·5	1·55	0·57	0·55	0·39
15 × 15	2·03	0·45	0·44	0·33
× 7·5	1·44	0·61	0·59	0·40
7·5 × 7·5	1·27	0·68	0·65	0·43
3 × 3	0·75	1·05	0·99	0·56

(1) assuming $R_s = 0·20$
(2) assuming $R_s = 0·26$
(3) assuming $R_s = 0·86$

Suspended ground floors

A suspended ground floor above an enclosed airspace is exposed to air on both sides but the air temperature below the floor is higher than the outside air temperature because the ventilation rate of the underfloor air space is very low. The U-values are therefore lower than was assumed in the past, when the underfloor temperature was taken as the temperature of the outside air.

Table 3 gives the basic thermal resistances of suspended ground floors excluding the resistance of the structure, which must be added in order to calculate the U-values. The figures in column 2 are the sums of the inside surface (ie the floor surface) resistance R_{si}, the resistance of an airspace R_a ventilated by 2000 mm^2 gaps per linear metre of boundary and the resistance of the earth R_e. To these basic resistances must be added the resistance of the proposed floor structure together with any added insulation R_s. The reciprocal of the sum of these resistances is the U-value.

The results of this calculation for timber floors, either bare or covered with a thin finish of low thermal resistance (R_s assumed to be 0·20), are given in column 3. U-values obtained with a covering of higher thermal resistance (R_s assumed to be 0·26) are given in column 4. U-values obtained with any surface finish and 25 mm quilt, as shown in Fig 4b (R_s assumed to be 0·86), are given in column 5.

Insulation of a suspended ground floor

Additional insulation of a suspended wooden ground floor is commonly provided either in the form of a continuous layer of semi-rigid or flexible material laid over the joists (Fig 4) or semi-rigid material between the joists (Fig 5). With concrete or hollow pot floors it will usually be more convenient to place the insulation above the structural floor.

The thermal resistance of boards or slabs laid over joists (Fig 4a) and of airspaces, or of overall insulation above the structural floor must be added to the basic resistance (column 2 of Table 3) and the resistance of the floor structure to calculate the U-value. Standard thermal resistances for airspaces between the various layers of the construction are given in Table 4. Blankets and quilts laid over the joists will be effective only over their uncompressed area between the joists; foils are effective only where they operate in conjunction with an airspace—again, therefore, only between the joists. This is unimportant, however, because the resistance of the joists compensates for the absence of insulation.

Table 4 Standard thermal resistance of unventilated airspaces

Type of airspace		Thermal resistance* $m^2 °C/W$
Thickness	Surface emissivity	
5 mm	High	0·11
	Low	0·18
20 mm or more	High	0·21
	Low	1·06
High emissivity planes and corrugated sheets in contact		0·11
Low emissivity multiple foil insulation with airspace on one side		1·76

* Including internal boundary surface

Vapour barriers

The introduction of an insulating layer on the underside of the floor raises questions as to the incidence of condensation on the colder faces of the construction and, consequently, of the need for a vapour barrier. A vapour barrier could set up a dangerous situation in a wood floor in the event of the space within the floor collecting water either by spillage or by leakage from plumbing or heating installations. The water could not then escape by draining or evaporating to the airspace below the floor and evaporation through the flooring would be so slow that dangerous conditions conducive to fungal attack could persist for a long period. The designer usually has no control over future treatment of the floor, particularly as to the nature of any finish that may be laid later. An impervious or nearly impervious finish such as pvc, rubber or linoleum sheeting would give good protection against spillage but in the event of leakage within the thickness of the floor it would further aggravate the situation. The safest course is not to provide a vapour barrier to a floor that is constructed of or incorporates wood.

References

1 Journal IHVE Nov 51 19 (195) pp 351–372

2 IHVE Guide, Book A: 1970. The Institution of Heating and Ventilating Engineers, 49 Cadogan Square, London SW1. £6·00

3 BRS Digest 108: Standardised U-values

4 Thermal insulation of buildings; C C Handisyde, D J Melluish: HMSO London 1971. £1·75

Table 5 Thermal properties of some materials

Material	Density kg/m^3	Conductivity (k) $W/m°C$	Resistivity (1/k) $m°C/W$
Asbestos insulating board	750	0·12	8·3
Carpet		0·055	18·2
—wool felt underlay	160	0·045	22·2
—cellular rubber underlay	400	0·10	10·0
Cork flooring	540	0·085	11·8
Eelgrass blanket	80–215	0·039–0·049	25·6–20·4
Fibreboard	300	0·057	17·5
Glasswool quilt	25	0·04	25·0
Linoleum to BS 810	1150	0·22	4·6
Mineral wool			
—felted	50	0·039	25·6
	80	0·038	26·3
—semi-rigid felted mat	130	0·036	27·8
—loose, felted slab or mat	180	0·042	23·8
Plastics, cellular			
—phenolic foam board	30	0·038	26·3
	50	0·036	27·8
—polystyrene, expanded	15	0·037	27·0
	25	0·034	29·4
—polyurethane foam (aged)	30	0·026	38·5
PVC flooring } Rubber flooring		0·40	2·5
Straw slab, compressed	350	0·11	9·1
Wood			
—hardwood		0·15	6·7
—plywood		0·14	7·1
—softwood		0·13	7·7
Wood chipboard	800	0·15	6·7
Woodwool slab	500	0·10	10·0
	600	0·11	9·1

For a more comprehensive list, see IHVE 'Guide' [2]

3 Walls, Windows and Doors

Strength of brickwork, blockwork and concrete walls

This digest gives briefly the basis of some of the main recommendations of CP 111:1964, 'Structural recommendations for loadbearing walls', to assist designers especially where a particular construction calls for stresses approaching limiting values allowed in the Code of Practice.

Strength relationships between walls and mortars

CP 111 : 1964 recommends permissible stresses for load-bearing walls, including brickwork, blockwork and various types of concrete wall. From these stresses the thickness of walls can be determined in relation to the loads to be carried. Permissible loads are now generally higher than formerly. This means that the wall thickness for, say, a two-storey dwelling is often governed by considerations other than strength.

For a considerable range of bricks the optimum strength of brickwork is obtained with mortar mix proportions of (cement and lime)-sand of 1 : 3 by volume. There is, however, little advantage in using a very strong cement-sand mortar in most brickwork and blockwork walls, and reference should be made to Digest 160, 'Mortars for bricklaying,' for an account of factors, in addition to strength, which affect the choice of mortars.

Figure 1 illustrates the comparison between strengths of mortar and brickwork for a number of mortar mixes when a medium-strength brick is used. As the proportion of lime in the mortar is increased, although the mortar loses strength the effect on the brickwork is not nearly as marked.

For brickwork with high-strength bricks the fall in strength of the brickwork is much more pronounced as progressively weaker mortars are used.

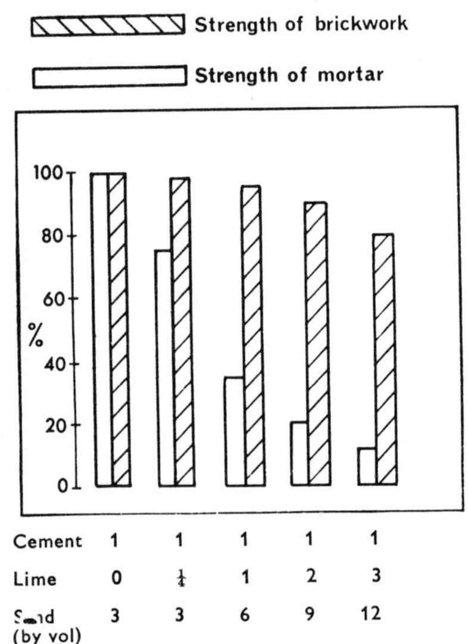

Fig. 1. Effects of mortar mix proportions on the crushing strengths of mortar and brickwork built with medium-strength bricks. Strengths are shown relative to the strength of a 1 : 3 cement-sand mortar and the brickwork built with it.

Table 1. Suitable mortar mixes for various brick strengths

Brick strength (MN/m²)		Mortar mix (by vol.)
Low	10	1 : 2 : 9
Medium High	20–35	1 : 1 : 6
(Class B)	48–62	1 : ¼ : 3
Very high (Class A)	70 or more	1 : 0 : 3

Thus in order to utilize the full capacity of high-strength bricks (70 MN/m² or more) a 1 : 3 cement-sand mix is needed. For lower strength bricks, mortars with an increased proportion of lime can be used without any great loss in brickwork strength (Table 1).

Strength relationships between walls and building units

Strength relationships between walls and building units are shown in Fig. 2. The walls were tested in the air-dry condition but the crushing strengths of the units were taken as the wet strengths, as required by the British Standards dealing with testing building units.

Brickwork

The strength ratio for brickwork : brick varied from slightly less than half to slightly less than one-fifth, the latter being that sometimes found for high-strength (Class A) bricks used with a weak mortar. The trends indicated in Figs. 1 and 2 are of particular importance for calculated brickwork. In CP 111 : 1964 the permissible compressive stresses for walls made of various bricks and mortar mixes are based on relationships of this kind.

Blockwork*

Loading tests on walls of one block thickness built with low to medium strength blocks gave a ratio of wall strength to block strength that was usually much higher than that for low to intermediate strength units of brick shape. It ranged from ½ to 1 (see Fig. 2) and was usually greater than ¾. The Code of Practice, based on limited evidence of this kind, allows increases in permissible compressive stress for walls built with blocks whose height : thickness ratio is between ¾ and 3. (A ratio of ¾ corresponds to the common brick.) This allowance for block shape is greatest for a height to thickness ratio of 2 to 3 and a crushing (wet) strength of up to 5·5 MN/m² (on gross area). For clay blocks of greater strength, tests indicate that the allowance should diminish as block strength increases, and should disappear when the latter exceeds about 20 MN/m².

*An amendment ('Amendment No. 2, published 22 May 1967, to CP 111 :1964') now deals in more detail with the allowance for the shape of the unit

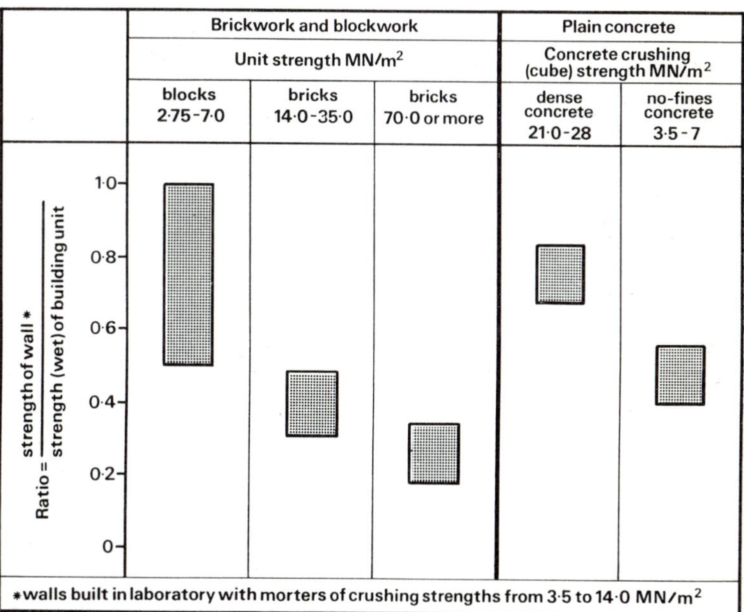

Fig. 2. Ratio of strength of storey-high wall to strength of building unit (block, brick or cube specimen)

Plain concrete walls

At present, the permissible stress of plain concrete walls is related to the crushing (cube) strength of the concrete with which the wall is made. The ratio of wall strength to cube strength in Fig. 2 is required when calculating the load-factor or the ratio of wall strength to design strength. This latter ratio can range from $2\frac{3}{4}$ to 3 for dense and no-fines concrete.

Slenderness ratio

Walls of a given thickness and material strength tend to fail at a lower load as the height increases. This loss of strength applies to brickwork and blockwork as well as concrete walls. It is caused primarily by elastic instability and limitations in workmanship. In design, therefore, the slenderness is important and for most purposes is defined as the ratio of effective height to effective thickness. These effective dimensions are used to take into account features that affect the behaviour and load-bearing capacity of the member.

CP 111 : 1964 recommends that, for walls set in other than lime mortar, the slenderness ratio should not exceed 18, except for dwellings of not more than two storeys when it should not exceed 24. Where lime mortar is used the corresponding limits of slenderness ratio are 12 and 18 respectively.

Effective height and lateral support

The effective height is related to the support provided to restrict lateral movement in the member. Rules governing the effective height to be assumed under various conditions of support are set out in CP 111. For example, with a simple column member supported top and bottom an effective height equal to the actual height between the lateral supports can be taken. For columns supported only at the bottom an effective height of twice the actual height must be used. Walls are similarly regarded but it is accepted that the factors responsible for loss of strength with increasing slenderness are less serious for walls than columns. A distinction is also made in the effective heights to take account of the lateral support provided by different forms of floor or roof system at the boundaries of the wall. For a timber floor to be accepted as a lateral support, metal anchors are required to tie the wall securely to the floor and so transmit any lateral forces to it. For small houses of conventional layout, the stiffening effect of partitions and return walls is such that metal anchors are normally unnecessary. However, in some recent forms of house construction, such stiffening features are absent and the omission of metal anchors is therefore not justified. In such cases the Code recommendation for the provision of metal anchors should be complied with.

Effective thickness

Normally the effective thickness assumed for calculating the slenderness ratio is the nominal thickness of the member. The two exceptions are :

(*a*) A loadbearing wall bonded into piers or intersecting walls. The wall is stiffened by the additional features and an allowance for this is made by taking an effective thickness greater than the nominal thickness. With suitably spaced and sized piers an effective thickness can be used of up to a maximum value of twice the nominal thickness.

While this increase is permitted for loadbearing walls stiffened by piers, where a loadbearing pier is bonded into adjacent walling any beneficial interaction is usually small so that no increase in the effective pier dimension can be assumed. In such cases it is preferable

Table 2. Stress reduction factors for slenderness, for walls axially loaded (CP 111:1964)

Slenderness ratio	Reduction factor	
	Brick wall	Plain concrete wall
6 or less	1·00	1·00
12	0·76	1·00
15	0·63	1·00
18	0·50	0·90
24	0·44	0·70

to assume that the load is applied centrally on the pier and to ignore the presence of the walling.

(*b*) A cavity wall. This is dealt with in a later section under 'Cavity walls'.

Effective length

As an alternative in a few types of construction the slenderness ratio of a loadbearing wall can be based on an effective length instead of an effective height, the effective length being the length between vertical supports, e.g. return walls and piers, into which the wall is bonded. When this concept is used the designer must carefully consider the adequacy of these lateral supports for the particular wall. In many practical cases vertical supports in the form of piers could not be considered to have sufficient stiffness to allow the loadbearing wall to be carried to any height.

Slenderness ratio and reduction factor

Slenderness is one of the several factors governing the permissible stress to be used in design. The permissible compressive stresses for a wall loaded axially are obtained from the product of a basic stress (which is related to the strength of the building unit, its shape and the mortar mix used) and a stress reduction factor (Table 2) related to the slenderness ratio of the wall.

The weakening effects associated with slenderness are most pronounced midway between lateral supports, so that reduction factors for slenderness do not apply when determining critical values for sections of the wall within one-eighth of the height of the wall from the centres of the lateral supports.

Effect of eccentric loading

Walls of brickwork and blockwork have a low tensile strength and are therefore particularly sensitive to eccentric loads. Table 3 illustrates the reduction in strength that occurs for walls of varying slenderness ratio as the eccentricity of vertical loading increases.

Using conventional theory, an eccentricity of $t/6$ theoretically halves the strength as compared with axial loading. Also, as no tensile strength is assumed (see Note 2 to Table 3), for eccentricities greater than $t/6$ the working thickness of the wall is reduced. Thus in Table 3 the load reduction factors for eccentricities of $t/4$ and $t/3$ are obtained

Table 3. Effect of eccentricity of loading on strength

Eccentricity of vertical loading as a proportion of the thickness of the wall	Load reduction factors according to conventional theory[2] for members of slenderness ratio 6 or less	Load reduction factors[1] for slenderness and type of loading. CP 111:1964			
		Slenderness ratio			
		6 or less	12	18	24
Axial loading	1·00	1·00	0·76	0·50	0·44
$\frac{1}{24}$	0·80	1·00	0·75	0·48	0·42
$\frac{1}{6}$	0·50	0·63	0·45	0·26	0·21
$\frac{1}{4}$	0·38	0·46	0·32	0·17	0·12
$\frac{1}{3}$	0·25	0·31	0·20	0·09	0·06

Notes

1. In some cases lateral load would modify these load reduction factors.

2. The assumptions are a limiting compressive stress equal to the permissible stress in axial compression, a linear stress distribution and no tensile strength.

from a working thickness of $3/4t$ and $\frac{1}{2}t$ respectively. These facts together with considerations of economy will tend to control the limits of eccentricity acceptable in design. Where supervision ensures good quality work a maximum eccentricity of $t/4$ for masonry is sometimes accepted.

CP 111:1964 gives *maximum* permissible compressive stresses for axial loading on members of varying slenderness and provides for modifying these stresses when the loading is eccentric. The reductions in strength recommended by the Code are in general less marked than if conventional theory is used, as indicated in Table 3. The extreme reduction in load that occurs as eccentricity and slenderness ratio approach limiting values should, however, be noted.

Assessment of eccentricity

The assessment of eccentricity of vertical loading at a particular junction in a brickwork or blockwork wall is made complex by the many possible combinations of wall and floor systems. The structural behaviour of the junction will be affected by the magnitude of the loads, the rigidity of the interconnected members and the geometry of the junction. It is not possible to determine the eccentricity of load at a junction by theory alone, and the designer's experience must be his guide. Some help in assessing the eccentricity of loading on walls in domestic construction is given below.

Where reinforced concrete and prestressed concrete floor and roof systems of normal span bear on external walls built with bricks or blocks, it is usually assumed that the point of application of the floor load is at the centre of the bearing of the wall. Thus when the bearing is over the full thickness of the wall, the floor load is considered to be axial, and where this is not the case, for example, with cover tiles, there is some eccentricity of load.

For long spans, say when the floor span is more than thirty times the thickness of the wall, and also for timber and other lightweight floors, larger deflections are involved and in such cases the point of application of the load is considered to be displaced from the centre of the bearing towards the span of the floor. The displacement cannot be assessed precisely, but a value of one-sixth of the bearing width is perhaps an acceptable assumption for most junctions.

Interior walls carrying continuous floors are assumed to be axially loaded except when carrying very flexible floor or roof systems. The assumption is valid also for interior walls carrying independent floors spanning from both sides, provided that the span of the floor on one side does not exceed that on the other by more than 50 per cent. Where the difference is greater, the displacement of the point of application of each floor load is taken as one-sixth of its bearing width on the wall.

For a wall corbelled out to support a floor or roof, the point of application of the load is assumed to be at the centre of the bearing on the corbelling. If the floor or roof is supported on a metal corbel or hanger, the load is assumed to be applied 25 mm away from the face of the wall.

Some typical examples of loading on walls in two or three-storey house construction are shown in Fig. 3.

Concentrated loading

Where floors or roofs are supported without interruption along the whole length of a wall, it is assumed that the load is uniformly distributed on the wall. Although not strictly in this category, floor and roof joists, filler joists and ribs of hollow tile floors are regarded similarly.

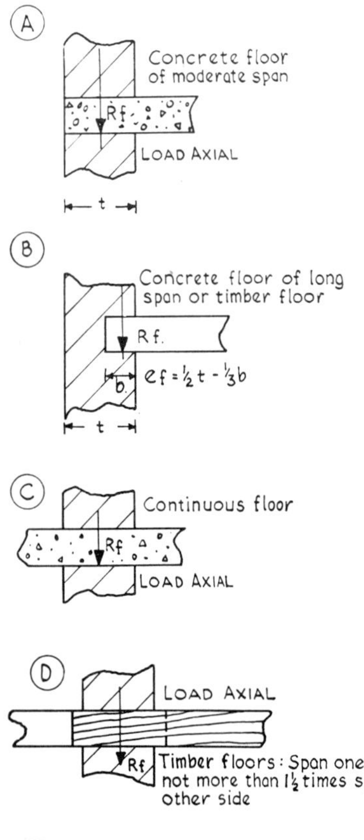

ef is the eccentricity of the reaction of floor load or loads

Fig. 3. Axial and eccentric loading on solid walls.

In practice, design calculations for load-bearing walls must also be capable of dealing with concentrated loads imposed, for example, by roof trusses, spine beams or lintels, in order to prevent failure from crushing or splitting under localised loading.

It is well known that allowable bearing stresses for concentrated loads can safely exceed the permissible stress allowed for uniformly distributed loading. This is because of both the strengthening effect provided by the lateral restraint of material directly under the local load, and the distribution of the load throughout the depth of the wall.

The maximum permissible stress allowed in CP 111 : 1964 for a combination of uniformly distributed and concentrated loading is 50 per cent greater than that for distributed load alone, and the effect of slenderness on permissible stress need not be considered within distances equal to one-eighth of the height of the wall above or below the position of the lateral support. Thus the maximum stress permitted under the contact area is greatest when the load is near a lateral support.

Occasionally concentrated loads cause high local stresses, and for hollow brick or block walls it may be necessary to increase bearing areas or to use padstones or spreaders. The latter may also be needed for any type of wall where indeterminate but very high local stresses occur, e.g. at the outer edge of a wall supporting a cantilever.

Cavity walls

Some structural aspects of cavity construction require careful consideration. CP 111 : 1964 specifies an effective thickness in design of two-thirds of the sum of the actual thickness of the two leaves. This is then used to calculate the slenderness ratio and obtain a corresponding stress reduction factor.

A comparison of wall strengths is given in Fig. 4. It will be noted that the strength of 260 mm cavity construction with both leaves loaded is 20 per cent less than that of the 220 mm thick solid wall. When the load is carried concentrically on one leaf only, the cavity construction is about one-third stronger than a loaded leaf standing alone, owing to the stiffening provided by the other leaf tied to it.

Wall ties are intended to share lateral forces and deflections between the two leaves. The butterfly type metal wall tie is the weakest in this respect and its use should be restricted to two-storey work.

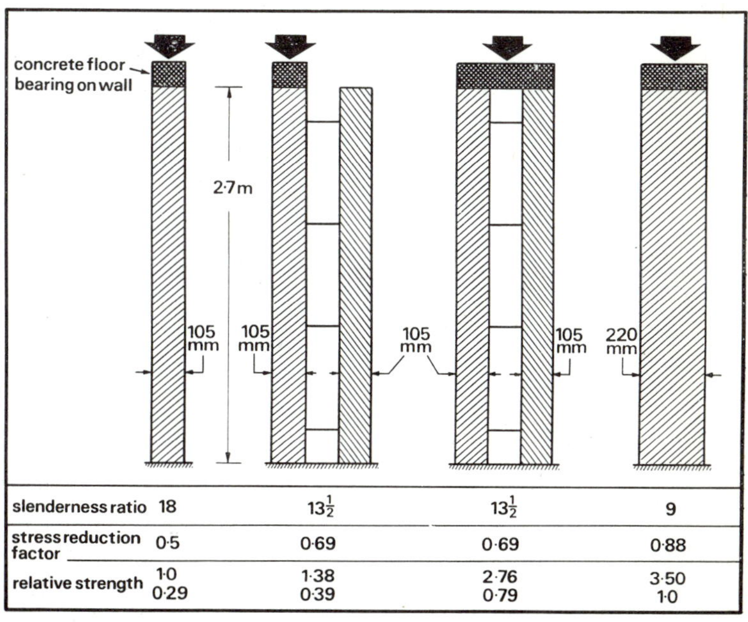

slenderness ratio 18	13½	13½	9
stress reduction factor 0·5	0·69	0·69	0·88
relative strength 1·0 / 0·29	1·38 / 0·39	2·76 / 0·79	3·50 / 1·0

Fig. 4. Design strengths of solid and cavity brickwork walls.

With any type of loadbearing wall it is structurally desirable for loads to be applied with minimum eccentricity. In cavity construction, concrete floors or joists that bear on one leaf only should have a width of bearing equal to the full thickness of this leaf, or at least a 75 mm bearing on any wall thicker than 75 mm. Metal hangers carrying floor joists to one leaf of a cavity wall can cause considerable eccentricity of load and their use should therefore be avoided if possible. It is good practice to stabilise the outer leaf of a cavity wall by loading it, preferably at roof level. When the inner leaf is built with a weaker building unit than the outer, some sharing of load is a definite advantage.

Concrete walls

Many forms of concrete are used for loadbearing walls. Dense and light-weight blocks, *in-situ* dense and no-fines concrete, precast wall units made with lightweight aggregate concrete, are examples. Recent amendments to CP 114 : 1957 specifically refer to lightweight aggregate concrete and in particular to reduction factors for wall slenderness which differ from those used with dense concrete walls.

Another form of lightweight wall is the unit made with aerated concrete. There are at present no *specific* recommendations for design stresses for precast wall units of this type in British Codes of Practice, though some guidance may be taken from CP 111 : 1964. In this Code the recommended permissible compressive stress for walls of slenderness ratio 15 or less and built with relatively weak material is one-fifth of the compressive (cube) strength of the saturated material, with a further reduction for walls of slenderness ratio above 15. The cube specimen is taken, after autoclaving, from the same cake of material as the wall specimens. Alternatively, permissible stress could also be obtained from tests on walls of storey height as allowed in CP 111 : 1964 for brickwork and blockwork walls. The ultimate strength of the wall obtained from the tests is divided by a specified load factor to give a permissible compressive stress.

For autoclaved aerated concrete walls the permissible compressive stress obtained from wall tests might, for example, be based on load factors ranging from 4 for walls of slenderness ratio of 10 or less, to 6 for walls of slenderness ratio of 24.

In CP 111 : 1964 the permissible stress for concrete walls shows a considerable increase compared with the earlier (1948) edition of this Code. This, considered in conjunction with the relatively small loads encountered in the small house, shows that strength considerations alone will not normally decide the thickness of walls in two-storey housing. For instance, if strength were the only consideration, a 100 mm thick single leaf wall built with blocks having a crushing strength equal to the lowest (wet) strength recommended for load-bearing walls, i.e. $2 \cdot 75 \, MN/m^2$ or $2 \cdot 00$ to $2 \cdot 75 \, MN/m^2$ where wall test data are available, might just be adequate for a small, well-built house. A 100 mm thick wall would not, however, normally meet the other functional requirements.

Wall panels of concrete now being manufactured range from 75 mm to 300 mm in thickness but again the minimum wall thickness for the small house will usually be determined by functional requirements other than strength. *Overall* stability, thermal and sound insulation, fire resistance and resistance to rain penetration, all need to be considered.

Damp-proof courses

Damp-proof courses are barriers so placed in a building that the passage of moisture between the parts they separate is negligible. Such barriers are needed in floors (see Digests 18 and 54) but this Digest is limited to the other parts of the structure where damp-proof courses are necessary. Components which shed rain directly, such as roof coverings and flashings, are mentioned only when used as, or in conjunction with, damp-proof courses.

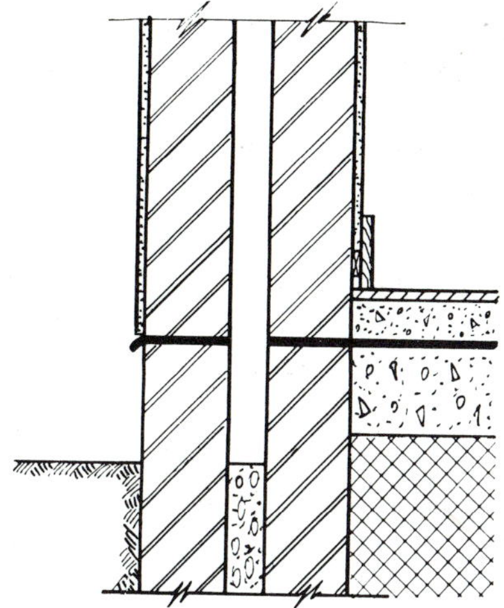

Fig. 1

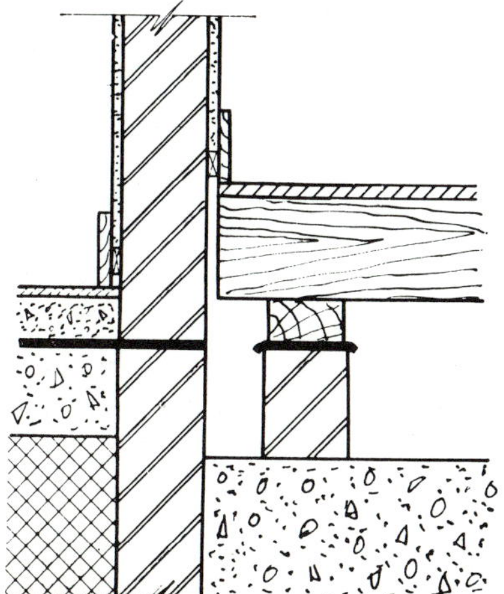

Fig. 2

Movements of moisture

Upward movement of moisture is normally prevented just above ground level; prevention of downward movement is needed at many levels, eg in parapets, chimneys and above lintels in cavity walls. Horizontal movement of moisture must occasionally be prevented, eg where the outer leaf of a cavity wall is returned at an opening to close the cavity, and where wall dpc's are at different levels from those of damp-proof membranes in abutting floors. Examples follow of the main position for dpc's.

Dpc's near the ground

Dpc's are required at the foot of a wall (solid and cavity construction) and should be at least 150 mm above ground level. Cavities must be kept clear. Stopping the dpc short of the external wall face and pointing the joint with mortar is not recommended. External render coats are frequently and incorrectly taken to ground level or below for the sake of appearance. The rendering should be stopped at dpc level (Fig. 1); the weaker the rendering the more necessary this precaution. Materials heaped against an outside wall can bridge the dpc, making it ineffective (*see* Digest 27).

Where an internal division wall has a solid floor on one side and suspended timber floor on the other (Fig. 2), the intermediate cavity must be kept clear.

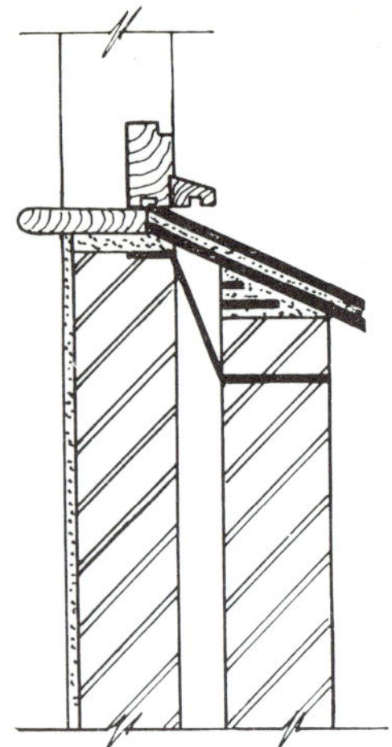

Fig. 3

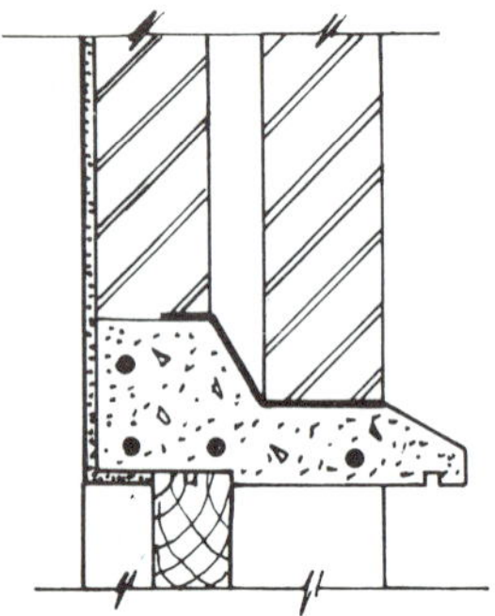

Fig. 5

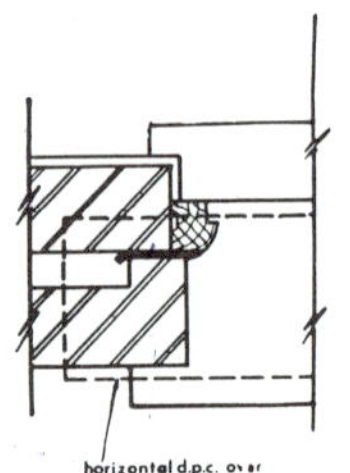

Fig. 6

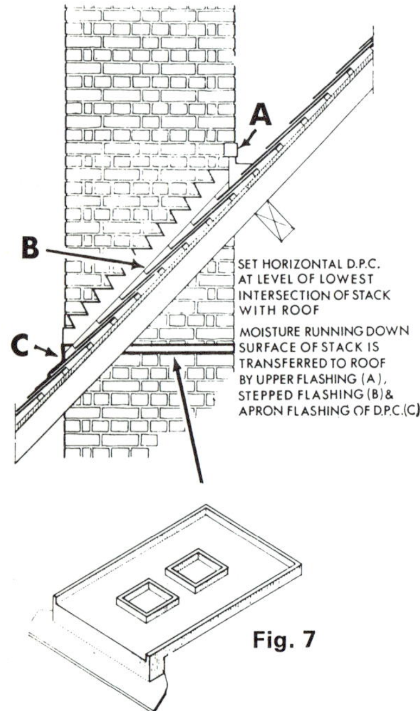

Fig. 7

LEAD OR COPPER DAMP-PROOF
COURSE TRAY WITH APRON FLASHING

Fig. 4

Sills

A sill of porous material or of jointed lengths of impervious material should be bedded on a flexible or semi-rigid dpc if it is part of a solid wall or bridges a cavity. In Fig. 3 the flexible dpc extends to the back of the sill. If the sill or its covering is impervious, no dpc is normally needed unless there is a danger of cracking, as in Fig. 4.

Lintels

Similar considerations apply to lintels. Fig. 5 shows a flexible dpc over a boot lintel. Sometimes a precast concrete lintel supports the inner leaf only and the outer is carried on a flat arch. However, the principle of protection against moisture is the same in all cases. A flexible dpc is sloped across the cavity to drain any moisture away from the inner leaf. It is tucked into a joint above the lintel or into a groove in the lintel. It is then brought down to the top of the opening to lie between the outer leaf and the window or door frame. The dpc should extend at least 100 mm beyond the ends of the lintel.

Openings in cavity walls

Where the outer leaf is returned to close the cavity a vertical dpc should be provided (Fig. 6). The dpc may be of two courses of slate in mortar but more usually a flexible sheet material is used.

Chimney stacks

For a chimney at or near the ridge and extending only slightly above it, good weather protection at the top is sufficient in all but very exposed positions. When the exposed part of the stack is more than about 1 m high, a dpc is advisable where the stack emerges from the roof. The simplest and most effective form is a combined dpc and flashing of lead or copper as shown in Fig. 7. Good protection at the top of the stack is also needed.

Table 1 Materials for damp-proof courses

Material	Minimum weight kg/m²	Minimum thickness mm	Joint treatment to prevent water moving: upward	downward	Liability to extrusion	Durability	Other considerations
Group A: Flexible							
Lead to BS 1178	Code No 4		Lapped at least 100 mm	Welted	Not under pressures met in normal construction	Corrodes in contact with fresh lime or Portland cement mortar	To prevent corrosion apply bitumen or bitumen paint of heavy consistency to the corrosion-producing surface and both lead surfaces
Copper to BS 2870, Grade A, annealed condition	Approx. 2.28*	0.25	Lapped at least 100 mm†	Welted†	Not under pressures met in normal construction	Highly resistant to corrosion	Where soluble salts are known to be present (eg sea salt in sand) similar treatment to lead is recommended. Copper may stain external surfaces, especially stone
Bitumen to BS 743: A—hessian base B—fibre base C—asbestos base D—hessian base and lead E—fibre base and lead F—asbestos base and lead G—sheeting with hessian base	3.8 3.3 3.8 4.4 4.4 4.9 5.4	— — — — — — —	Lapped at least 100 mm	Lapped and sealed	Likely to extrude under heavy pressure but this is unlikely to affect water resistance	The hessian or felt may decay but this does not affect efficiency if the bitumen remains undisturbed. Types D, E and F are most suitable for buildings that are intended to have a very long life or where there is risk of movement	Materials should be unrolled with care. In cold weather the rolls should be warmed before use
Polythene, black, low density (0.915–0.925 g/ml)	Approx 0.5	0.46	Lapped for distance at least equal to width of dpc	Welted	Not under pressures met in normal construction	No evidence of deterioration in contact with other building materials	Welted joints must be held in compression. When used as cavity trays exposed on elevation these materials may need bedding in mastic for the full thickness of the outer leaf of walling, to prevent penetration by driving rain
Pitch/polymer	Approx 1.5	1.27	Lapped at least 100 mm	Lapped and sealed			
Group B: Semi-rigid							
Mastic asphalt to BS 1097 or BS 1418	—	—	No joint problems		Liable to extrude under pressures above 65 kN/m². Material of hardness appropriate to conditions should be used	No deterioration	To provide key for mortar below next course of brickwork, grit should be beaten into asphalt immediately after application and left proud of surface. Alternatively the surface should be scored whilst still warm
Group C: Rigid							
Epoxy resin/sand	—	6.0	No joint problems		Not extruded	No evidence of deterioration in contact with other building materials	Resin content should be about 15%. The appropriate hardener should be used
Brick to BS 3921; max water absorption 4.5%	—	—	—	Not suitable	Not extruded	No deterioration	Use two courses, laid to break joint, bedded in 1:3 Portland cement:sand
Slates at least 230 mm long; should pass wetting and drying test and acid-immersion test of BS 3798	—	5.0	—	Not suitable although they may sometimes be used to resist lateral movement of water	Not extruded	No deterioration	

* Where copper is to form in one piece a dpc and a projecting drip or flashing it should weigh at least 4.12 kg/m²

† Copper may be welded, in which case phosphorus-deoxidised copper to BS 1172 should be used

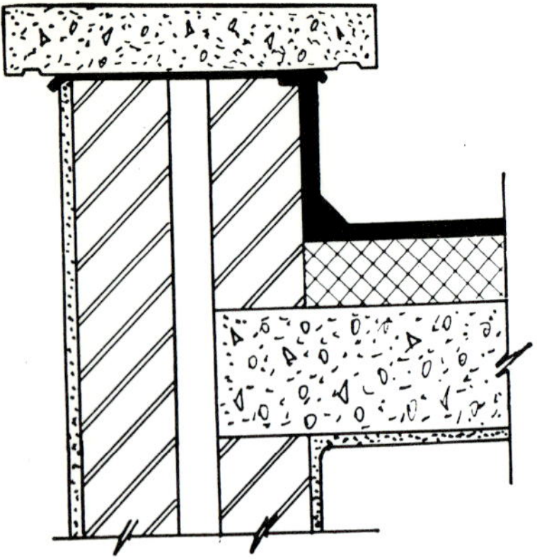

Fig. 8

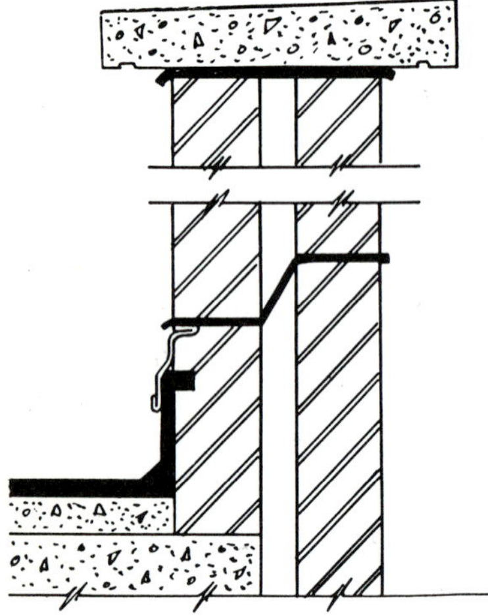

Fig. 9

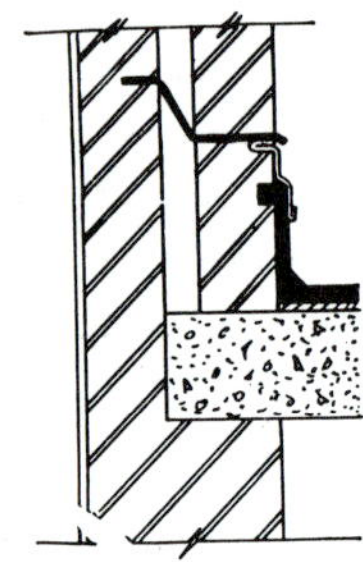

Fig. 10

Parapets

With a very low solid parapet wall there is nothing to be gained by adding a dpc at or near roof level but one under the coping is recommended. The roof covering is extended up the face of the wall and tucked in under the dpc (Fig. 8).

High parapets should contain a dpc at or near roof level which is lapped with the roof covering, as well as a dpc below the coping (Fig. 9).

Roofs at different levels

Where the roofs of adjacent parts of a building are at different levels, the wall exposed between them must have a dpc joined to the lower roof covering to give complete protection against downward penetration of moisture (Fig. 10).

Properties of damp-proof courses

As yet, no Standard lays down performance requirements for dpc materials. BS 743 specifies the quality of some of them and their properties are summarised in Table 1. The table also includes notes on pitch/polymer and epoxy resin/sand not included in the Standard. Group A materials are suitable for most situations and they alone should be used over openings and for bridging cavities. Flexible dpc's over openings in hollow walls should as far as possible be in one piece so as to avoid the sagging that can occur at lapped joints. Flexible materials are suitable for situations where a stepped dpc is necessary; some are also capable of being dressed to complex shapes without fracture.

Group B material is particularly suitable for very thick walls and where there is high water pressure.

Group C materials can withstand heavy loads but are more likely to be fractured by building movement than materials in the other groups. In addition to the materials listed, new types of dpc are likely to come from developments in plastics sheet materials, synthetic resin mortars and synthetic rubbers.

Semi-rigid asbestos bitumen sheet is already used for cavity gutters and flashings and has advantages in rigidity and ease of jointing.

High quality concrete will resist the passage of *liquid* water. As some transmission of water *vapour* through damp-proof courses is normally permissible, dense, quality-controlled concrete might, therefore, come to be accepted without the further protection of a damp-proof course. It is, however, difficult to specify a mix and its compaction to achieve the quality needed.

Methods of damp-proofing existing walls are discussed in Digest 27.

Rising Damp in Walls

This Digest examines the circumstances in which rising damp is likely to occur and suggests possible remedies. The measures considered include providing a complete moisture barrier by inserting a damp-proof course, reducing the amount of water in the wall by drainage and evaporation, and concealing the dampness by an impermeable wall lining. Some recent developments in the technique of damp-proof course insertion are described. The treatment of rising dampness in floors is not considered here, as it is the subject of Digest 54.

Rising damp

Building materials are porous and therefore have the capacity to soak up water in much the same way as a wick. If ground water is allowed to reach the foot of the wall it will tend to rise and, unless stopped by a barrier such as a damp-proof course, will reach a height depending on a variety of factors, such as the supply of water, the pore structure of the wall materials, and the rate of evaporation from the wall surfaces. The picture is further complicated by the fact that ground water almost invariably contains dissolved salts which tend to concentrate at the wall surfaces where the water evaporates (Fig. 1). Precisely how these salts affect the ultimate height of water in the wall, or its rate of rise, is at present not clear, but their presence in the wall surface means that, even if the further rise of water is prevented, the decorations are still likely to be spoiled. This is because some of the salts originally drawn up into the wall are hygroscopic, that is, they can absorb moisture from the air, so that the surface tends to become damp whenever the air in the building is humid. Ways of dealing with plasterwork affected by such salts are described later in this Digest. It should

not be thought that the provision of a damp-proof course is likely by itself to be a complete remedy. Replacement of salt-contaminated plasterwork is nearly always necessary as well.

Identification and occurrence

Apart from water used for construction purposes, which causes only temporary inconvenience whilst drying out, dampness can result from rain penetration and condensation, in addition to rising damp. The typical effect of rising damp is to produce a roughly horizontal 'tide-mark' on the wall, above which there is no damage but below which the plaster may be disrupted and the decoration destroyed by efflorescence, or bleached or caused to lift and peel under the damp conditions (Fig. 2).

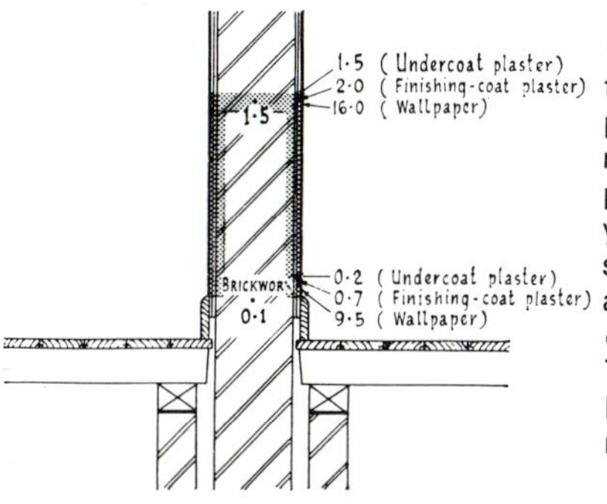

Fig. 1. Concentration of salts in a party wall in which rising damp has persisted for 80 years; the figures show the percentages, by weight of chloride plus nitrate. The shaded area is heavily contaminated.

Fig. 2. A wall affected by rising damp.

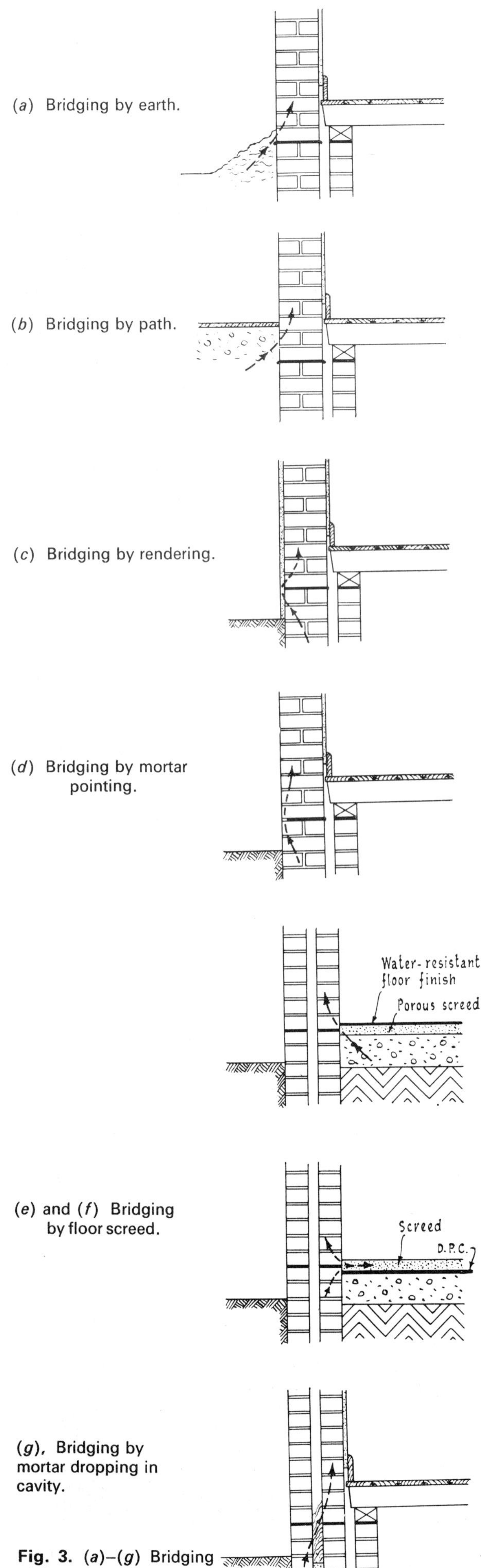

(a) Bridging by earth.

(b) Bridging by path.

(c) Bridging by rendering.

(d) Bridging by mortar pointing.

(e) and (f) Bridging by floor screed.

(g), Bridging by mortar dropping in cavity.

Fig. 3. (a)–(g) Bridging of damp-proof course.

Where hygroscopic salts have accumulated the surface appears damp and mould growth may be present; it is liable to occur on any persistently damp surfaces. Pictures and furniture placed against the wall, and clothes in wall closets, may suffer from the damp and become mouldy.

Rising damp commonly affects old houses having no damp-proof courses or ones which are faulty; it was not until the Public Health Act of 1875 that the incorporation of a dpc became compulsory. Since then, damp-proof courses in many old houses have given trouble because they were either inefficient when laid or have since become so with age. BS 743 specifies materials acceptable for damp-proof courses, including asphalt, bitumen sheet, lead, copper, slate, dense burnt clay brick and polythene. During the next few years the use of other plastics materials may be permitted. There is a good case also for including dense quality-controlled concrete, provided that means are found for ensuring the high standard that is needed.

The use of these materials in buildings constructed nowadays can easily prevent rising dampness, but, unfortunately, it is still seen in many modern houses; this is because the dpc has been bridged in some way, thus providing an alternative path along which moisture can rise up the wall. Bridging can occur in a number of ways, for instance, by solid fuel or garden earth being piled against the wall (Fig. 3a), or a path raised above dpc level (Fig. 3b). It is not at all uncommon, with buildings butting on the street, for pavements to be placed above the level of damp-proof courses—and they are likely to be bedded in ash or clinker containing soluble salts. Co-operation between architects and local authorities is needed to ensure that the pavement is below dpc level. Bridging of the dpc can also be caused by a porous rendering (Fig. 3c), by covering the exposed edge of the dpc with pointing mortar (Fig. 3d), or by mortar that has been allowed to drop between the two leaves of a cavity wall (Fig. 3g).

A common form of bridging that is not easily recognized is that caused by a porous floor screed. Although building regulations require the inclusion of a dpc in walls, there is no such regulation for solid floors. Some solid floors incorporate a damp-proof membrane beneath the floor screed, others rely on the floor finish to serve as a moisture barrier. Unless care is taken, it may be possible for the screed, which is usually porous, to transmit moisture from the wall below dpc level to that above (Figs. 3e and f). With this form of construction the position of the moisture barrier must be carefully planned and a vertical dpc may be needed to join the wall dpc to the membrane in or on the floor.

Remedial measures

No hard and fast rules can be laid down for the treatment of rising damp; the measures decided on will depend on the cost and estimated effectiveness of the treatment, as seen against the value, condition and probable life of the property. Some of the causes of rising dampness described above can be dealt with fairly simply—for instance, earth or paths above dpc level, or rendering or pointing bridging the dpc—but where there is no damp-proof course, or where it is defective, more radical measures are necessary. There are three possible approaches to the problem.

First, a complete moisture barrier may be provided, by inserting a dpc; this should lead to a permanent cure. There are also various proprietary methods that aim to produce a dpc by the injection of water-repellent substances; these are probably more appropriate for walls too thick or too unstable for a conventional dpc to be inserted. Secondly, it may be possible to reduce the amount of moisture in the wall, by hindering the access of water to it and by increasing evaporation from it. With any of these measures, it will also be necessary to replace any decayed plaster and remove accumulated salts.

Thirdly, many of the methods available for dealing with walls heavily contaminated with salts can be used alone to *conceal* the rising damp. These involve battening out or lining the inside of the wall, or using various barrier layers which prevent the damp and salts affecting the decorations. The rising damp is not stopped by such measures, and will often be driven higher; but where it is causing no structural damage such steps are useful where the property has a limited life or more drastic treatments are too costly. It may, of course, be appropriate to use measures of the second category—reducing the moisture content of the wall—in conjunction with such a lining treatment.

Methods to prevent rising damp

Inserting damp-proof courses

The traditional method of inserting a damp-proof course is to cut out a course of bricks, a short length at a time, and replace it by a course of dense engineering bricks or slates. A much less costly method is by cutting or grinding a narrow slot through a bed-joint. New tools have now been developed for this work and new materials have become available; this Digest brings the description of the newer method up to date.

Method of work

The cut may be made either by a hand-saw or by one of several types of power-driven saw*; these are described below. The cut is started in the selected course at a jamb or corner, preferably the latter as this avoids knocking out a brick to start the cut; once the cut reaches the inside it is more convenient to continue the work from inside the building. With a new chain saw developed at the Station it is possible to mortice into a joint and so work entirely from one side of a wall.

The cycle of cutting the slot and inserting the membrane may then proceed without interruption along the wall. The membrane must be inserted immediately the cut is made, generally in lengths of about 0.5m; in heavily loaded sections of the wall, e.g. at jambs and junctions, work must proceed in shorter lengths.

The sheets forming the membrane should be wide enough to project about 5mm on each side of the wall. The membrane may be of any impermeable and durable sheet material; the choice will depend in part on the width of the slot, and this in turn depends on the type of mortar and on the saw that is used. A narrow slot is produced in a wet and sticky mortar by the thin blade of a handsaw or reciprocating saw; in this case a thin rigid sheet must be driven into the slot. More usually, in a dry and crumbly mortar, a wider slot will be cut; and the new chain saw cuts a slot about 7mm wide in any mortar. This will require two layers of a thicker material to fill the slot; alternatively, a thin sheet may be used and the joint packed, by spreading a bed of mortar on the membrane before inserting it

* Since this Digest was first issued, a new method has been evolved using a combined cutting disc and dust extractor, details of which are given on p. 83.

Fig. 4. Damp-proof course insertion: mortar spread on the membrane to pack the joint.

Fig. 5. Driving a metal sheet into a narrow slot with the help of a guide frame.

and wedging slips of slate into the joint at intervals of about 0.3m (Fig. 4).

Even when the membrane is a close fit into the slot there is a slight settlement as the cut advances; measurements on a test house have indicated that 1.5mm may be expected. With a joint packed as described above, about half this settlement will occur. As a rigid sheet material, for driving into a narrow slot, half-hard copper may be used or, for a limited subsequent life of, say, 10–15 years, sheet zinc will give sufficient protection at lower cost. A guide frame may be used for driving in the sheets (Fig. 5), but this is often unnecessary.

With a wide slot, two layers of bitumen felt, each of 0.6m lengths butted together, may be inserted; they should be placed so that the joints are staggered, with at least a 75mm interval between the joints in the two layers. Copper or polythene sheet may be used and the joint packed. Soft copper sheeting can be inserted into the wide slot without fear of damage, and this is available more cheaply than the half-hard sheet required for driving into a narrow slot. High- or low-density polythene sheet, 0·02 in. (0.5mm) thick, is a cheaper alternative; high-density sheet is the more easily inserted. The polythene should be filled with carbon black, to retard deterioration where it is exposed to sunlight.

The rate of progress depends much more on the state of the wall than on the type of saw used. In regularly coursed brickwork, with a dry crumbly mortar, it may exceed 3m per hour in a 220mm wall. Where courses are irregular or obstacles have to be negotiated, the rate may be only 0.6m per hour, but corners and junctions are usually simple to negotiate. The average rate in brick walls of the type generally met with in old houses is about 1.3m per hour in 220mm walls, and 2m per hour in 105mm walls. These estimates cover the time spent in preparation, inserting the membrane and in overcoming obstacles, but not making good the plastering or repointing the brickwork externally.

Saws

The saws* available for cutting the slot include handsaws, which are operated by two men (Fig. 6),

* The names of suppliers of the saws and the proprietary treatments described in this Digest may be obtained from any Building Centre.

power-operated reciprocating saws, a circular saw, and the new chain saw.

The reciprocating saws, hand and power-operated, are of mild steel, with teeth tipped with tungsten carbide or stellite. The handsaw has a 42-in. (1066mm) blade; the 3-in. 75mm width is preferable to the 4-in. (100mm) wide blade normally supplied. Sawing so near the ground is exhausting but for short periods the work can be done as quickly as with a power saw. For a single job, the handsaw provides a cheap and practical method of cutting the slot; in any case, a handsaw is a most useful adjunct to a power saw.

Two types of power-operated reciprocating saws have been used: a light pneumatic model, weighing 2.5kg and designed to be held by the operator, and a more powerful type which requires a supporting frame for more continuous work. Other forms of saw may well be suitable; larger and more powerful tools are used by specialist contractors, and one has developed a circular saw claimed to be highly efficient.

A disadvantage of reciprocating saws is that the swarf from the slot is not removed as the cut advances, and after a time this slows down the rate of working. With a chain saw (as well as with the handsaws and circular saws), the swarf is removed by the sweep of the blade, but the chain saws ordinarily used for cutting and dressing stone cut a slot that is too wide for dpc insertion—at least 12mm wide, that is, wider than a mortar joint.

Fig. 6. Handsaw.

A new chain saw blade has been developed at the Station for use with a lightweight petrol-driven chain saw. In the modified chain (Fig. 7), the blade tips, have been replaced by tungsten carbide cutting blocks brazed into the links and supported by mild-steel buttresses; the chain is $\frac{1}{4}$ in. (6.35mm) wide and it cuts a slot only slightly wider. The saw (Fig. 8), which

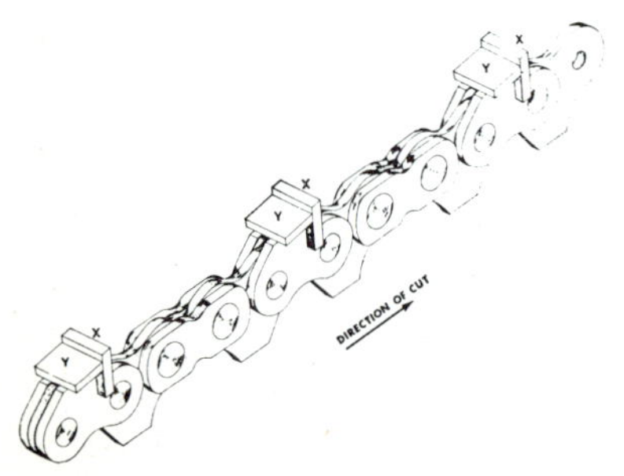

Fig. 7. The modified saw chain, with tungsten carbide cutting tips (X) supported by horizontal buttresses (Y).

Fig. 8. Chain saw.

weighs 3kg, is fitted with a guide bar 14 in. (355mm) long and will satisfactorily cut a 220mm solid wall. It is driven through a flexible shaft by a lightweight 1.675kw air-cooled petrol engine. This unit can be carried and operated by one man. The expected life of a chain and guide bar (which must have a hardened nose) is about 50m of 220mm wall.

This new development has overcome several earlier limitations of the method of dpc insertion. The chain saw is convenient to use, the rate of work is faster than with other power saws, and, when commencing a cut, it is possible to mortice into a joint—this avoids knocking out a header brick. The possibility of working entirely on one side of a wall brings with it several other advantages which are discussed below.

Practical details

The level at which the cut is made must be decided after careful inspection of the site, taking account of ground and floor levels.

For solid floors, the cut should be made in the first bed joint in which it is possible to work, above the level of the slab and at least 150mm above the outside ground level. If an impermeable flooring such as pitchmastic is to be laid over the concrete base, it must be linked with the new dpc. If a permeable flooring that is not sensitive to moisture is used, the finished level of the floor should preferably be at the same level as the dpc or, if this is not practicable and the floor level is below the new dpc, the brickwork between the dpc and the top of the floor should be treated with two coats of bitumen paint and the skirting should be protected against rot. If a moisture-sensitive floor, for instance of timber, is to be laid on the concrete, a damp-proof membrane should be provided and linked up with the new dpc; the protection of timber floors is dealt with in Digest 18.

For suspended timber floors, for proper protection the dpc should be inserted below plate level. Provided that the outside ground level permits it, this can be done with the chain saw working from the outside only, without the need to take up and relay the floor. Where this is not practicable, inserting a dpc in the first bed joint above floor level is very much a second-best solution; if it is adopted, it would be prudent to ensure that the wooden skirting will not be affected by damp and to increase the ventilation in the under-floor space.

An effective barrier against rising damp will be afforded only when the damp-proof course is taken around all internal and external walls, including chimney breasts if possible. With the chain saw, chimney breasts can be cut without removing the firebacks, except where fireplaces are back-to-back; in that case, or if other tools are used, it is possible to insert narrow widths of membrane around the chimney breast with the fireback in position, and this protection combined with the heat from the fire may suffice to keep the chimney breast fairly dry.

The technique of inserting damp-proof courses has been applied both by contractors and by the Building Research Station to a wide variety of dwellings, some of which were in poor repair. It has been shown that if care and common sense are exercised, the method is safe to employ providing:

(i) Vertical joints are filled with mortar; otherwise, the brickwork may not bridge the slot.

(ii) Walls are reasonably plumb and are stiffened by returns or by partitions; this implies that a damp-proof course should not be inserted in, for example, long parapet walls which sometimes occur in buildings such as factories.

Of course all brickwork should be inspected before the work begins, especially as many old walls were built in two leaves, with the outer leaf much more carefully laid than the backing and with a broken bed joint. When the brickwork is in a very decayed condition, or there is little mortar in the vertical joints, many bricks in the course immediately above the saw may become loose and drop as the saw passes under them. Sometimes one or more courses of bricks must be completely removed in sections as the cut advances and, after the membrane is inserted, replaced as headers bedded in mortar and wedged with slate. Despite the very poor conditions, settlements after inserting damp-proof courses in such walls did not exceed 0.8mm. It is convenient for the bricklayers to have several sandbags filled with dry cement-sand mix which can be prepared quickly to meet emergencies.

A problem that may be encountered in this work is that of removing cement mortar. Frequently old brickwork has been repaired—quite mistakenly—by raking out the joints and repointing with a cement-sand mortar; elsewhere cement-sand plinths have been added in an attempt to combat rising damp. In both instances the cement mortar must be removed before cutting a slot, otherwise progress is slow and the saw blades are quickly damaged. Usually this is done with a cold-chisel, a slow process which often damages the brickwork. Both jobs may be tackled more easily with a grinding disc, either fitted in an angle grinder or as a separate unit driven by the portable engine, which would in any case be required to drive the chain saw (Fig. 9). A batten

Fig. 9. Grinding disc for removing cement mortar.

aligned with the bed joint provides a convenient method of supporting and guiding the tool; a neat cut without ragged edges, about 7mm wide, is produced when two carborundum-faced grinding discs are clamped to the same spindle. Slots up to 40mm deep are cut at a rate of about 0.3m per min. When a wall prepared in this way is subsequently cut, the slot is cut more rapidly and the cost of making good the plinth or rendering is avoided.

Injecting damp-proof courses

There are several proprietary techniques which are claimed to produce a moisture barrier in a wall. These generally involve injecting water-repellent substances into holes drilled at regular intervals along the base of the wall. The wall to be treated is almost certainly damp and may be nearly saturated. Consequently, it is difficult to form a continuous barrier by injection of water-repellent substances. The most successful techniques appear to be those employing a siliconate solution in water or a siliconate/latex mixture. These materials are miscible with water at the time of injection but change, after injection, to form a water-repellent band within the wall. The siliconate solution is allowed to permeate the wall under the influence of gravity; the siliconate/latex mixture is forced into the wall under high pressure.

The success of either technique depends on the care taken by the operatives to ensure continuity in the silicone barrier.

Electro-osmosis

Several systems are in use in this country and on the Continent which aim to prevent the rise of water above a certain position in a wall (or reduce it to an insignificant amount) by electrical means. In one system, electrodes of similar metal are placed in the wall and in the ground and connected together. In another system an electrical potential from an external source is applied between such electrodes. A related system has both sets of electrodes in the wall, one on the outside and the other at a higher level inside. Yet another system has electrodes of one type of metal in the wall connected to those of another type in the ground, the dissimilar metals producing an electrical potential by galvanic action. It has been reported that some of the systems employing dissimilar metals become inoperative after a few years because of electro-chemical corrosion of the electrodes. It would be expected that similar corrosion would occur when metal electrodes are used in systems employing an external electrical potential.

The scientific basis for these systems is a subject of controversy and the Station is investigating the efficacy of the system widely used in this country.

Drainage and drying

There are some types of walling where it may be impracticable to insert or inject a dpc, for instance natural stone walls having a rubble core. Or, in other cases, the provision of a new dpc may not be feasible economically. In these cases there are several measures which can be taken to reduce the amount of ground water reaching the wall or to increase evaporation from it. These may be effective alone or they may be decided on as preliminaries to a lining treatment.

One such measure is to lower the ground water level near the wall by site drainage; another is to expose an area of wall below ground or floor level so as to encourage evaporation there, for example, by laying corrugated asbestos cement sheeting against the wall. A different approach is to provide a moisture barrier, for instance, by providing impermeable surrounds such as a pathway, draining away from the wall.

Increased heating in the building will, of course, help to dry out a wall but its effectiveness greatly depends on how bad the rising damp is and what air circulation is available in the rooms affected by it. Neither heating nor ventilation is likely to be adequate where the decorative finish is impermeable, e.g. a gloss painted surface, and restricts the evaporation of water from the inner surface of the wall. Whatever form of heating is used needs to be maintained if the effects of rising damp are to be avoided permanently, and this measure will therefore involve a fairly high and continuing expense.

According to a proprietary system designed to increase evaporation, porous tubes are inserted into the wall to encourage evaporation at the outer surface. Removing some bricks from the wall may have a similar effect, though in both these methods it is difficult to see how the increase in evaporation is sufficient to be of real benefit. Better ventilation of the external wall surface will also increase evaporation, and clearing the nearby trees and shrubs may assist in some cases.

Methods to conceal rising damp

Wall linings

Various treatments are available that will ensure that the decorations are kept out of contact with the moisture and salts in a damp wall. None of these does anything to lessen the dampness in the wall and as already mentioned they may, if they impede evaporation from the wall, actually drive it higher. It is therefore recommended that the wall surface up to 0.5m above the highest affected parts should be treated, and suitable precautions must be taken if there is any risk of the dampness reaching the first floor timbers. To limit this rise it is best to use porous decorations, such as wallpaper, distemper or emulsion paint.

The cheapest and in some ways the best method, which also provides a measure of thermal insulation, is to line with plasterboard on timber battens. To prevent rotting, the battens should have been

Fig. 10. Combined grinding disc and dust extractor in use.

pressure-impregnated with a non-staining inodorous preservative. To prevent mould growth in the cavity, the wall surface and back of the lining board should be treated with a fungicide.

Two other methods introduce an impervious layer between the wall and the new plaster and decoration. They both involve stripping off the old plaster first. One uses a backing of corrugated pitch or bitumen lathing which is mechanically fixed to the wall and plastered over. This is an effective method, provided that the fastenings are made efficiently. In the other method the wall is coated with a rubber/tar or bitumen preparation. The old plaster is first stripped, and the wall is coated with the preparation. This also is an effective treatment but the preparation may lose adhesion after about ten years. In both these cases plastering with an undercoat that does not shrink on drying is necessary; a mixture of one part browning plaster to one-and-a-half parts clean sand is usually recommended.

As a temporary expedient only, a cheaper alternative exists where the old surface is fairly sound; it may be possible to fix a barrier layer on the old plaster and to decorate over that. Suitable layers are pitch paper fixed with bitumen, lead foil fixed with a mixture of red lead and gold size, or with old fat varnish, or aluminium foil backed with a heat-sensitive adhesive which can be ironed on to the old surface.

Plastering

Apart from the proprietary treatments described above that provide an impervious backing to the new plaster and decoration, there are several plastering methods that may effectively impede damp penetration.

Where a dpc has been inserted in a wall, dampness, as already mentioned, will persist as long as hygroscopic salts remain in the wall surface; replastering will therefore be necessary to a height of, say, 0.5m above the affected area. This, however, should be delayed as long as possible to allow soluble salts to move from the masonry into the plaster.

Replastering may also be effective, for a while, where no dpc has been inserted, in preventing the dampness from reaching the wall surface. For this purpose the old plaster is stripped, the wall surface is keyed by raking out the mortar joints squarely to a depth of 10mm and an undercoat of 1 : 3 cement/sand, possibly with an integral water-proofer, or an undercoat of aerated 1 : 6 cement/sand, is applied, followed by a plaster finish coat. As a further precaution, decorations should be porous as mentioned above.

Many old walls of soft brick or porous stone are not sufficiently strong to restrain the shrinkage of a dense 1 : 3 cement/sand rendering, and in these circumstances the use of a weaker mix of 1 : 6 cement/sand plasticized by means of an air-entraining agent is preferred to a cement/lime/sand undercoat of similar strength. The aerated mix is likely to be the more effective in impeding the passage of moisture and salts.

Insertion of damp-proof courses

Since this Digest was first published, there has been a further development in slot cutting technique prior to dpc insertion, resulting in increased speed and economy. A patent has been applied for.

In this method the wall is cut with a grinding disc, driven by a portable motor via a flexible drive. The glass-fibre discs impregnated with carborundum are up to 24 in. (0.6m) in diameter and are provided with a shield connected to a vacuum extractor (Fig. 10). Because almost complete extraction of dust is ensured operatives can work without masks and in confined spaces. The machine is hand-held and the rate of cut is about 0.3m per minute for a 105mm wall and about half this rate for a 220mm wall. The method has been shown to be satisfactory both with irregularly coursed brickwork and with brickwork set in cement mortar.

Further details of contractors using this method may be had on application to any Building Centre.

Choosing specifications for plastering

The greater range of plaster finishes now available and of backgrounds able to receive them, and also the development of mechanised plastering techniques, have all tended to complicate the choice of suitable plastering specifications. The purpose of this Digest is to show, with the help of tabulated summaries, how the main factors affecting this choice can be assessed in logical sequence.

Plastering is intended to conceal unevenness in the background, and to provide a finish that is smooth, crack-free, hygienic, resistant to damage and easily decorated. It may also be required to improve fire resistance, to provide additional thermal and sound insulation, to modify sound absorption or mitigate the effects of condensation; but where these are of secondary importance, as is usually the case, the selection of a plastering specification will depend first upon the surface finish desired. The stages in choosing plastering specifications will therefore be considered from this starting point.

Stage 1

Choosing the finish

The quality of the plaster finish, whether it should be hard or soft, smooth or textured, and also the choice of the decorative finish and the time likely to be available for drying out the plaster before it can be applied, will usually be the main considerations. The finish may have to meet special requirements concerning sound absorption or condensation as well. Table 2 lists various plaster finishes and sets out the more important properties likely to determine their choice. For ease of reference in Table 1, some of the finishes have been designated as Categories I, II and III in Table 2.

Stage 2

Number of coats

Until recently, three-coat plasterwork was applied to most backgrounds: a first, levelling undercoat, followed by a second undercoat to provide uniform suction for the finishing coat. Nowadays, the surface of brickwork and blockwork is such that two-coat work is often possible, though it is more important than ever, given a regular surface, to avoid extreme variations of suction. Some backgrounds, such as plasterboard and smooth concrete, have level surfaces and uniform suction which make single-coat plastering possible. In this context, the references in Table 1 to 'sufficiently level' backgrounds require explanation. Different plasters are able to tolerate surface irregularities to different extents; this is because of the different thicknesses of one-coat plasters normally applied. Thin-wall plasters based on organic binders are usually no more than 1·5 mm thick to be economic. Holes and scratches in the background must therefore be filled in by hand before a finish of this thickness can be applied. Thin-wall plasters based on gypsum are thicker (up to 5 mm) and so are more tolerant of blemishes. Single-coat board finish gypsum plasters may range from 3 mm thickness on solid backgrounds (*see* Table 1) to 5 mm on plasterboard or plaster lath.

It is sometimes felt that two-coat work is superior to single-coat work, but this is not necessarily true. The use of single-coat plastering on plasterboard tends to increase the risk of cracking at the joints, as compared with two-coat plastering, but it is better on surfaces of smooth concrete or painted surfaces treated with a bonding agent—the thinner plaster is less likely to fail in adhesion as a result of differential thermal movement. Another advantage of single-coat plastering on dry backgrounds such as precast concrete slabs or boards, is that it considerably reduces the quantity of water introduced during the later stages of construction. The thin plaster dries very quickly and there is no restriction on the type of decoration.

Stage 3

Plaster undercoats

So long as proper attention is given to the selection of plaster undercoats to match the suction of the background, its surface irregularities and other important properties, a wide range of finishes is possible, whatever the background. Rarely does the plaster finish required determine the background.

Table 1 sets out the properties of backgrounds that determine which undercoats should be used, or whether single-coat work is possible. Notes on the use of Table 1 are given below, describing these properties in more detail; *the reference numbers that precede these notes correspond to the reference numbers in the table.*

(1) Strength of background

This is particularly important where undercoats based on cement are to be used; the strength of the background must be adequate to restrain their drying shrinkage. Undercoats based on gypsum plasters do not shrink appreciably on drying, so that unless the background itself is likely to shrink after these undercoats are applied considerations of strength are unimportant.

(2) Suction of background—strength and adhesion of plasters

Strength

(*2a*) The ratio by volume of plaster to Type 1 sand (BS 1198–1200 : 1955) in sanded browning plasters will be seen from Table 1 to vary from 1 : 3, for backgrounds of moderate to high suction, to $1 : 1\frac{1}{2}$ for low-suction backgrounds. The higher the sand content of the plaster, the greater the amount of water needed to make the mix workable. The 1 : 3 mix would give too weak a plaster unless the backing removes some of the water before the plaster sets. A high-suction backing reduces the water/plaster ratio—and hence increases the strength of the set material—to much the same level as that of a $1 : 1\frac{1}{2}$ mix applied to a low-suction backing.

(The alternative volume proportions of sand in sanded browning plasters given in Table 1, e.g. $2\frac{1}{2}$–3, relate to the variations in bulk density of proprietary plasters; recommendations of the appropriate manufacturer should be closely followed.)

(*2b*) When Type 2 sands are used instead of Type 1, the proportions of sand in all sanded browning plasters should be reduced by one-third below the figures given in Table 1. This is because finer sands require more water to obtain a workable mix. Too much water leads to a loss of strength in the plaster, hence the need to reduce the sand content if finer sands are used.

(*2c*) The proportions of gypsum plaster in ready-mixed lightweight gypsum undercoat plasters have been similarly designed to match the suction of particular backgrounds.

Adhesion

(*2d*) A background of very high suction, e.g. aerated concrete or insulating bricks, may absorb water so rapidly from the plaster mix that its adhesion to the background is weakened; also, there may be too little time in which to level the coat. To deal with

this problem by excessive wetting of the background before plastering is to invite risk of trouble later on from drying shrinkage, efflorescence, or spoiled decorations. It is better instead to use mixes which retain their water in contact with backgrounds of high-suction. Plasters based on lime can be improved by using a fat lime putty rather than dry hydrated lime, or by mixing in a high-speed mixer. Alternatively, proprietary plasters are available which derive their improved water-retaining properties from cellulose additives, and such additives may also be used to improve the adhesion and 'workability' of aerated cement-sand mixes.

(3) Key or bond

While high-suction backgrounds can have an immediate effect upon the adhesion of plaster, as discussed above, the continuing adhesion of the plaster is dependent upon the key or bond provided. The adhesion is threatened by the stresses which can develop between the plaster and the background as a result of differences in thermal expansion or drying shrinkage. The key or bond must therefore be adequate to contain these stresses, which can be such as to cause failure in plaster otherwise able to withstand a tensile load of 0.7 MN/m². The background, the key or bonding agent, and the plaster, can all be chosen to minimise risk of failure from this cause.

Thus, the greater elasticity of lightweight plasters helps to reduce stresses; the use of thin plaster coats is recommended because shear stress is directly proportional to thickness; and the good mechanical key of metal lathing or wood-wool should be fully utilised by working the plaster well into the interstices with the trowel.

With backgrounds offering little or no key, such as smooth concrete, the choice of undercoat is more restricted, but bonding agents applied to the background surface before plastering can improve adhesion in these instances considerably. The bituminous type of bonding agent is advantageous on walls, where it resists damp penetration, but it is not recommended for use on soffits. The PVA emulsion type is of value for dense concrete containing brick, granite or limestone aggregate, and for some structural lightweight concrete. (This type of bonding agent is also useful in controlling the suction of backgrounds or dry undercoats.)

(4) Drying shrinkage and thermal movement

These movements have been mentioned above, but some points of detail are better dealt with separately. As noted in Table 1, the shrinkage of clay brickwork or blockwork is low compared with that of concrete blockwork and other products based on cement. With the latter, therefore, it is particularly necessary to allow them to dry out as completely as possible before plastering.

The high drying shrinkage of wood-wool slabs causes little trouble when the slabs are fixed dry and

Table 1 Suitable plastering systems for various backgrounds
The numbers in parentheses correspond to the numbered notes in the text

Background					Suitable alternative undercoat plasters					
	Properties						Sanded browning plasters (2a) mix proportions—Class B plaster: Type 1 sand (2b)		Undercoats based on cement (7) or of lime-sand gauged with gypsum plaster*		
Type	Suction (2)	Key or bond (3)	Drying shrinkage (4)	Type of finish (6) and category (see Table 1)	Lightweight gypsum plasters (2c)		By wt.	By vol.	Cement: lime: sand (by volume)	Cement: sand (aerated by means of mortar plasticizers) (by volume)	Single-coat work
Solid background											
Normal clay brick- and blockwork	Moderate–high	Good, if joints well raked or bricks keyed; otherwise spatter dash or other bonding treatments may be needed	Negligible	I II III	— — Browning or multi-purpose	1 : 4½ 1 : 4½ —	1 : 2½–3 1 : 2½–3 —	1 : 1 : 6 1 : 1 : 6 or 1 : 2 : 9 —	1 : 6 1 : 6 or 1 : 8 —	Not suitable for this type of background
Dense clay brickwork (other than engineering brickwork) and blockwork, calcium silicate or concrete brickwork or blockwork	Low–moderate		Low–high for materials other than clay brickwork	I II III	— — Browning or multi-purpose	1 : 3 1 : 3 —	1 : 2 1 : 2 —	1 : 1 : 6 1 : 1 : 6 or 1 : 2 : 9 —	1 : 6 1 : 6 or 1 : 8 —	As above
No-fines concrete, open-textured lightweight aggregate concrete or blocks	Low–moderate	Good	Low–high	I II III	— — Browning or multi-purpose	1 : 3 1 : 3 —	1 : 2 1 : 2 —	1 : 1 : 6 1 : 1 : 6 or 1 : 2 : 9 —	1 : 6 1 : 6 or 1 : 8 —	As above
Aerated concrete slabs and blockwork	Moderate–very high	Good with suitable plasters	Low–high	I II III	— — Bonding or multi-purpose	1 : 3 1 : 3 —	1 : 2 1 : 2 —	1 : 1 : 6 1 : 2 : 9 —	1 : 6 1 : 8 —	Sufficiently level surfaces may be plastered with thin-wall plasters; or with board finish plasters on low-suction concrete
Clay engineering brickwork, dense concrete or closed surface lightweight concrete and blocks	Low–moderate, but concrete needs wetting or a bonding agent	Poor, bonding treatments may be necessary (3)	Low–high for materials other than clay brickwork	I II III	— — Bonding or multi-purpose	1 : 2 1 : 2 —	1 : 1½ 1 : 1½ —	Not normally suitable for this type of background (1)		Sufficiently level concrete surfaces may be plastered with thin-wall plasters or board finish plasters; a bonding treatment may be necessary for the latter
Surfaces such as glazed bricks, tiles or painted surfaces, all treated with bonding agents	Low	Adequate with suitable plasters	Usually negligible	I II III	— — Bonding or multi-purpose	1 : 2 1 : 2 —	1 : 1½ 1 : 1½ —	Not suitable for this type of background (1)		Sufficiently level surfaces may be plastered with board finish plasters

Background					Suitable alternative undercoat plasters					
	Properties					Sanded browning plasters (2a) mix proportions—Class B plaster: Type 1 sand (2b)		Undercoats based on cement (7) or of lime-sand gauged with gypsum plaster*		Single-coat work
Type	Suction (2)	Key or bond (3)	Drying shrinkage (4)	Type of finish (6) and category (see Table 1)	Lightweight gypsum plasters (2c)	By wt.	By vol.	Cement: lime: sand (by volume)	Cement: sand (aerated by means of mortar plasticizers) (by volume)	
Slabs										
Wood-wool	Low	Good	Usually fixed dry, but may be high when used as permanent shuttering	I II III	— — Metal lathing or multi-purpose	1 : 3 1 : 3 —	1 : 2 1 : 2 —	— 1 : 2 : 9 —	— 1 : 8 —	Not suitable for this type of background
Strawboard	Low	Poor, requires treatment with bonding agent, or metal lathing	Fixed dry	I II III	— — Bonding or multi-purpose on p.v.a. bonding coats	When metal lathing is stapled to this background to provide a key, the mixes recommended below for metal lathing may be applied				Sufficiently level surfaces trated with p.v.a. bonding agents may be plastered with board finish plasters
Boards										
Plasterboard, plaster lath and insulating fibre-board	Low	Adequate with suitable plasters	Fixed dry, but moisture movement of fibre-board may be high	I II III	— — Bonding or multi-purpose	1 : 2 1 : 2 —	1 : 1½ 1 : 1½ —	Not suitable for this type of background (1)		Sufficiently level surfaces may be plastered with board finish plasters but in the case of expanded plastics the boards should be fully bonded to a firm background
Expanded plastic boards	Low	Adequate with suitable plasters	Fixed dry	I II III	— — Metal lathing or multi-purpose	1 : 2 1 : 2 —	1 : 1½ 1 : 1½ —			
Lathing										
Metal lathing	Low	Good	None	I II III	— — Metal lathing or multi-purpose	1st 2nd 1 : 2 1 : 3 1 : 2 1 : 3 —	1st 2nd 1 : 1½ 1 : 2 — 1 : 1½ 1 : 2 1 : 2 : 9 —		— 1 : 8 —	Not suitable for this type of background
						Mixes for first undercoat should contain hair				

* Undercoats of lime-sand gauged with gypsum plaster (1 volume of Class B browning plaster; 3 volumes of lime; 9 volumes of sand) are suitable to receive weak lime finishes and may be substituted whenever a 1 : 2 : 9 cement-lime-sand undercoat is indicated in the table.

Table 2 Characteristics of plaster finishes

Category in Table 1	Plaster finish	Surface hardness and resistance to impact damage	Other surface characteristics	Restrictions on early decoration	Shrinkage or expansion	Remarks
	Lime plasters Lime	Weak and very easily indented	Level, open-textured (depending on sand), absorbs condensation	Initially only suitable for permeable finishes that are unaffected by alkali	Shrink on drying, but shrinkage is reduced by addition of fine sand	Slow hardening. Only apply on dry undercoats
II	Gauged lime	Resistance to damage increases with proportion of gypsum plaster	Similar to above, but smoother finishes obtainable		Shrinkage is restrained by the gypsum content provided that over-trowelling is avoided	Only apply on dry undercoats
	Gypsum plasters Class D (Keenes)	Very hard and resistant to damage	Very level and smooth. Particularly suitable for low-angle lighting conditions	None, except on undercoats containing cement or lime, or unless lime is added to the finishing coat	Expand during setting. Subsequent movements usually small, but too rapid drying can lead to delayed expansion	Sets slowly and so allows ample time for finishing to a smooth surface. Should not be allowed to dry too quickly
	Class C (anhydrous)	Hard and resistant to damage	Slightly less smooth than Class D			
I	Class B (hemihydrate)	Sufficiently hard and resistant for most normal purposes, but weakened by additions of lime	Sufficiently smooth and level for most purposes		Expand during setting, though extremely slightly with board finish plasters. Subsequent movements are small	Sets quickly, should be allowed to dry as soon as possible
III	Lightweight	Surface hardness similar to Class B plasters. Ease of indentation varies with the type of lightweight undercoat, but resilience tends to prevent serious damage	Sufficiently smooth and level for most purposes	None, but the higher water content of lightweight undercoats makes these somewhat slower to dry than sanded gypsum plaster undercoats	Expand during setting. Subsequent movements usually small and easily restrained by background	Maximum fire resistance Lightweight plaster surfaces warm up more quickly than others and so help to prevent temporary condensation
	Cement-lime-sand 1 : 0–¼ : 3	Very strong and hard	Wood float finish	Initially only suitable for permeable finishes that are unaffected by alkali	Shrink on drying, but surface cracking can be minimized by avoiding over-working	Suitable for damp conditions
	1 : 1 : 6	Strong and hard	Wood float finish			
	Single-coat finishes Board finish gypsum plasters (Class B)	Surface hardness similar to Class B above, but resistance depends on background	On suitable backgrounds, similar to Class B above	None; finish dries very quickly	Extremely small expansion on setting. Subsequent movements small	
	Thin-wall finishes Based on gypsum	Softer than board finishes	Smooth and level on sufficiently level backgrounds	Dries very quickly. No restrictions when dry	Extremely small expansion on setting. Subsequent movements small	
	Based on organic binders	Moderately hard. Resistance depends on background	Matt surface, closely following the level of the background	Dries very quickly. No restriction when dry	The very thin coats are restrained by the background	

are restricted by their fixing. But if used as permanent shuttering for casting concrete the slabs get wet, and subsequent drying shrinkage can cause serious cracking of the plaster unless this is strong enough to restrain the movement.

Thermal movements of backgrounds are rarely large enough to cause serious defects, except where concrete roofs and heated concrete floors are concerned. Precautions are discussed later to prevent cracking of plaster at the junctions of walls and ceilings.

Some backgrounds of smooth-surfaced concrete (*see* the previous note) have negligible coefficients of expansion compared with that of gypsum plaster. Differential thermal movement is then not only much greater than normal, but the smooth surface of the concrete offers little key to the plaster to restrain the movement. Bonding agents of the PVA emulsion type will strengthen the bond in these instances and also control the suction of the background if this is excessive. Alternatively, plasters with improved water-retaining properties can be used.

(5) Efflorescence

Ideally, backgrounds should be dry before plastering, or, if plastered, before decorations are applied. Any efflorescence can then be brushed off before the decorations can be spoiled. So long as the background remains dry there should be no further efflorescence.

Backgrounds are, however, frequently plastered before they have dried out, and if they are likely to contain efflorescent salts (as may clay brickwork and blockwork, for example) it is advisable to choose an undercoat that will present a barrier to them. For this purpose, undercoats based on cement are preferred to those based on gypsum plaster. Aerated cement/sand undercoats are more effective than cement/lime/sand undercoats. *See also* Note (7).

Stage 4

Plastering specification

Tables 1 and 2, with the assistance of the above Notes (1)–(5), should enable a particular type of finish from Categories I, II and III to be obtained on a given background. Some further notes are given here on the constitutions of these plasters and on the application of the other finishes described in Table 2.

(6) Type of finish

Category I. Any Class 'B' and 'C' plasters, neat, or gauged with not more than a third of their volume of lime.

Category II. A weak finish of lime putty gauged with a quarter to half of its volume of gypsum plaster, to which up to one volume of fine sand may be added.

Category III. Any of the proprietary finishes usually recommended for use on lightweight gypsum plaster undercoats.

Keenes. This is stronger than other gypsum finishes

and undercoats must be strong enough to take it. Traditionally, a 1 : 3 Portland cement/sand undercoat is used, but this is only suitable for strong backgrounds with good mechanical key, e.g. brickwork or no-fines concrete. A Class 'B' browning plaster : Type 1 sand undercoat gauged 1 : 2 by weight (1 : 1½ by volume) may be used on other backgrounds, but manufacturers of the plaster should be consulted as to its suitability for this purpose.

Cement-based finishes. These finishes should be applied only to *cement-based* undercoats of similar strength. Cement-lime-sand finishes (1 : 1 : 6) may be applied to similar undercoats on any of the backgrounds specified in Table 1 as being suitable for these, but the stronger 1 : 0–¼ : 3 mixes require correspondingly stronger backgrounds to restrain the shrinkage, e.g. brickwork or no-fines concrete.

Thin-wall plasters. These include the so-called Scandinavian plasters commonly applied by spraying or with a broad spatula, and the finishing plasters of high water retention designed for high-suction backgrounds.

(7) Alternative specifications for gypsum plaster or lime finishes (Categories I and II)

Table 1 shows that for several backgrounds there is a choice between two undercoats based on cement. and a third of sanded gypsum browning plaster, as alternative specifications with gypsum plaster or lime finishes. (In addition, undercoats of lime-sand gauged with gypsum plaster, in a volume ratio Class 'B' browning plaster to lime to sand of 1 : 3 : 9, can receive weak lime finishes and may, subject to the following precautions, be substituted for a 1 : 2 : 9 cement-lime-sand undercoat wherever this is indicated in the table.)

If the background is salt-free and likely to remain dry, gypsum has the advantage over cement-based undercoats in drying out and hardening quickly, and enabling application of the finishing coat to be made in less than 24 hours. However, the advantage of speed in plastering is lost if efflorescence delays or spoils decorations, and in these circumstances, as explained under 'Efflorescence' above, it is better to use cement-based undercoats. The choice then lies between cement undercoats plasticized by lime or by air-entraining mortar plasticizers. An aerated mix does not develop the high suction of lime-plasticized mixes and is a more effective barrier to efflorescent salts; but if aerated mixes are applied in cold and humid conditions they may be slow to develop the suction needed to ensure satisfactory adhesion of the finish. Further, aerated mixes are less easily pumped for spray-plastering purposes than cement/lime/sand mixes.

Design of plasterwork to prevent cracking

The causes and prevention of cracking on plaster have been largely covered in the preceding notes. To prevent cracking over the joints between board and slab backgrounds other precautions are necessary, however. The plaster there can be reinforced with jute scrim or metal mesh and should be of adequate strength and thickness. Where boards are fixed in buildings likely to experience large fluctuations in humidity it is better to isolate the boards from any movement in their timber supports by means of special clips.

Discontinuities in backgrounds, e.g. where brickwork abuts concrete, are likely to cause cracking because of differential thermal expansion and drying shrinkage. At these positions the plaster should, if possible, be cut through, and the cut may then be masked by a cover strip fixed to one side. Alternatively, the plaster can be isolated from the background for some distance on both sides of the discontinuity and held instead on metal mesh or lathing. For pier and panel construction this isolation may be provided by means of building paper, and the reinforcing metal lath is fixed to the panels on both sides.

Junctions between walls and ceilings can be covered by cornices that will tolerate slight movements without cracking. Alternatively, a straight cut can be made through the wall plaster to isolate it from the effects of shrinkage of the joints.

Standards and Codes

The various factors that should be considered when selecting a plastering specification are dealt with in greater detail than in this Digest in the British Standard Code of Practice, CP 211 : 1966, Internal plastering.

Further information may also be obtained from the following publications:

British Standards

BS 12	Portland cement
BS 146	Portland-blastfurnace cement
BS 405	Expanded metal (steel)
BS 890	Building limes
BS 1142	Fibre building boards
BS 1191	Gypsum building plasters
BS 1198–1200	Building sands from natural sources
BS 1230	Gypsum plasterboard
BS 1317	Wood laths for plastering
BS 1369	Metal lathing (steel) for plastering
BS 2552	Polystyrene tiles for walls and ceilings

Several Building Research Station Digests also give information on plastering, and reference should be made to the Digest Index for these. Digest 95 (first series), which contains a section on flaking of plaster finishing coats, is now out of print, but a note on this subject is available from the Station on request.

Double glazing and double windows

Two diverse environmental requirements—one, a desire for improved heat insulation, the other, the need for protection against external noise—have stimulated the production and use of various forms of double glazing.

The essential difference between double glazing designs to meet these two needs lies in the width of the air space. For heat insulation, air-space widths as small as 6 mm are of some value; for sound insulation, air-space widths of not less than 100 mm, and preferably more, are essential, but there are other factors which are also discussed in this digest.

Heat insulation

In response to the general desire for improved heat insulation in heated buildings, various double glazing and double window systems are available, ranging from simple 'do it yourself' sealed *in situ* systems to the more sophisticated factory produced hermetically sealed double-glazing units and from double-rebated frames to openable coupled casements and sashes or separate secondary windows (see Figs 1–5). Some of the practical considerations peculiar to the various types are discussed later.

Width of air space

For effective heat insulation, the optimum width of air space for vertical double glazing varies according to the mean temperature of the air space and the temperature difference across it, but it is usually taken as 20 mm. Figure 6 shows that, for all practical purposes, the optimum is maintained down to an air-space width of about 12 mm. Below this width, the heat transmission becomes progressively greater until it approaches the figure for single glazing. For air-space widths greater than 20 mm, the heat transmittance remains practically constant.

Table 1 Thermal transmittance (U) in W/m² °C through glazing (without frames)

Glazing system	Degree of exposure:		
	sheltered	normal	severe
Single	5·0	5·6	6·7
Double air space 3 mm wide	3·6	4·0	4·4
6 mm	3·2	3·4	3·8
12 mm	2·8	3·0	3·3
20 mm (or more)	2·8	2·9	3·2
Triple each air space 3 mm wide	2·8	3·0	3·3
6 mm	2·3	2·5	2·6
12 mm	2·0	2·1	2·2
20 mm (or more)	1·9	2·0	2·1

Comparative performance of double glazing

Given air-space widths of not less than about 12 mm, the heat insulation of single-glazed windows can be improved appreciably by double glazing; in general terms, an air-space between two layers of glass halves the thermal transmittance. Table 1 shows the comparative thermal transmittance ('U') values for double glazing for four widths of air-space and three grades of exposure. For comparison, the U-values for single glazing and triple glazing are included.

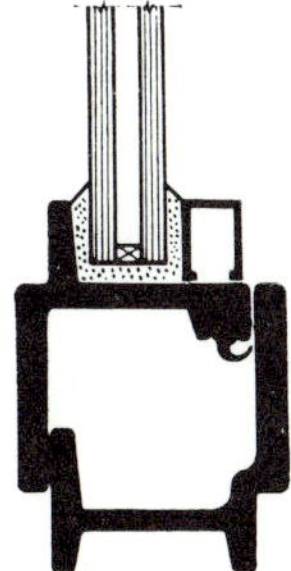

Fig 1 Typical factory-sealed double glazing units

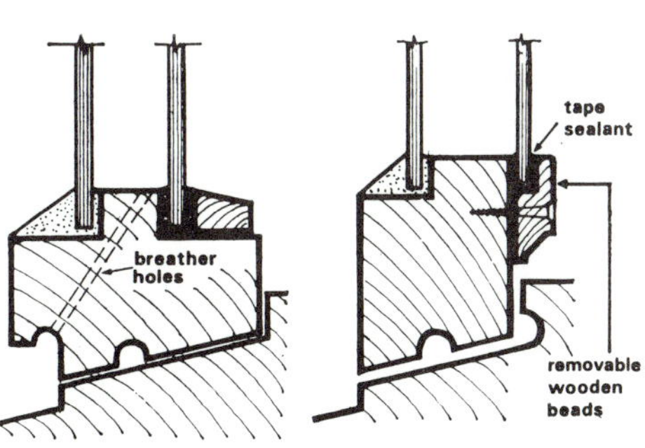

Fig 2 Typical glazed *in situ* double glazing

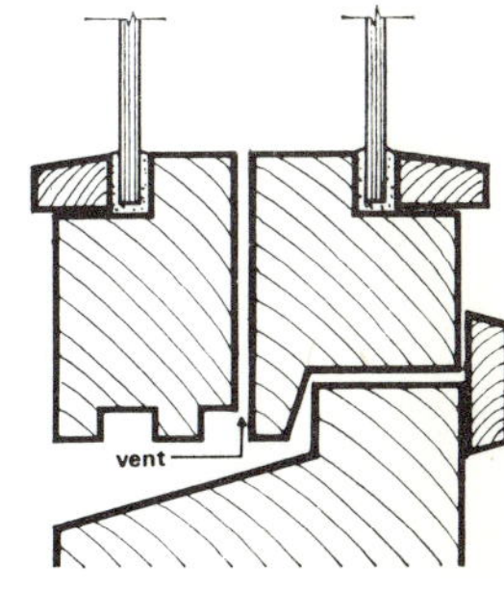

Fig 3 Typical double windows, coupled type

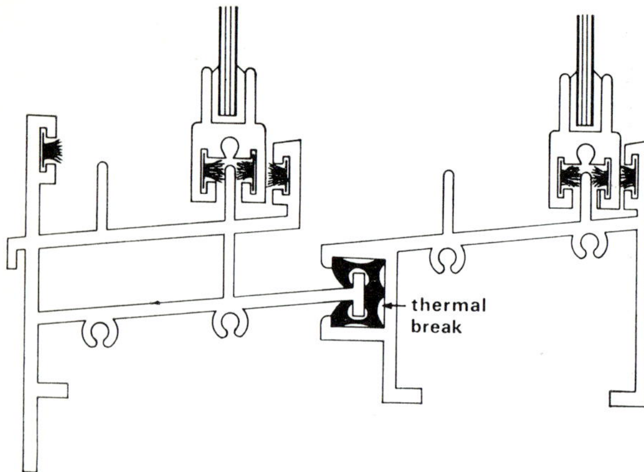

Fig 4   Horizontal sliding type, metal

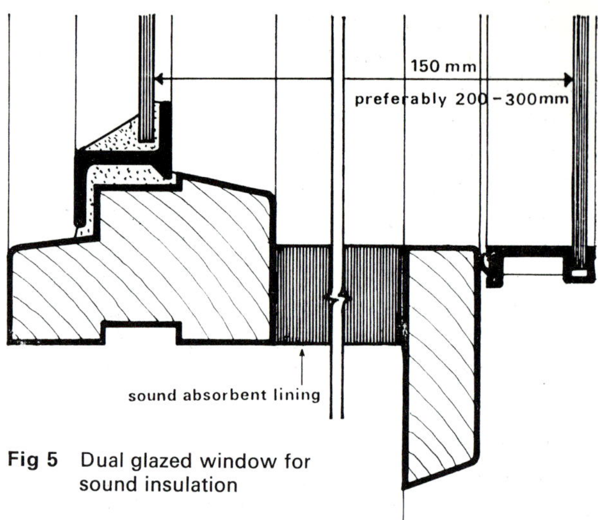

Fig 5   Dual glazed window for sound insulation

The above data relate to transparent and translucent glasses. For glazing systems using heat-reflecting glasses with metallic surface coatings (not laminates) of low heat emissivity, the metallic coating improves the insulation against heat loss. For double-glazing units, the improvement in insulation provided by the unit may be as much as 30–40 per cent.

Effect of window frames

The U-values in Table 1 are for the glass only and do not allow for the effect of the surrounding frame and glazing bars. In wood windows, the thermal resistance of the frame can contribute usefully to the insulation value of the whole window but the reverse is true of metal windows (because of the relatively high conductance of the metal frame) unless thermal breaks are introduced to offset the 'cold bridge' effect. Some proprietary metal double windows now incorporate insulating plastic insets in the frame section to break the thermal continuity (Fig 4). For heat loss calculations, it is often useful to know the average U-value for the window as a whole, and unless the frame section is complex in profile, its effect on the average thermal transmittance can be worked out on a proportional area basis. Table 2 gives comparative U-values for a number of typical wood and metal windows.

Tables 1 and 2 are derived from the IHVE 'Guide'.[1] The three categories of exposure are defined as:

sheltered Up to third floor of buildings in city centres

normal Most suburban and country premises: fourth to eighth floors of buildings in city centres

severe Building on the coast or exposed on hill sites: floors above the fifth of buildings in suburban or country districts: floors above the ninth of buildings in city centres.

Loss through windows in relation to the whole

To assess the value of double glazing in terms only of heat saving under steady-state conditions, the heat losses through the glazing must be considered in relation to the overall heat losses. A broad comparison can be made by taking the average thermal transmittances for the different parts of the fabric, multiplying them by the appropriate areas to give the comparative rates of conduction heat loss per degree C difference of temperature between indoors and outdoors and then adding the ventilation heat loss. In a two-storey semi-detached house of 100 m² gross floor area, the component rates of heat loss might be:

Conduction heat loss	Average U-value W/m² °C	Area m²	Rate of heat loss W/°C
Roof	0·50	× 50 =	25
Ground floor	0·76	× 50 =	38
Unglazed walling	1·50	×100 =	150
Windows (single-glazed)	4·30	× 17 =	73

Total rate of conduction heat loss through fabric 286

Ventilation heat loss (assuming 1 air change/hour)

$$= 240 \text{ m}^3 \times \frac{\text{W/m}^3 \text{ °C}}{3} = 80$$

Total   366

Thus the single-glazed windows account for about 20 per cent of the total heat loss including ventilation, or about 50 per cent of the conduction loss through the unglazed external walls. If double-glazed windows with an average U-value of 2·5 W/m² °C are substituted for the single-glazed windows, the rate of heat loss through the windows would be reduced to about 42 W/°C, and the corresponding total rate to about 335 W/°C. The heat loss through the windows would thus be reduced to about 13 per cent of the total.

Table 2 Thermal transmittance (U) in W/m² °C through typical windows (including frames)

Window type	% total window area occupied by frame	Degree of exposure: sheltered	normal	severe
Single-glazed:				
metal casement*	20	5·0	5·6	6·7
wood casement	30	3·8	4·3	4·9
Double-glazed:† metal horizontal sliding window with thermal break	20	3·0	3·2	3·5
wood horizontal pivot window	30	2·3	2·5	2·7

* Metal frame assumed to have a similar thermal transmittance to that of the glass
† With 20 mm air space

Double glazing and comfort

Apart from the reduction in heat loss gained by using double glazing, the increase in surface temperature of the glass facing the room may improve the comfort particularly of persons sitting or working near the windows; the discomfort caused by radiation losses and cold convection currents from the cold surface of single glazing is well known. In general, the surface temperature of the glass facing a heated room in a double glazing system in winter will be some 4–7°C higher than that of single glazing for the same internal and external temperature conditions (Fig 7).

Condensation

An explanation of the conditions which lead to condensation is given in Digest 110.[2] Condensation can occur on the surface of the glass and also on the frame, particularly if it is metal. The condensation on the glass is more troublesome because of the relatively large surface area and because it interferes with the view out. Where condensation on the glazing persists for long periods the run-off of excess moisture can lead to deterioration of the bottom member of the frame and spoil the appearance of the window reveals and areas of wall below the sill. Double

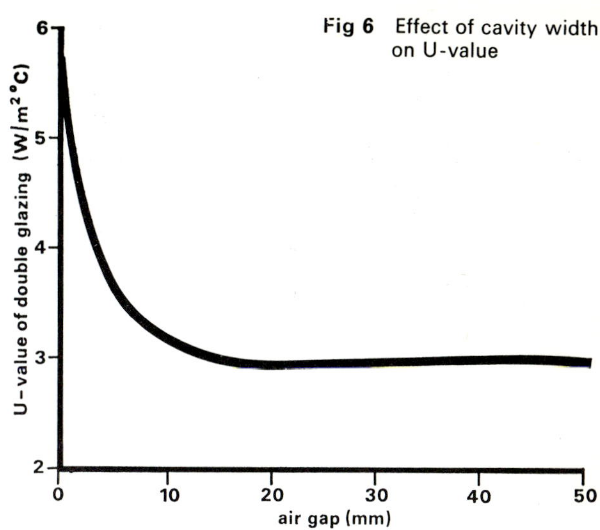

Fig 6   Effect of cavity width on U-value

glazing can reduce the risk of condensation on the glass because the surface exposed to the room is warmer than single glazing and has a better chance of being above the prevailing dew-point temperature. In practice, this will depend on the relative humidity, and in buildings where high humidities occur, coupled with low standards of heating, there is no guarantee that double or even triple glazing will

Fig 7   Thermal gradients   Temperature conditions referred to in text :   (a) outside   0°C   (b) outside   0°C
                                                                                       inside   12°C   inside   20°C
                                                                                       dewpoint   6°C   dewpoint 14·5°C

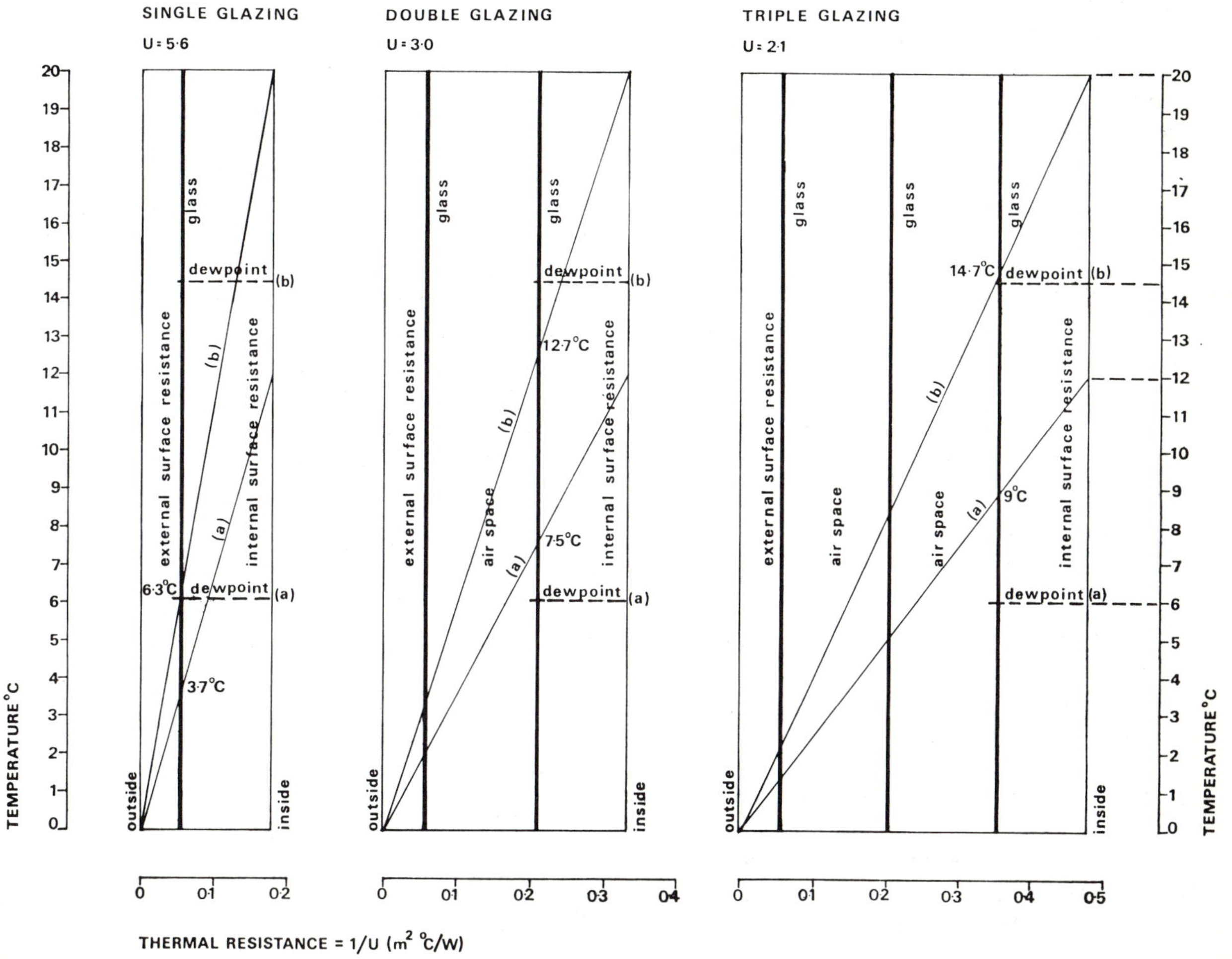

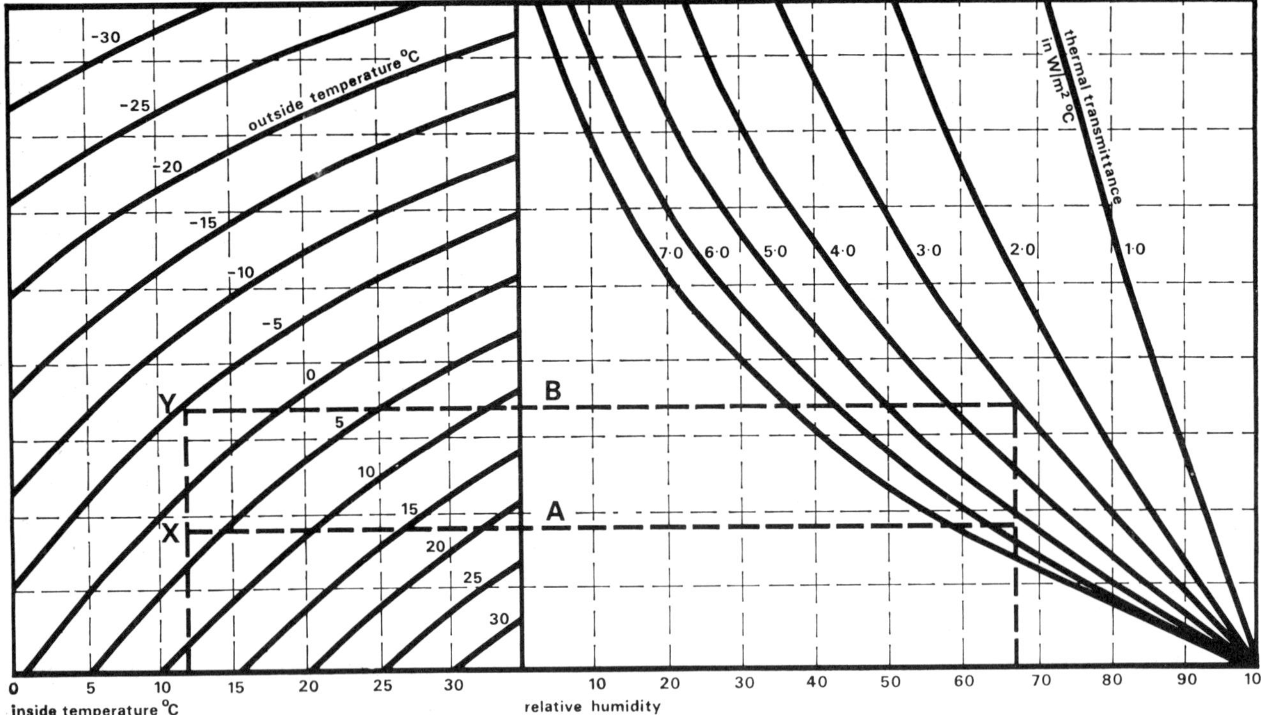

Fig 8 Condensation prediction chart

always avoid condensation on the glass. For instance, assuming an internal temperature of 12 °C, an outside temperature of 0 °C and a dew-point temperature of 6 °C (typical of conditions in a bedroom), condensation would occur on the surface of single glazing, but not on double or triple glazing (see Fig 7). Raising the room temperature to 20 °C and hence the glass surface temperature, for the same humidity conditions, there would still be a risk of condensation on single glazing. For conditions of high humidity, such as might occur in a domestic kitchen, assuming an internal temperature of 20 °C, an outside temperature of 0 °C and a dew-point temperature of 14·5 °C, condensation would occur even on double glazing.

If the relative humidity and indoor and outdoor temperatures are known, the possibility of condensation on glazing can be predicted from Fig 8. This gives comparative dew-point curves for a range of U-values which embraces those given in Table 1 for single, double and triple glazing. The chart applies only to steady-state conditions but it can be used for a first check on the likelihood of condensation. As an example, the broken line **A** on the chart indicates the conditions of relative humidity 67 per cent, thermal transmittance 5·6 W/m²°C (for single glazing) and an inside temperature of 12 °C; condensation is likely to occur when the outside temperature drops to about 3 °C (point 'X' on the chart). The line **B** shows the same temperatures and relative humidity plotted against a thermal transmittance of 3·0 W/m² °C (for double glazing); the outside temperature must now drop to nearly –5 °C (point 'Y') to cause condensation on the glass.

Sound insulation

The simplest way of increasing the sound insulation of an element of structure is to increase its mass; in a general sense this is true for windows, but with such lightweight components a substantial improvement can be obtained by double-leaf construction, as shown in Table 3 (taken from Digest 128[3]), provided that the windows are sealed, for example by weather-stripping, and that the air space is wide enough to give the required insulation at the lower frequencies.

Even small gaps in double windows can impair the sound insulation and, therefore, if at all possible, one leaf of a double window should be a fixed light system. If both leaves have openable lights in them, all the openable lights should be weather-stripped. For rooms subject to high humidity, it may be necessary to seal the inner leaf of glazing to prevent condensation in the air space as discussed on page 97.

Table 3 Sound insulation of windows

Description	Sound Reduction (av 100–3150 Hz)
Any type of window when open	about 10 dB
Ordinary single openable window closed but not weather-stripped, any glass	up to 20 dB
Single fixed or openable weather-stripped window, with 6 mm glass	up to 25 dB
Fixed single window with 12 mm glass	up to 30 dB
Fixed single window with 24 mm glass	up to 35 dB
Double window, openable but weather-stripped, 150–200 mm air space, any glass	up to 40 dB
Double window in separate frames, one frame fixed, 300–400 mm air space, 6–10 mm glass sound-absorbent reveals	up to 45 dB

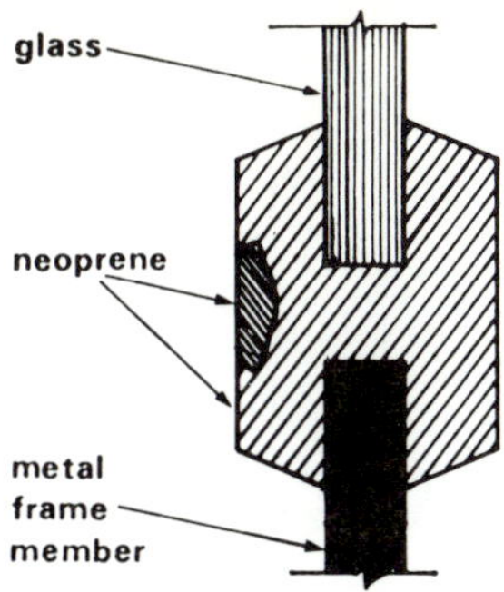

Fig 9 Flexible mounting of glass in neoprene gasket

Width of air space

The optimum air-space width for thermal insulation is about 20 mm, but this is too small to be of any practical advantage for sound insulation. For protection against road traffic noise, in which the low frequency components predominate, a minimum air-space width of 150 mm is recommended and preferably more—say 200–300 mm whenever it is economically obtainable. Unless an air space at least 100 mm wide can be provided, practical insulation against road traffic noise may well be obtained more effectively by heavy single glazing selected to give the same average performance.

Room ventilation

Table 3 shows that to obtain a useful improvement in sound insulation windows must be closed. A room with permanently closed double windows will usually need some form of mechanical ventilation through a duct system. The ducts should be lined with acoustic absorbent material to give sound attenuation comparable with that of the insulation provided by the building enclosure. For the ventilation of individual rooms there are fan-operated ventilating cabinets with sound attenuation designed to match the sound insulation provided by double windows.

Absorbent linings and flexible mountings

The provision of sound-absorbent linings, eg acoustic tiles, on the reveals between double windows can give marginal improvements in sound

Table 4 Effect of comparative area of window on the sound insulation of external walls rated at *40 dB and †50 dB respectively

	Average sound insulation (dB)			
	20 dB window		40 dB window	
Percentage of glazing area to wall area	40 dB wall	50 dB wall	40 dB wall	50 dB wall
0 (windowless wall)	40	50	40	50
10	30	30	40	47
25	26	26	40	45
33	25	25	40	44
50	23	23	40	43
75	21	21	40	41
100 (fully glazed wall)	20	20	40	40

* Equivalent of walls weighing about 120 kg/m², for example, a 50 mm dense concrete composite panel backed with thermal insulation material
† Equivalent of walls weighing about 480 kg/m² for example, a 215 mm solid or a 255 mm cavity brick wall

insulation. Flexible edge-mounting of the glass, for instance in neoprene gaskets, see Fig 9, can also help by promoting resonance damping. Recent experiments have shown that the latter treatment can improve the performance of double windows by at least 5 dB over most of the frequency range 100–3150 Hz, but the extra sound insulation can only be realised if the windows are well sealed; air leakage would tend to mask any benefit from the use of flexible mountings.

Windows in relation to the whole building

The improvement in sound insulation obtainable from double windows depends to some extent on the standard of insulation of the rest of the fabric and on their respective areas, although there is no simple relationship. The net insulation for windows and wall for a range of relative areas is given in Table 4, taken from CP153 : Part 3.[4] If the standard of the walling is of the same order as the windows, the percentage area of glazing will have no significant effect on the net sound insulation; if it is appreciably higher, the net insulation of the wall and windows will be less than the wall insulation. The converse is also true as the standard of insulation of the wall or other parts of the building fabric may be lower than that of the windows. For example, in an ordinary house, a tiled roof plus ceiling may have an average sound insulation of about 35 dB as against 40 dB or more for double windows. Thus if it is required to improve the protection against external noise the restricted performance of the roof could inhibit the improvement sought by double glazing.

Effect on light transmission

The transmission of light through glass varies with the angle of incidence and the properties of the glass, but for categorising different types of glazing, a percentage representing the light transmission at normal incidence is usually given by manufacturers. The light transmission through clear glass up to about 6 mm thick could be of the order of 85–90 per cent for single glazing, and about 70 per cent for double glazing. For daylight calculation purposes, it is usually convenient to calculate the daylight factor on the assumption of a single thickness of clear glass and then if necessary to apply a correction factor to allow for the reduced transmission of double glazing or other types of glazing material. Table 5 gives correction factors for a range of typical glazing materials in double glazing, the factor for single clear glass being taken as unity.

For comparison purposes, the transmission of radiant solar heat through various glasses is also given by manufacturers on a percentage basis assuming normal angles of incidence and taking into account the proportion of incident radiation reflected at the outer surface of the glass and the proportion absorbed by the glass and re-radiated.

Table 5 Recommended daylight correction factors and solar gain factors to allow for effect of various types of double glazing

Glasses	Recommended correction factor to apply in daylight calculations	Solar gain factor (S)	Alternating solar gain factor (Sa)	
			Heavy-weight building	Light-weight building
4–6 mm clear single glazing	1·00	0·76	0·42	0·65
4–6 mm clear both panes	0·90	0·64	0·39	0·56
6 mm surface-tinted bronze outer pane 4–6 mm clear inner pane	0·45	0·47	0·32	0·44
6 mm selectively absorbing green outer pane 4–6 mm clear inner pane	0·70	0·42	0·30	0·40
6 mm body-tinted grey outer pane 4–6 mm clear inner pane	0·35	0·39	0·28	0·37
6 mm heat-reflecting laminate (gold) outer pane — 4–6 mm clear inner pane		0·15	0·12	0·14

When calculating cooling loads in air-conditioned buildings or temperatures in uncooled buildings, it is necessary to know the solar gain through the windows; this varies with the angle of incidence and the properties of the glass. A simplifying concept, adopted in the IHVE Guide[1] allows for these varying characteristics under representative conditions and employs a *solar gain factor* to allow for the effect of a specified glass and any shading devices. In a broad sense, solar gain factors can be used as an approximate measure of the relative effectiveness of different types of glass and sun controls in combating solar gains. For calculating daily mean solar cooling loads or indoor temperatures, the solar gain factor **S** is used; for calculating the fluctuations of solar cooling load or indoor temperature about the mean, the factor S_a is used. Values of **S** and S_a for five types of double glazing and for single clear glazing are given in Table 5.

Practical considerations

Installation

Factory-sealed units Recommended glazing procedures for various types of factory-sealed double-glazing units are given in a booklet[5] issued by the Insulation Glazing Association. The booklet limits its advice to new work; for installing units in existing frames it recommends consultation with the manufacturer.

The glazing instructions relate to minimum depth and width of rebate to accommodate the thickness of the double-glazing unit, fixing of beads, type of bedding compounds, provision of setting blocks and distance pieces to ensure that the unit is correctly positioned in the frame and preparation of the frame. In addition, the booklet gives 'step by step' site glazing procedures appropriate to various kinds of bedding systems, and recommendations for the handling and

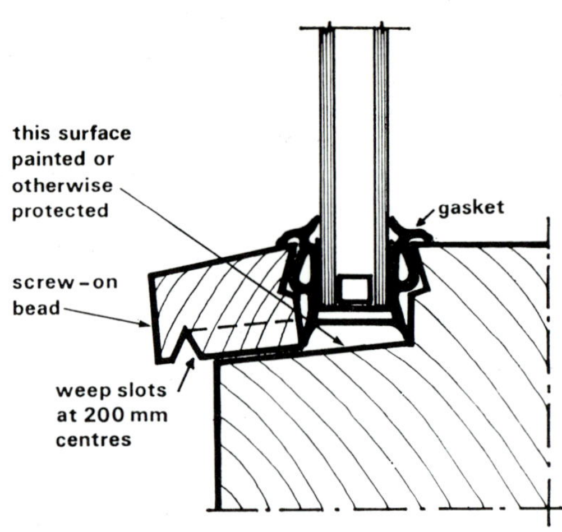

Fig 10 Drained glazing system

storage of units. Some manufacturers provide lists of approved glazing compounds.

Manufacturers' standard instructions are generally in line with the IGA recommendations, supplemented in some cases by additional guidance on the application of the bedding compounds and routine pre-glazing checks.

Failure to follow the glazing instructions of individual manufacturers may invalidate their warranty, which is usually for a period of ten years, but may be longer, and covers failure of the unit to function properly due to deterioration of the seal. If the seal fails, the unit cannot be repaired and can only be replaced by a new one.

Sealed units with fused all-glass edges are being increasingly used and a type using a welded glass-to-metal seal is also available, but there is still considerable use of the form in which flat sheets of glass are bonded to spacing strips and sealed. With the latter, it is important that the edges are kept dry, as continual wetting tends to destroy the bonded joints. For this reason, special care is needed to ensure that the glazing method and compound will protect the edges of the unit. The compound must also be compatible with the bonding materials of the unit. Attention has recently been directed to the value of drained glazing systems because of the difficulty of ensuring, particularly at the bottom edges, that no

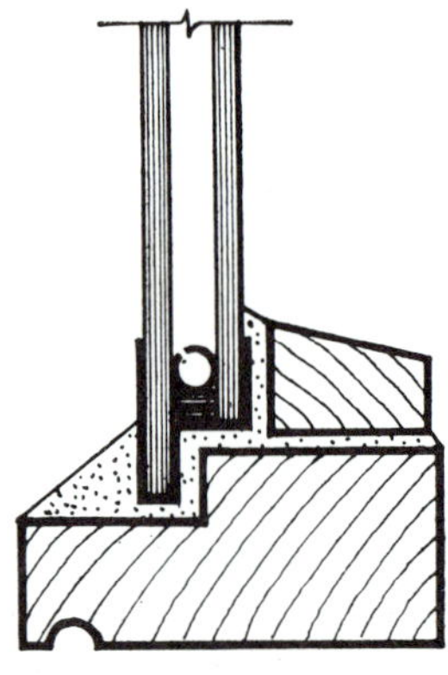

Fig 11 Typical stepped unit

crevices are formed where water may lodge between the unit and the bedding compound; a typical system is shown in Fig 10.

Some firms produce double-glazing units to fit into the rebates of standard sections, wood and metal, without the use of face beads. For frames that are too small, or unsuitable for enlarging to accept the thickness of standard units, stepped units are available (Fig 11).

Sealed double-glazing units with dark coloured, heat-absorbing or heat-reflecting glasses are liable to be heated strongly by the sun. If during a sunny spell the glass is put in partial shade, for example by other buildings or parts of the building, differential thermal movements may cause fracture. There is some evidence also of an increased likelihood of thermal stresses in double-glazed units when tinted glasses are backed by internal blinds of high reflectance. If these conditions are likely to arise the manufacturers should be consulted.

Single-frame double glazing systems sealed *in situ* Many attempts have been made to provide sealed double glazing by glazing to double rebated frames, or by fixing a second line of glazing with wood or plastic face beads, generally to existing frames (see Fig 2). Experience has shown that however well the glazing seals are made, the cavities cannot be expected to remain air-tight indefinitely; under the various movements and shrinkages that occur the seals may break down, water vapour (and dust) then finds its way into the air space and at some times condenses on the inside of the outer glazing.

if the inner panes are bead glazed it may be fairly easy to remove them for cleaning but it is often difficult to reseal the windows effectively and the 'slip-on' or 'press-on' type of added glazing, using proprietary compressible plastic edging strips, may be easier to reseal.

For wood windows, double-glazed *in situ*, the wood exposed to the air space should be painted or varnished to reduce the evaporation of moisture from the timber into the air space and breather holes to the outside should be provided on the basis of one 6 mm diameter hole for 0.5 m^2 of window. The holes should be plugged with material such as glass fibre insulation or nylon to exclude dust and insects. The frame material of metal or plastic windows will not contribute to a build-up of moisture in the air space. The double glazing should be completed under relatively dry conditions (preferably in cold weather) to prevent trapping humid air in the air space and the room-side glazing should be sealed as effectively as possible.

Coupled or sliding double sashes These may be pivoted, with openable coupled sashes, or they may slide, with pairs of sliders operating in the same frame; the former may be wood or metal but the latter

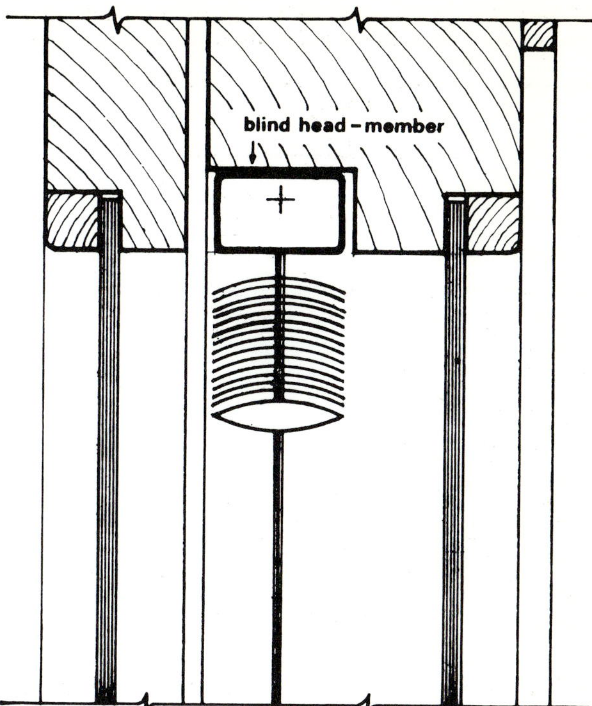

Fig 12   Coupled type double sash with venetian blind

are usually metal (Figs 3 and 4). To reduce the risk of condensation, the inner sashes should be well sealed when the sashes are closed together. For the same reason, in the coupled pivoted type a ventilating slot is commonly left round the periphery of the outer sash to ventilate the air space to outdoors when the sashes are closed. This does not seriously reduce the insulation of the window. In the sliding types, the air space is not usually ventilated because the inner surfaces are readily accessible for cleaning. The air space in the double-sash types is usually sufficient to allow retractable blinds to be installed between the inner and outer glazing (Fig 12). The blinds, usually venetian or pleated, are controlled by cords or rod action, but fully automatic control of the venetian types is available.

Double windows with separate frames When wide cavities are required for sound insulation, or it is not practicable to double glaze single-frame windows, secondary windows can be fixed to the existing frames or to ancillary frames. With the latter, a space can be left between the two panes, depending on the depth of the reveals, for fixing acoustic absorbent material to give additional sound insulation. There are proprietary secondary windows available in aluminium or PVC. They may be either hinged or sliding, to give access for cleaning; fixed lights can also be incorporated. Units can be supplied for screw fixing to the original window by householders, but specialist firms will supply and fix. The casement types usually have a compressible strip to provide a seal against the existing window or the ancillary frame; the sliding types usually have a wool pile seal in the track. Both types can be removed for cleaning or storage but large glazed units are awkward to handle. Where the secondary window has to provide access to the opening lights of the outer frame for cleaning it is necessary to ensure that the opening lights in both frames work conveniently together. In this respect, difficulties can arise with pivoted windows because of the unavoidable projection of their opening lights across the cavity.

If tinted glass is used to reduce solar heat gains, blinds of high reflectance in the inter-space may give rise to increased thermal stresses as already mentioned.

Maintenance

General recommendations on the durability and maintenance of windows are given in CP 153 : Pt 2.[6]

The glazing joints of factory-sealed double-glazing units should be checked at intervals recommended by the manufacturer and when repainting the window frames. For low-cost schemes using self-hardening compounds and without beads, a maintenance check every three years may be required; for silicone acrylic-based or polysulphide sealants a check every ten or twelve years may be sufficient. Where there are surface cracks, extrusion or retraction of the original compound, it should be raked out to a depth of 3 mm and replaced with a compound preferably of the same type. The joint should be shaped to direct water away from the glass on the outside and room side. Glazing beads should be re-secured as necessary and drainage holes in drained joint and gasket systems should be cleared of any dirt or other obstructions. For gasket glazing, maintenance checks at one-year intervals are recommended by specialist firms.

With sealed *in situ* systems it may be necessary to remove the glazing on the room side, clean the glass and reseal at intervals of two years or more depending on the effectiveness of the seal. The appearance of the inner surface of the outer glazing will show whether it is necessary to remove the inner glazing. The joints of wood frames, specially the bottom joints, should be checked for cracks, and when the glass is removed the opportunity should be taken to re-paint and re-varnish the wood exposed to the air space. The glass should be replaced under dry conditions, preferably a cold dry day, with the room amply ventilated to avoid entrapping warm moist air from the room.

For coupled or sliding double sashes or double windows with separate frames, maintenance will be as required for single windows but the fit of the inner opening lights should be checked to ensure that the seal on the room side is maintained as effectively as possible and if necessary weather-strip or flexible seals replaced by new.

Tests

There are currently no British Standard tests that are specific to sealed double-glazing units or double windows, but the IGA booklet [5] includes performance tests for glazing compounds for use with double-glazing and multiple-glazing units. BS 4254[7] includes methods of test for two-part polysulphide-based sealing compounds.

In general, for glazed *in situ* double-glazed single-frame windows and coupled or sliding dual sashes, tests related to the performance requirements for single-glazed windows would apply. The provisional recommendations published by the British Standards Institution in DD4[8] for resistance to wind loads, air infiltration and water penetration are relevant.

The thermal transmission of double windows is usually calculated, taking into account the width of the air space, the type of frame and the proportional areas of frame and glazing. A description of the method of calculation and the concept of standardised U-values is given in the IHVE Guide.[1]

For sound insulation measurements of windows, the standard methods given in BS 2750[9] apply.

References

1 IHVE Guide : Book A, Design Data (1971) ; Institution of Heating and Ventilating Engineers, London

2 BRE Digest 110 Condensation

3 BRE Digest 128 Insulation against external noise—1

4 British Standard Code of Practice CP 153 Part 3 : 1971 Windows and roof lights—Sound Insulation

5 Glazing requirements and procedures for double glazing units ; Insulation Glazing Association, London 1968

6 British Standard Code of Practice CP 153 Part 2 : 1970 Windows and roof lights—Durability and maintenance

7 British Standard BS 4254 : 1967 Two-part polysulphide-based sealing compounds for the building industry

8 British Standard Draft for Development DD4 : 1971 Recommendations for the grading of windows

9 British Standard BS 2750 : 1956 Recommendations for field and laboratory measurement of airborne and impact sound transmission in buildings

Further Reading

Beckett, H. E. and Godfrey, J. A.—*Windows*, Crosby Lockwood Staples : London (1974)

Prevention of decay in external joinery

An increasing number of reports of 'wet rot' in external joinery, particularly in new houses, suggests that the factors responsible are not as widely understood as they should be. This Digest examines the causes of decay and makes recommendations as to its prevention.

Incidence

Decay in exterior windows and doors has always occurred sporadically, but the number of reported cases of decay has increased appreciably over the past few years and is found even in joinery complying with current Codes of Practice and Specifications. Window joinery in newly built houses has given most cause for complaint; decay has sometimes become a serious problem within five or six years from the time of construction. Usually, decay has been noted earlier in the wetter, western areas of the British Isles though it has by no means been confined to these regions.

While decay may occur anywhere in opening lights and in frames permanently in contact with brickwork or blockwork, it is particularly marked in ground floor windows, especially kitchen and bathroom windows. The lower parts of these windows, ie the cills, the bases of jambs and mullions and the lower rails of opening lights, are particularly susceptible.

Type and symptoms

Almost invariably decay is of the wet rot type. This means that the fungi responsible, unlike the dry rot fungus, will not spread the rot to other woodwork in the building. Discoloration of the paintwork and 'cupping', or softening of the underlying wood, are the usual first symptoms of decay. Later, cross-cracking and dark discoloration of the wood itself may be observed. Sometimes in the later stages the wood becomes soft and stringy. There is no evidence of Baltic redwood having been infected prior to manufacture by the fungi responsible for wet rot.

Causes and prevention

Decay is caused by timber of low natural resistance becoming sufficiently moist to allow wood-destroying fungi to grow. Prevention of decay therefore depends on the selection of timber which is either naturally resistant to decay or which is otherwise protected by preservative treatment. Only heartwood has natural resistance to decay; sapwood * *must* be excluded where preservative treatment is not used.

---

* Sapwood cannot economically be excluded from joinery grade redwood.

Moisture control

Irrespective of the species of timber employed, adequate seasoning to remove moisture is essential before manufacture as joinery. A low moisture content restricts dimensional changes, permits satisfactory painting and contributes generally towards decay prevention. After seasoning, further access of moisture should be restricted:

by good design and careful fabrication of the joinery itself, particularly of the joints;

by storage indoors or raised clear of the ground and covered with waterproof sheets to protect the woodwork from exposure to the weather prior to final installation and painting;

by good detailing at the window/wall joint;

by thorough painting and regular maintenance;

by prompt attention to any necessary repairs.

Timber and its selection

Timbers for mass production joinery are chosen mainly because they are inexpensive and easily machined, not because they are particularly resistant to decay.

Baltic redwood (also known as red or yellow deal, red pine or fir) is the softwood most used for exterior joinery; its heartwood is more resistant to decay than its sapwood. The heartwood of Douglas fir is also used.

English oak is the traditional hardwood for exterior joinery, especially cills, in which greater resistance to decay is an asset. The hardwoods teak, utile, gurjun and agba have also proved acceptable. All these should be specified by name; merely to specify 'hardwood' can lead to the use of unsuitable timbers such as abura, beech, obeche and ramin (all with poor resistance to decay). The inclusion of perishable sapwood of an otherwise decay-resistant timber can be avoided by specifying 'Heartwood only'.

Suitable timber species and their properties are set out in Table 1 (on p. 105), extracted from CP 153 : Part 2 : 1970.

As Baltic redwood predominates in window joinery in the United Kingdom, the remarks which follow apply mainly to this timber. However, the principles of prevention and treatment of timber decay do not vary much between any of the different species used.

Preservative treatments

Care in design, construction, handling and maintenance can reduce but not eliminate the risk of decay. The obvious but often too costly remedy is to use naturally durable timber, excluding all sapwood. Alternatively, the risk of decay can be eliminated by adequate preservative treatment.

Recommended preservative treatments include:

Vacuum/pressure impregnation with water-borne preservative
Complete penetration of the sapwood of Baltic redwood can be achieved by this method. Treatment with an aqueous solution and the subsequent redrying that is necessary, however, cause the wood to swell and then to shrink; this can result in some raising of the grain and a risk of distortion.

Double-vacuum treatment with organic-solvent type preservative
The degree of treatment obtained can be controlled by varying the treating cycle used and it is possible to ensure that the net retention of the relatively high-cost preservative is adequate but not excessive. Complete penetration of sapwood is not usually obtained, but the average absorption and penetration achieved is about twice that from immersion treatment below.

Diffusion treatment This is carried out at the sawmill; after processing, seasoned treated timber can be supplied to the joinery manufacturer already penetrated throughout with a water-soluble preservative. This stock can then be converted and assembled in exactly the same way as untreated wood. Because the preservative is water-soluble, there is a risk that some may be leached out if the joinery is exposed to rain for months without paint or other protection.

Immersion treatment with organic-solvent type preservative Machined components or assembled units are submerged in a tank of preservative for three minutes. This provides satisfactory protection even though sapwood penetration is incomplete. Preservatives used for immersion, and for double-vacuum treatment, often contain water repellents: these improve the dimensional stability of the wood and hence benefit the performance of the joinery.

For further information on preservative treatment see PRL Technical Note 24, *Preservative Treatments for External Softwood Joinery Timbers*; this includes a list of the commercial processes and preservatives suitable for the treatment of external joinery.

Design and fabrication of joinery

Entry of water is most likely to occur at joints. Some modern joints loosen more easily than the old-style mortice-and-tenon joints with wedged tenons, and are therefore more vulnerable to water penetration. Joints are usually made with the grain of one member at right angles to that of another. Quite small fluctuations in moisture content stress these joints severely and once cross-grain movements begin, more water can enter and threaten decay.

The animal and casein glues permitted by some joinery specifications are likely to fail in persistently moist conditions. When this happens the joints in doors and opening sashes are less able to resist sudden stresses; the joints loosen, the paint film is broken, and penetration by moisture follows. The modern, synthetic resin glues are preferred because of their greater resistance to moisture. (See Forest Products Research Bulletin 38, *The efficiency of adhesives for wood.*)

The sectional sizes of timber scantlings accepted for windows have become smaller in recent years. It is bad practice to employ flimsy sections in window joinery, particularly where weather conditions are severe, as in some parts of Scotland. Such sections are prone to distort, permitting ingress of moisture, and they may not provide adequate space for the rebate. In opening lights the strength may be insufficient to prevent racking and consequent water penetration at the joints.

Window designs with horizontal surfaces which do not effectively shed water, still more those that actually entrap it, should be avoided. This applies both to rain-water on the outside of the window and to condensation on the inside.

The present-day omission of condensation channels and adequate means of drainage from them is to be deplored. Often water stands on the sill against the bottom of the frame and seeps into the joint between the two.

Cills present problems because they require timber of larger dimensions, and hence of greater cost, and also because they tend to present a higher decay risk. Whilst decay risk can be met by using more durable timber, increased cost is involved even if such timber is restricted to the cill alone. The current trend is to reduce costs by building up the cill from two smaller dimension timbers; the increased complexity of the two-piece unit calls for greater care in design and manufacture. However, even when Baltic redwood is used, a properly designed jointed cill effectively glued with suitable adhesives and preservative treated as necessary should give adequate service.

Fig 1 Complexity of joint design in a modern frame and cill. Result—joints expose a large surface area to risk of moisture penetration and facilitate decay

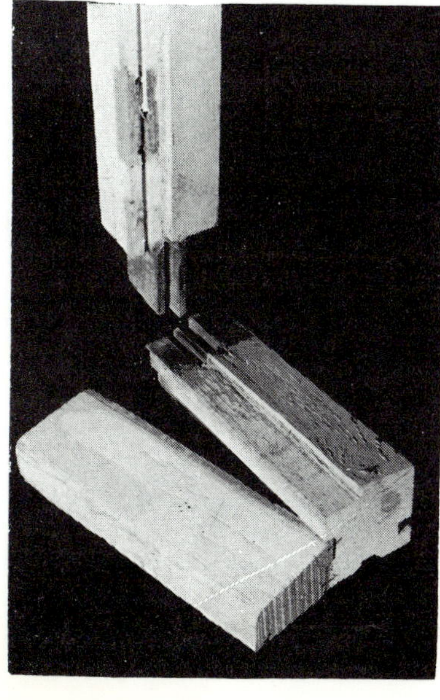

Weather protection before installation

Joinery is often to be seen exposed to all weathers while in transit or awaiting installation on building sites. This is thoroughly bad practice. All woodwork should be fully protected from the weather and stacked clear of the ground. Pink primer may look protective but it is rarely effective in preventing penetration of moisture—moisture which may later cause trouble if trapped by relatively impermeable coats of paint.

The window/wall joint

A further contributory cause of moisture penetration, and therefore of decay, may arise from contact between window frames and wet brickwork. There is rarely any damp-proof barrier separating them, nor is it usual to protect the backs of the frames with extra paint before installation. Even when the joinery is set in the inner leaf of brickwork there is risk of dampness continuing in the frames for as long as it takes the brickwork to dry out.

The more usual practice, however, except in Scotland, is to set the frame in the outer leaf of brickwork; not only is this brickwork often permanently damp, the window itself is much more directly exposed to the weather. Clearly this practice carries the greater risk of timber decay. Thorough priming—two full coats of an aluminium based primer—is needed on all timber surfaces that are to be in contact with external walls.

External and internal climates

The direct exposure of windows to driving rain and to rain-water run-off will depend on their regional location and position in the building, ie storey height and orientation (see Digest 127). That decay tends to be most marked in the lower, external parts of ground-floor windows is only to be expected. However, too little thought is given to the effects of internal climate on window joinery; this can be as damaging to the woodwork of single-glazed windows as the weather outside. In bathrooms and kitchens especially, temperature and humidity are often high, and in cold weather condensation on these windows is troublesome. Changed living habits and heating methods have aggravated the problem (see Digest 110). Often there is a pool of water along the lower edge above the back putty. This back putty is rarely as free from defects as that on the face, and water finds an easy entry. (The need for condensation channels at the base of the frame has already been mentioned.)

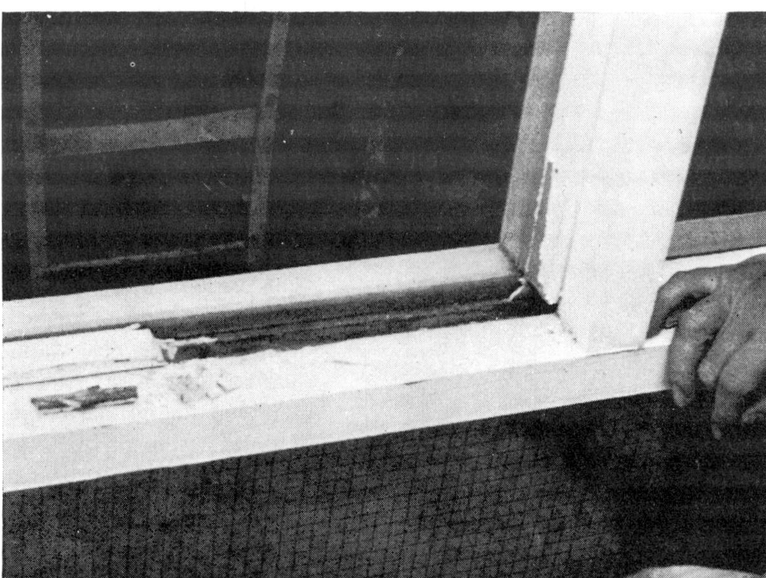

Fig 2

Outer glazing bead nailed—not glued or sealed. Result—free moisture penetration.

Grooves machined for outer glazing bead run right through mullion/transom joint. Result—void for collection of moisture

Fig 3 External part of mullion/transom joint held by nail. Result—water penetrates freely once the paint film is broken. Failure through decay began after three years' service

Painting and maintenance, primers
(*see also* Digest 106)

As noted, the ability of a single coat of primer, even a good quality lead-based type, to prevent ingress of water is greatly over-estimated. 'Pink' shop primers are of variable quality and many do little more than give the joinery a uniform appearance; after quite a short exposure they become weak or powdery and unfit to take further coats of paint. Many paint failures can be traced back to poor quality primers. Thus it is most important to store joinery under cover; if this is considered impossible much better priming than usual should be specified. A water-repellent preservative plus a good quality primer, or two coats of aluminium primer, would be suitable. (Since aluminium primers will also keep moisture in, they should only be used on correctly seasoned timber, or after the moisture or solvent from preservative treatment has dried out, otherwise blistering may occur.)

Lead-based pink priming paints (BS 2521:1966) are among the best for durability, both as part of a full paint system and if left exposed on site. They are expensive and their toxicity makes their

Fig 4 Gap left beneath tenon in joint between stile and cill or, in this case, between lower rail of main frame and base of mullion. Result—void for collection of moisture and eventual decay

Fig 5 Cupping of paintwork and putty failure associated with decay in lower rail of an opening light

use indoors inadvisable where children could contact them (even under further paint). Acrylic emulsion-based primers are coming into use, because they offer convenience to the joinery manufacturers. Although their water repellency may be no greater than that of many oil-based primers, they appear, on the limited evidence available, to have adequate durability before and after painting.

If a thin, weak or powdery shop coat is present, it should be sanded and a further coat of good quality primer applied. End grain and surfaces in contact with brickwork or masonry should always receive an extra coat of primer. Poor quality putty cracks or loses adhesion easily and allows water to enter at the bottom rail: all putty should be completely painted, with a slight overlap on to the glass. Hardwood cills should have the grain filled with a knifing stopper or filler (not a water-mixed type), preferably between two coats of primer.

Shellac knotting, even if it holds back resin, usually permits adhesion failure of the paint coats; an aluminium pigmented knotting is preferable, but good practice requires knots to be cut out and the cavities to be stopped.

Treatment of decay

It is important that all decayed wood should be cut out and paint stripped from adjacent areas so that damp, but not rotten, timber can be given a chance to dry out. If decay is extensive the entire window unit may have to be replaced. Wood exposed by cutting should be treated liberally with preservative, again allowed to dry, then given a coat of primer. Depending on their size, holes should be made good with stopper (see above) or primed timber, suitably shaped. Another coat of primer should be applied to the new surface, followed by undercoat and finishing coats.

Recommendations

1   When making use of Specifications and Codes of Practice bear in mind that their requirements may not be adequate in every circumstance to ensure satisfactory durability of the timber.

2   To ensure adequate service from Baltic redwood and other timbers of low decay resistance, insist on preservative treatment.

3   If durable hardwood is to be used specify the timber by name and insist on the exclusion of sapwood.

4   Protect joinery from the weather during delivery to the site, and on the site before installation.

5   Refuse to accept low quality shop primers. Ensure an adequate painting system.

Table 1 Timber species and properties

Standard name	Sapwood distinct	Heartwood durability†	Resistance to impregnation with preservatives	Woodworking properties		Amount of movement on re-wetting	Paint performance in service	Remarks
				Resistance in cutting	Blunting effect			
Afrormosia	Yes	Very durable	Extremely resistant	Medium	Moderate	Small	Good, provided paint is not affected by oily exudations	Non-ferrous fitments necessary
Afzelia				High	Moderate	Small	*	May discolour other building materials
Iroko				Medium	Fairly severe	Small	*	
Kapur				Medium (variable)	Moderate (occasionally severe)	Medium	*	Non-ferrous fitments necessary
Makoré				Medium	Severe	Small	*	Non-ferrous fitments necessary
Teak					Fairly severe (variable)	Small	Good, provided paint is not affected by oily exudations	Has been known to cause discoloration of granite
Western red cedar	Yes	Durable	Resistant	Low	Mild	Small	Good	Strength relatively low. Non-ferrous fitments necessary to avoid marked discoloration of the wood under clear finishes
Agba	Moderate			Medium	Moderate	Small	Moderate, but experience is limited	Log core may contain brittleheart. Resin exudation may occur with some parcels, especially if the timber has been air-dried
European oak, American white oak, Japanese oak	Yes		Extremely resistant	Medium (variable)	Moderate (variable)	Medium	Poor	Non-ferrous fitments necessary. Occasionally disfigures other building materials
Idigbo	Moderate				Mild	Small	*	Log core may contain brittleheart. May cause staining of other building materials
Sweet chestnut	Yes			Medium	Mild	Small	*	Non-ferrous fitments necessary
Utile					Moderate	Medium	Good	
Douglas fir	Yes	Moderately durable	Resistant	Medium	Medium	Small	Moderate	Non-ferrous fitments advisable
African mahogany			Very resistant	Medium (variable)	Moderate (variable)	Small or medium	*	Log core may contain brittleheart
Gurjun Yang			Resistant	Medium (variable)	Moderate (occasionally severe)	Large	Moderate	Resin exudation occurs with some parcels of these timbers (especially keruing); select material free from resin for painted work
Keruing						Medium	Moderate	
Red meranti Red seraya	Moderate					Small	*	Log core may contain brittleheart
Sapele	Yes		Extremely resistant	Medium	Moderate	Medium	Good	
Redwood	Yes	Non-durable or perishable	Sapwood permeable Heartwood moderately resistant	Low	Mild	Medium	Good	
Western hemlock	No		Sapwood and heartwood resistant	Low	Mild	Small	Good	
Whitewood								
American red oak	Yes		Sapwood permeable Heartwood resistant	Medium	Moderate	Medium	Poor	Non-ferrous fitments advisable
Beech	No		Permeable	Medium (variable)	Moderate (variable)	Large	Good	
Ramin				Medium	Moderate	Large	Moderate, but experience is limited	

† The sapwood of all species is either perishable or non-durable.
* Seldom painted; there is therefore little painting experience.

Timber fire doors

A well-designed door will fulfil its normal functions of isolation, insulation and security without causing too much hindrance to the movement of people and goods. Delaying the spread of fire and smoke is one important aspect of security.

The precise role to be played by doors in various locations in buildings, the methods of assessing performance and the criteria to be applied are topics under active discussion in this country and abroad. This digest gives the current situation and will be revised when these activities have resulted in changes in specifications and requirements.

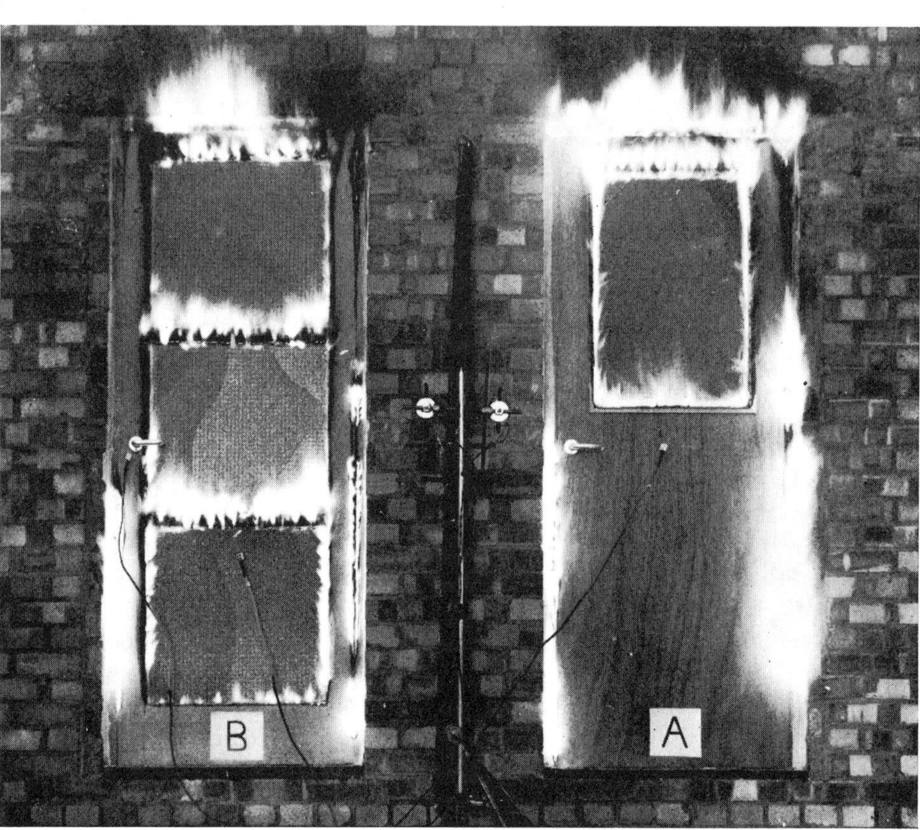

Fig 1 Ignition of glazing beads and door edges

Different parts of a building may be separated from each other into compartments by a fire-resisting construction; any openings leading to and from them will be closed by doors which will have a precise function to fulfil in case of a fire. First, the door should maintain the effectiveness as a fire barrier of the wall in which it is located; secondly, it should prevent excessive transmission of products of combustion which can interfere with the safe use of escape routes. Every fire door is therefore required to act as a barrier to the passage of smoke and fire to varying degrees dependent upon its location in a building and the fire hazard associated with the building.

Functions

The majority of fires start from a small source and develop quantities of smoke in the early stages. The rise in temperature is accompanied by an increase in pressure which tends to force the smoke and hot gases to other areas through any available openings and gaps in the fire compartment. A door in the enclosing walls will be subjected to these pressures and smoke will be forced through any gaps and orifices. As the fire progresses both the gas temperature and the pressure will increase; there may also be some damage to the door which may become deformed and consequently allow greater quantities of smoke to pass through. The collapse of windows will result in some relief of the pressure but the resulting increase in ventilation can lead to more rapid combustion of the contents. The door is then required to withstand exposure to the high temperature conditions without losing its integrity (See Table 1).

The functions of fire doors may therefore be summarised as:

1 To provide adequate resistance to the passage of smoke and other combustion products during the early stages of a fire.

2 To provide a barrier to a well-developed fire without permitting excessive quantities of smoke to pass.

Some doors may be required to fulfil only the first function as they may not be subjected to the full severity of a fire because of their location; others may have the main aim of resisting fire penetration as indicated by the second function. Some may have to meet both requirements. At present, fire doors are specified as smoke-stop doors when required to fulfil the first function and fire-check and fire-resisting doors to fulfil the second.

Assessment of performance

The performance of timber doors is judged by subjecting them to the standard test procedure specified in BS 476: Part 8. This Standard has recently replaced the specification given in BS 476: Part 1 which was the basis of extensive work on timber doors. As the revised Standard makes a few fundamental changes in the procedure for testing doors reference will be made to both procedures.

Tests are made on complete door assemblies, ie the door and frame with the necessary hardware, mounted in a wall representing its use in practice. By testing a door in one type of frame and using it in another, no guarantee can be given of its behaviour under fire conditions. The test procedure is fully described in the Standard and consists of exposing one face to the standard heating condition whilst observing the door for the maintenance of stability and integrity. The revised Standard requires the tests to be carried out with the upper part of the door under a small positive pressure, to simulate the conditions likely to occur in a fire. It also provides a more objective method of establishing the loss of integrity of a door by the use of a combustible fibrous pad; this was previously established by visual observation. The earlier Standard did not require any measurements to be made of the insulation provided by a door against the transfer of heat but the revised version does so.

Fig 2 Deformation of door

A door should be tested from each side to establish its performance with either face exposed to fire conditions; this requires two specimens but if it can be established that in practice fire is likely to attack only from a specified direction, only that face needs to be exposed in the test.

Fire behaviour

On exposure to high temperatures, charring of the exposed face occurs leading to surface ignition. Burning of the timber from the exposed face and the consequent movement of moisture inside the section causes deformation of the door panel with a tendency to form a concave shape towards the heat source (Fig 2). Timber doors are usually provided with two or three hinges along one vertical edge and a latch at about mid-height along the other. The triangular section at the top and the bottom of the door are free to deform and the maximum movement is likely to occur at the two corners on the closing edge. If the closing face of the door is exposed to the fire conditions, the stops on the frame will restrain this movement but this will stress the hinges and latch. When the opening face is exposed there is no restraint at the corners which are then free to deform. The gaps between the door edges and the frame provide an area which the flames can exploit and the extent of charring is greater than through the solid section. The amount of damage is significantly influenced by the initial size of the gaps, by any increase in gap that may occur by shrinkage and warping of the door and by the flow of hot gases through the gaps.

Fig 3 Fire penetration along edges

Gradual deterioration of the door assembly along the edges between the door and frame leads to the transmission of hot gases, widening of gaps, appearance of glowing on the unexposed face (Fig 3) and finally flaming. Initial failure is frequently in the upper half of the door, either along the top edge or the vertical sides close to the hinges. Failure in these areas usually occurs before penetration of fire through the door.

Over the last few years a technique has been developed for minimising the susceptibility of door edges to early penetration by fire. It consists of applying special materials having intumescing

characteristics along the edges so that a rise in temperature will cause the material to swell and close the gaps. Intumescent paints have been used but the most successful and reliable technique is to insert an intumescent strip, about 3 mm thick by 10 mm wide, into a groove cut in the door or the frame edge. As soon as the temperature in the vicinity of the strips exceeds 200° C, usually about 10–15 minutes after the start of a fire, the strip swells and seals the gaps (Fig 4). The intumesced material is soft and cellular in structure and will not prevent deformation of the door. For the increased protection needed with one-hour doors, a number of these strips, both in the door and the frame edge, may be necessary. The protection must be continuous and not interrupted at hinges and latch plates.

Specifications

The data in this section are based on the actual performances of a large number of timber doors and frames and are intended to highlight the important features and measures that are necessary. The performance of a given type can be established categorically only by test but an assessment is possible by a close study of the design features.

Smoke-stop doors (Fig 5)

There are no precise definitions available for this type of door and work is in progress nationally and internationally to define criteria and test procedures which may be used for classification. Single-swing doors must be close fitting in a rebated frame with 25 mm deep stops so that despite a pressure difference across the two sides only small quantities of smoke will pass. Additional seals must be provided around the edges if stops are smaller. Normal draught excluders can be satisfactory for a time but only a few types, employing metallic strips, can be relied on to remain effective. The non-metallic types will also become less effective as the temperature rises. Doors to corridors and lobbies which swing in both

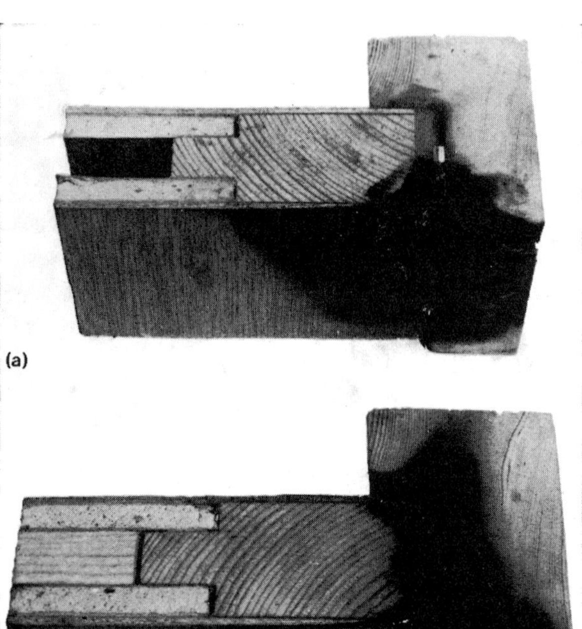

(a)

(b)

Fig 4 Exposure to fire on door (a) with intumescent strip and (b) without

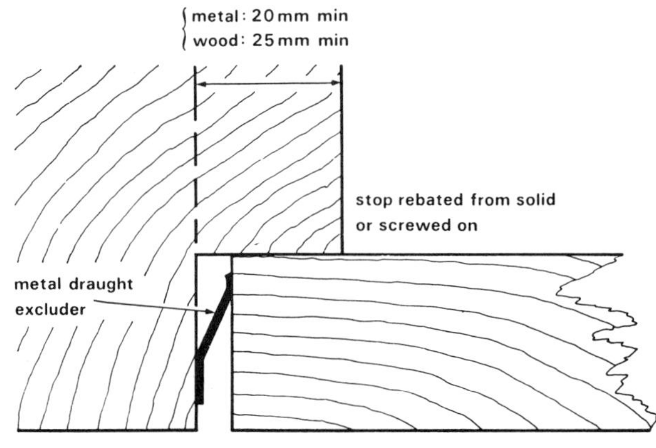

Fig 5 Smoke-stop door

directions and therefore cannot be rebated, are likely to offer less resistance to the passage of smoke. Smoke-resistance can be improved by fitting smoke seals such as the felt brush or flexible polyurethane foam type; intumescent strips will not become effective until the later stages of a fire.

The thickness of smoke doors is not necessarily critical, nor is the size of glazing.

Fire-check and fire-resisting doors (Fig 6)

The distinction between these two types of fire-door is not often fully appreciated. Fire-check doors are those which either comply with the construction specification given in BS 459: Part 3 or have given an equivalent performance in fire tests; fire-resisting doors have a higher standard of integrity. Table 1 illustrates the difference between the two types of door and Table 2 summarises some of the design differences.

A fire-check door has lower integrity than the corresponding fire-resisting door; its use in place of a fire-resisting door will not provide equivalent protection particularly against the passage of smoke during the later stages of a fire.

Half-hour fire-check door The specification given in BS 459: Part 3 is widely used and even with doors of a different design it is customary to use the frame specified in the Standard. The doors are 45 mm thick but some 40 mm doors have been found successful by test: in all cases the frame has 25 mm deep stops which may be glued and pinned or screwed to the frame. The gap between the door edge and the frame should not exceed 3 mm; two or three hinges and a latch are required to hold the door in the closed position (Fig 7). Alternative door constructions which have been shown to be successful are solid timber, compressed straw, extruded chipboard, cork, flaxboard and compressed fibre-board strips. Glazed panels, if any, should consist of wired glass not less

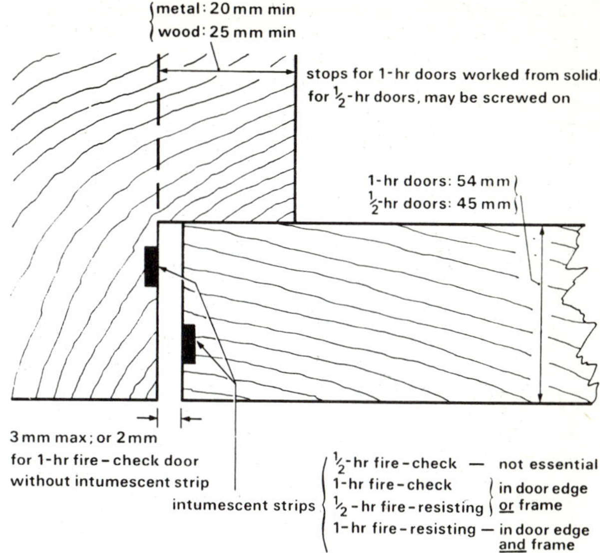

Fig 6 Fire doors

than 6 mm thick and no more than 1·2 m² in area; the glass should be held in position by timber beading not less than 13 mm square in cross-section, securely pinned or screwed.

Half-hour fire-resisting door As half-hour fire-check door, 45 mm thick, but with three hinges and an intumescent strip in the door or the frame edge. Glazing beads should be protected against ignition preferably by encasement with metal cover strips.

One-hour fire-check doors A specification is given in BS 459: Part 3 for 54 mm thick doors of composite construction with asbestos board mounted in an impregnated timber frame with 25 mm stops cut from the solid or in a steel frame with 20 mm

Table 1   Fire-check and fire-resisting doors

Door type	Integrity[1] *Minutes*	Stability[2] *Minutes*
Half-hour fire-check	20	30
Half-hour fire-resisting	30	30
One-hour fire-check	45	60
One-hour fire-resisting	60	60

(1) Integrity   Failure is deemed to occur when cracks or other openings exist through which flames or hot gases can pass or when flaming occurs on the unexposed face.

(2) Stability   Failure is deemed to occur when collapse of the specimen takes place

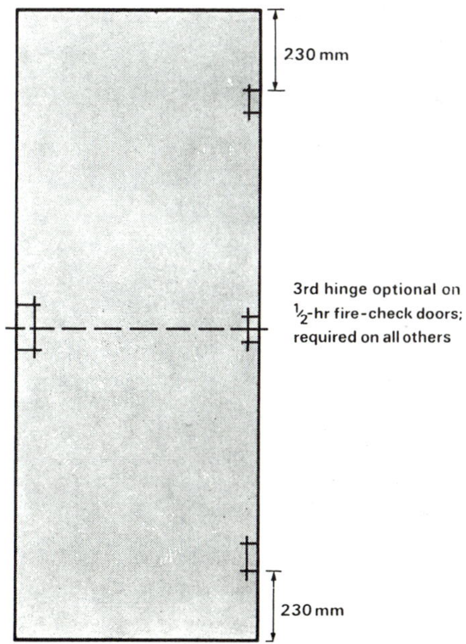

Fig 7 Recommended hinge and latch positions

Table 2 Fire doors: design features (see text for more detail)

Door type	Door thickness	Stops[1]	Glazed panels: max. size	glass	beads	Hinges[2]	Door/frame max. gap	seals
Smoke-stop	Not critical	Solid re-bated or screwed on	Not critical	6 mm wired		2	Not more than 6 mm	Draught excluder
$\frac{1}{2}$-hour doors: fire-check	45 mm to BS 459 Pt 3 (but see text)	solid re-bated or screwed on	1·2 m²	6 mm wired	not less than 13 mm × 13 mm wood	2 or 3	3 mm	—
fire-resisting					Ditto but encased in non-combustible cover strip	3	3 mm	Intumescent strip in door edge *or* frame
1-hour doors: fire-check	54 mm to BS 459 Pt 3.	solid re-bated	0·5 m²	6 mm wired	Non-combustible sub-frame	3	2 mm if no seal / 3 mm with intumescent strip	
fire-resisting						3 broad flap	3 mm	Intumescent strips in door edge *and* frame

(1) Stops to be at least 25 mm deep in wood frames, or 20 mm deep in metal frames

(2) All doors require, in addition, a substantial latch and latch plate

stops. The gaps between the door edge and the frame should not be more than 2 mm unless intumescent strips are used; the use of intumescent strips is desirable in 1-hour doors. At least three hinges and a substantial latch and latch plate are needed. The door constructions which have been shown to be successful include blockboard, compressed straw and cork cores with asbestos facings and veneer. Glazing of 6 mm wired glass in pane sizes not exceeding 0·5 m² should be mounted in a non-combustible sub-frame. The most successful method is to use profiled asbestos or glass-fibre reinforced cement framing members. Steel beading may be successful provided it is so fixed that it does not distort on expansion. The beading or framing to the door must not provide a point of weakness; the use of intumescent paints, cover strips or recessed joints will be necessary. The glazing frame should be fixed with long screws or nails from each side to ensure its retention after nearly 35 mm of timber has been charred from either side.

One-hour fire-resisting doors As for one-hour fire-check doors but with special design to prevent fire penetration at the edges. Double sealing with intumescent strips is essential as well as the use of extended flap hinges and latch plates. The frame should preferably incorporate a non-combustible membrane which may be faced with timber veneer. Very few one-hour fire-resisting doors are commercially available and there is scope for development by the use of intumescent sheet material embedded below the decorative facing of the door and the frame.

Existing doors Occasionally it is necessary to improve the performance of an existing door to a half-hour fire-check or fire-resistance standard, although in practice it may be more economic to replace the door with a fire-check or fire-resisting door rather than alter it. The doors are usually either panel type or a light core flush type about 35 mm thick: they require a facing on each side of quarter-

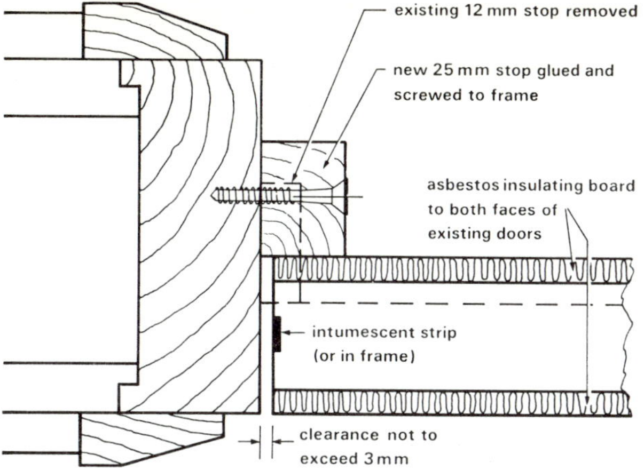

Fig 8 Conversion of an existing door

inch (6·4 mm) asbestos insulation board. The stops on the frame are likely to be 12 mm or less; these have to be removed and replaced by 25 mm stops, glued and screwed in situ. It is important that the finished door should not be deformed and that clearances do not exceed 3 mm. To meet the fire-resistance standard, it will be necessary to provide an intumescent strip in the door or the frame edge (Fig 8).

No advice is possible for improving the performance of existing doors to a one-hour standard but more elaborate protection will obviously be necessary and the biggest problem is likely to be the improvement necessary along the door and frame edges.

Glazing

Glazing may range from a small vision panel in a door to a glazed screen with hinges for maximum light transmission. Ordinary glass cracks when exposed to heat and is liable to fall out fairly early in a fire. Wired glass 6 mm thick can withstand exposure to the heating condition in a fire test for at least 60 minutes before it reaches a temperature high enough to soften it. The main reason for this is that nearly 50 per cent of the incident heat is transmitted through the glass by radiation.

The size of the glass and the method of its retention are important factors which influence its integrity. As the temperature approaches the softening point a large sheet will tend to collapse earlier than a smaller one. On the unexposed face, beading retaining the glass is subjected to radiant and conducted heat through the glass and to convection currents at the top of the pane. This can raise the temperature sufficiently to ignite timber beading after about 20 minutes (Fig 1). To delay the ignition of beading to 30 minutes it is necessary to provide protection by impregnation, surface coating or a surface covering of non-combustible material. For longer periods of fire protection, an improved retention system for the glazing is needed; so far only non-combustible glazing sub-frames have been shown to be satisfac-

tory. The glass panel should be small and the method of fixing it should ensure that no direct path can be created for the transference of hot gases.

Fittings and furniture

Hinges and latches have an important role in ensuring the integrity of the door. The hinges must remain adequately screwed, in spite of the charring of wood in the vicinity, to provide the necessary fixing to the edge of the door. It is common to use three hinges although tests have shown that with some doors two hinges may be adequate for the half-hour period. Steel and brass hinges are effective for a half-hour door, but only steel hinges will be satisfactory for a one-hour door. For the latter, it may be necessary to use hinges with extended flaps (broad butts) so that fixing is maintained even when severe charring has taken place.

The latch is the only holding device on the closing edge, hence it is important that it should be strong and that the nib of the latch should engage into the latch plate at least 12 mm to ensure that the closing edge will not spring open when the door deforms. Extended-flap latch plates are advisable for one-hour doors. Where mortice locks are provided, the cut-out in the door should be the minimum necessary to prevent any voids which fire can penetrate. After the edges, the mortice lock and latch areas represent the next zone of weakness. Filling the voids in the cut-out with intumescent paste will markedly lessen the weakness in this area.

Plastics and aluminium handles and knobs will be destroyed on the fire side but this may not have a serious effect on the integrity of the door if steel spindles are provided. Letter plates introduce a weakness in a door. No investigation has been carried out to establish a design for satisfactory performance, but those made of aluminium are likely to melt and even with steel flaps deformation can lead to integrity failure. A double flap of steel, one on each side of the door, is likely to be necessary to maintain integrity. Letter plates should ideally be positioned in the lower part of the door.

Self-closing devices can suffer damage and consequently become inoperative if they are surface mounted on the exposed face. If their sole function is to close the door with a latch, when they have done so any damage suffered need not be critical for the integrity of the door. Recessed self-closing devices will suffer less damage, but in the top of the door panel can produce a local weakness. The ideal place is within the sill where it will cause the least interference with the performance of the door. The Building Regulations set out the circumstances in which rising butts may be used: where they are fitted, compensation must be provided for the cut-out from the top edge of the door by increasing the depth of the stop (Fig 9).

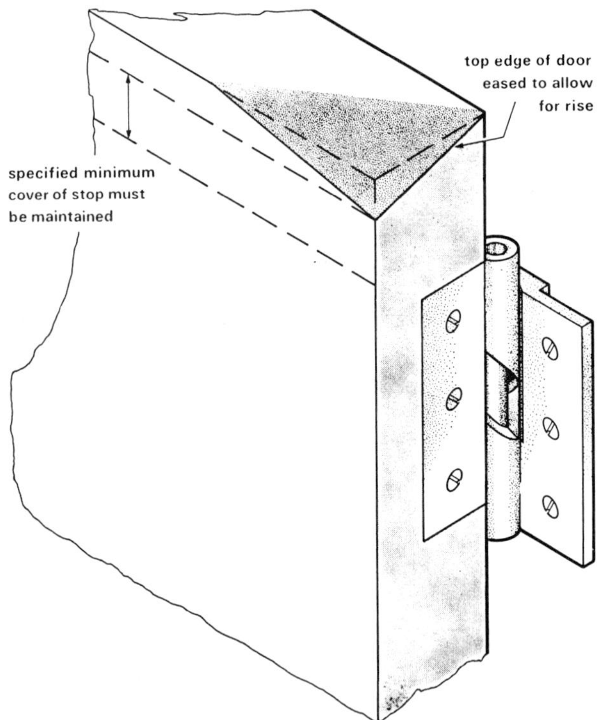

Fig 9 Rising butts need a deeper stop

A self-closing device used with double swing doors should not only close the door but also hold it firmly in the closed position. It needs to be strong enough to keep the door closed against any pressure to which the door leaf may be subjected in a fire. The only type of device which has been shown to be successful is a spring type mounted in the sill.

British Standards

BS 459 Part 3:1951 Fire-check flush doors and wood and metal frames (half-hour and one-hour types).

BS 476 Part 1:1953 Fire tests on building materials and structures.

BS 476 Part 8:1972 Test methods and criteria for the fire resistance of elements of building construction

4 Roofs

Permissible spans for trussed rafters

Conventional engineering principles applied to the design of trussed rafters may lead to an uneconomic use of material. Designs have therefore been developed on the basis of a simple analysis followed by tests on full-size prototypes and subsequent modification, if necessary.

The digest sets out the maximum recommended spans for various sizes of rafter and ceiling ties for the most widely used configurations of trussed rafter, the fink and the fan, for roofs of 15 to 35 degrees pitch, using stress-graded timber.

No deviation from established designs should be made without the advice of an engineer.

A most significant development in domestic roof construction in recent years has been the introduction of the trussed rafter. This is a light-weight timber truss which is used to replace the rafters, purlins and ceiling joists of traditional roofs. It is fabricated prior to erection and so eliminates in situ nailing other than for fixing purposes. Some common trussed rafter configurations are illustrated in the diagrams, and of these the *fink* and *fan* are the most widely used.

Joints in trussed rafters are made with metal plate fasteners or with glue or nailed plywood gussets. As the gussets or plates are fixed to both faces of the joint it is of course necessary that all members are of the same thickness. Metal plate fasteners are usually made from 18 or 20 gauge galvanised steel and may have either a pattern of holes punched in them through which nails can be driven, or they may have integral teeth. Several proprietary brands of each type are in common use in this country. The punched metal plate fastener with integral teeth requires a special press to form the joints and is only used when trusses are prefabricated in the factory. With the other type of metal plate and with plywood gussets

it is only necessary to drive in the requisite number of nails at each joint. Ideally this should be done in the factory, but if this is not possible it can be done on the site.

Apart from simplicity the plywood gusset and metal plate fasteners have advantages over other jointing methods. All the truss members are in the same plane thus avoiding lateral eccentricity of loading. The strength of a joint can also be varied simply by changing the size of the plywood gusset or metal plate or by varying the number of nails. In this way a proper balance can be obtained between joint strength and member strength.

Trussed rafters can be designed by conventional engineering principles, but it has been found that this may lead to an uneconomic use of material. Simple design methods are not really effective for this type of truss with semi-rigid joints, nor is there adequate information on the characteristics of metal plate or plywood gusset joints to enable these to be dealt with theoretically. Trussed rafters have therefore been developed on the basis of a simple analysis which is then followed by tests on full-size prototype units. Modifications are then made if necessary on the basis of the test results. The procedure adopted for such tests is that recommended in CP 112 : Part 2 : 1971 'The Structural Use of Timber'.

Trussed rafters are normally produced in spans which range from 5 m to 11 m with pitches between 15° and 35°. For the smaller spans the cross-section of members may be as small as 38 mm × 75 mm, increasing to 50 mm × 150 mm for the larger spans. Members within one trussed rafter, although of the same thickness, may be of different widths because of the different forces in the members. The advantages to be gained from varying the sizes of the internal members is obviously greatest in the longer spans, but may not be practical for the shorter span trusses.

For trussed rafter construction, as with any timber structure where strength and stiffness are important, it is necessary that the timber be stress graded in accordance with CP 112 : 1971 to ensure that it is of the strength assumed in the design and capable of withstanding the stresses which will be induced in the members under service conditions. Stress grading is therefore an essential part of trussed rafter construction and must not be neglected.

Compared with traditional types of roofs the trussed rafter has a number of practical advantages. Trussed rafters are generally fabricated under factory conditions and this should enable the quality of materials and workmanship to be controlled much more effectively than could normally be done on site. The volume of timber used in a trussed rafter roof is appreciably less than in its traditional counterpart and the erection time is also reduced. Using trussed rafters spaced at 600 mm centres it is claimed that roofs can be erected in about one quarter of the time taken to construct a traditional roof. Trussed rafters are normally designed to span between the outer walls of a building. This eliminates the need for load-bearing internal partitions, and the internal space can be divided as required and changed subsequently if necessary without structural alterations.

Each of the members in a trussed rafter performs a specific function and it is therefore important that the advice of an engineer is obtained before any deviation from established designs is made. It is equally important that no cutting, birds-mouthing, etc, of any member should be allowed other than at the positions specified by the engineer or designer.

Trussed rafters are intended to be used under dry conditions. The moisture content of the timber when fabricated should be 22 per cent or less and this should not be exceeded subsequently. Care must therefore be taken to protect the units from moisture during transit and site storage. Once trussed rafters are erected the roof covering should be laid with the minimum of delay.

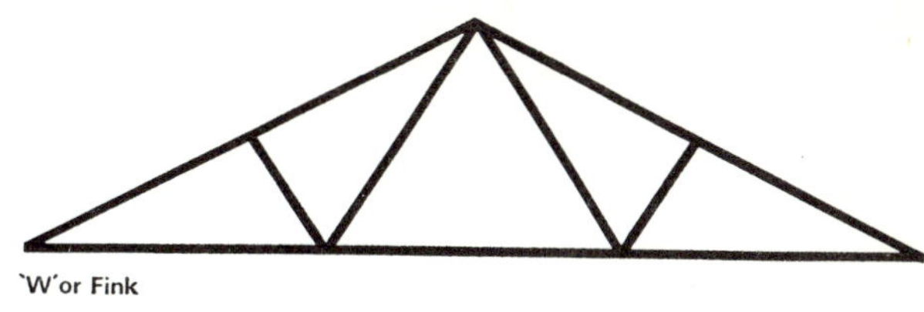

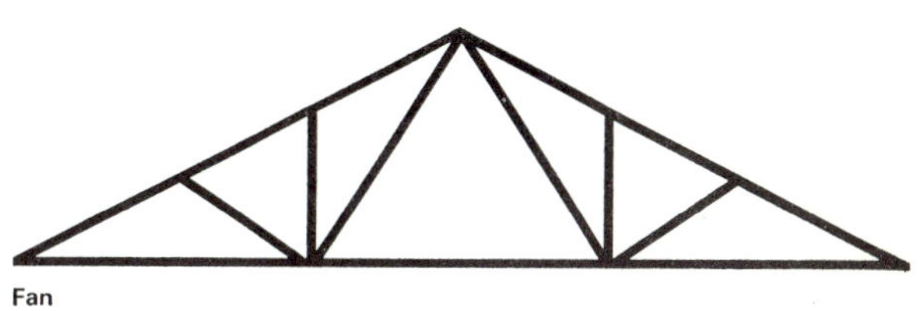

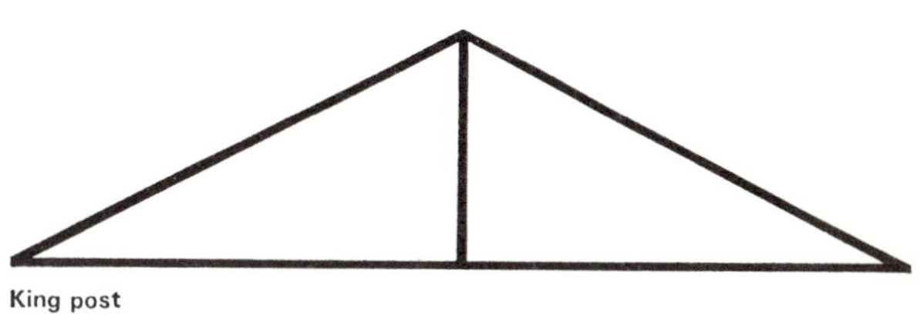

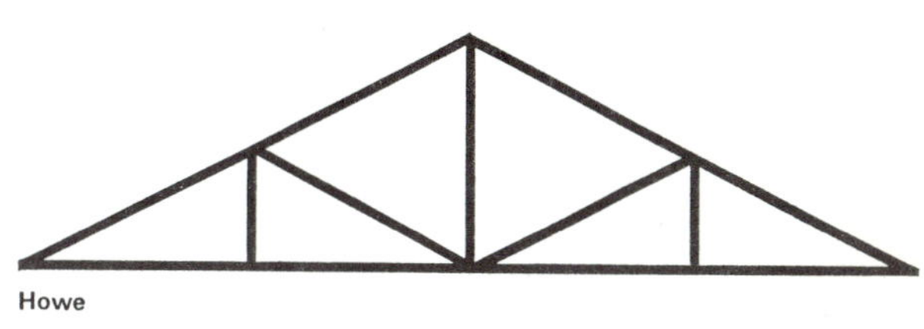

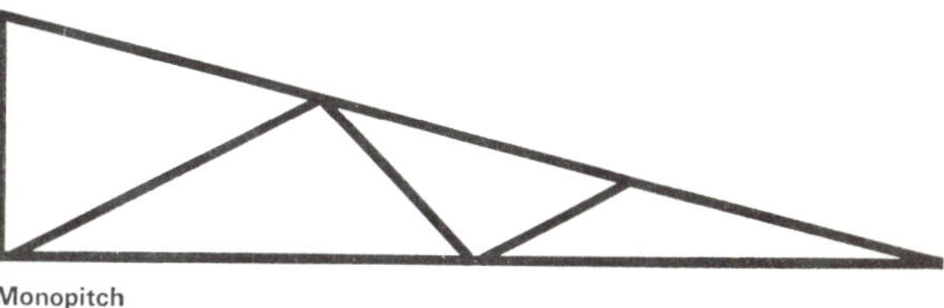

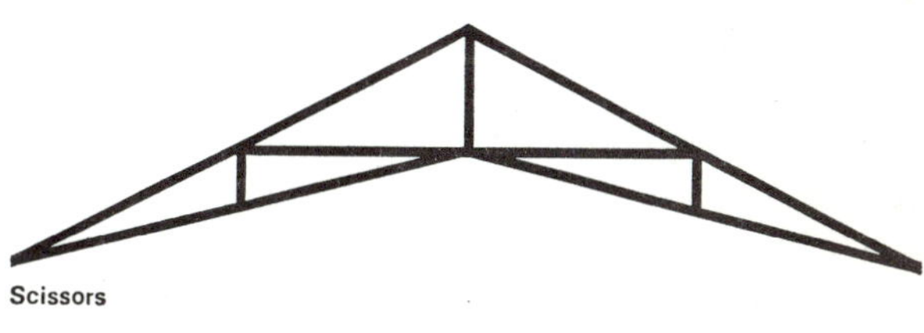

Common trussed rafter profiles

Maximum permissible spans (m) for rafters for fink trussed rafters (Timber to BS 4471 and of composite grade to CP 112)

Basic size (mm)	Actual size (mm)	Pitch (degrees)								
		15	17½	20	22½	25	27½	30	32½	35
38 × 75	35 × 72	6·03	6·16	6·29	6·41	6·51	6·60	6·70	6·80	6·90
100	97	7·48	7·67	7·83	7·97	8·10	8·22	8·34	8·47	8·61
125	120	8·80	9·00	9·20	9·37	9·54	9·68	9·82	9·98	10·16
44 × 75	41 × 72	6·45	6·59	6·71	6·83	6·93	7·03	7·14	7·24	7·35
100	97	8·05	8·23	8·40	8·55	8·68	8·81	8·93	9·09	9·22
125	120	9·38	9·60	9·81	9·99	10·15	10·31	10·45	10·64	10·81
50 × 75	47 × 72	6·87	7·01	7·13	7·25	7·35	7·45	7·53	7·67	7·78
100	97	8·62	8·80	8·97	9·12	9·25	9·38	9·50	9·66	9·80
125	120	10·01	10·24	10·44	10·62	10·77	10·94	11·00	11·00	11·00

Maximum permissible spans (m) for rafters for fan trussed rafters (Timber to BS 4471 and of composite grade to CP 112)

Basic size (mm)	Actual size (mm)	Pitch (degrees)								
		15	17½	20	22½	25	27½	30	32½	35
38 × 75	35 × 72	8·03	8·38	8·64	8·87	9·08	9·27	9·46	9·65	9·85
100	97	9·89	10·37	10·67	10·96	11·00	11·00	11·00	11·00	11·00
125	120	11·00	11·00	11·00	11·00					
44 × 75	41 × 72	8·65	9·00	9·25	9·48	9·73	9·89	10·08	10·30	10·48
100	97	10·71	11·00	11·00	11·00	11·00	11·00	11·00	11·00	11·00
125	120	11·00								
50 × 75	47 × 72	9·26	9·62	9·86	10·10	10·36	10·53	10·70	10·93	11·00
100	97	11·00	11·00	11·00	11·00	11·00	11·00	11·00	11·00	

Maximum permissible spans (m) for ceiling ties for fink and fan trussed rafters (Timber to BS 4471 and of composite grade to CP 112)

Basic size (mm)	Actual size (mm)	Pitch (degrees)								
		15	17½	20	22½	25	27½	30	32½	35
38 × 75	35 × 72	5·07	5·31	5·53	5·74	5·94	6·12	6·31	6·50	6·67
100	97	7·03	7·36	7·68	7·99	8·27	8·54	8·81	9·06	9·33
125	120	8·66	9·10	9·49	9·88	10·24	10·59	10·93	11·00	11·00
150	145	10·17	10·71	11·00	11·00	11·00	11·00	11·00		
44 × 75	41 × 72	5·53	5·78	6·03	6·26	6·48	6·69	6·89	7·08	7·28
100	97	7·53	7·90	8·25	8·59	8·90	9·19	9·48	9·75	10·04
125	120	9·13	9·60	10·04	10·46	10·86	11·00	11·00	11·00	11·00
150	145	10·52	11·00	11·00	11·00	11·00				
50 × 75	47 × 72	5·92	6·20	6·46	6·72	6·94	7·17	7·39	7·60	7·81
100	97	7·93	8·33	8·71	9·06	9·38	9·70	10·02	10·32	10·62
125	120	9·42	9·94	10·40	10·86	11·00	11·00	11·00	11·00	11·00
150	145	10·59	11·00	11·00	11·00					

Care is also needed in handling, particularly during erection, to ensure that the joints are not damaged by excessive lateral deflection. To assist in this, trussed rafters are frequently banded together in bundles before leaving the factory. Erection may be carried out individually, or a complete bundle lifted by crane on to the wall plates and the individual trusses spread to the required centres.

From the results of prototype tests and the experience gained in the use of trussed rafters in this country over the last seven years, permissible spans for the more common types of trussed rafter have been determined. These are given in the tables. The spans cater for the imposed loading requirements of the present CP 3 Chap V Part 1 : 1969 and the Building Regulations 1972 which makes CP 3 a mandatory requirement. They will also satisfy the requirements of the proposed amendment to CP 3 if and when it is adopted.

The spans are valid for the following conditions:

1 Timber to be to the dimensions and tolerances given in BS 4471 and of composite grade or better to CP 112:1971. Moisture content not to exceed 22 per cent.
2 Design dead loads not to exceed:
   Rafters: 685 N/m^2 measured in slope area
   Ceilings: 250 N/m^2
3 Design spacing of trusses not to exceed 0·6 m generally.
4 Joint design, fabrication and erection to meet the requirements of Timberlab Paper No 29.

For further information on the design, construction and use of trussed rafters reference should be made to Timberlab Paper No 29 which may be obtained free of charge from:

Building Research Establishment
Princes Risborough Laboratory
Princes Risborough
Aylesbury, Buckinghamshire

Built-up felt roofs

Built-up bitumen felt roofing is an efficient and durable form of roof covering provided it is properly laid. Many roofs of this kind have remained serviceable for 20 years or more with little need for maintenance or repairs. There are others, though, which for a variety of reasons, have become defective, sometimes quite early in their life, causing leakage and other troubles.

With due care, difficulties can be avoided; this digest explains the reasons for the defects and shows how they can be overcome. The recommended precautions are included in British Standard Code of Practice CP144 Roof coverings, Part 3:1970, Built-up bitumen felt.

Roofing felt may be used as a single thickness for temporary structures, such as sheds, but for permanent construction three layers of felt are usually bonded together on site with hot bitumen to form 'built-up' felt roofs.

Types of felt

Roofing felt consists of a mat of organic or inorganic fibres impregnated and usually coated with bitumen. BS 747 Roofing felts Part 2:1970 groups the felts into four classes, sub-divided as follows:

Class 1. Bitumen felts (fibre base)
1A Saturated
1B Fine sand surfaced
1C Self-finished
1D Coarse sand surfaced
1E Mineral surfaced
1F Reinforced

Class 2. Bitumen felts (asbestos base)
2A Saturated
2B Fine sand surfaced
2C Self-finished
2E Mineral surfaced

Class 3. Bitumen felts (glass fibre base)
3B Fine sand surfaced
3E Mineral surfaced
3G Venting base layer

Class 4. Sheathing felts and hair felts
4A(i)   Black sheathing (pitch)
        Black sheathing (bitumen)
4A(ii) Brown sheathing
4B(i)   Black hair
4B(ii) Brown hair

Durability

As already stated, built-up bitumen felt roofing of reputable quality, when properly used, can be expected to last for many years. Some of the felts made under the 'austerity' conditions of the last war gave poor service, but these are not representative of the general run of production before or since.

When felt roofing fails to give satisfactory service, the failure is generally due to incorrect use of the material, or neglect of the necessary precautions, rather than to poor quality felt. If, for example, no suitable provision is made against the movements of the sub-structure of the roof that result from drying shrinkage or thermal changes, the felt may be stretched beyond its capacity, and crack or tear.

Again, defects may arise through the entrapment of large amounts of water under the felt, or through excessive condensation on the underside. Effects such as these usually occur long before the felt shows any serious deterioration as a direct result of exposure to the weather.

The long-term effect of exposure is to cause a gradual loss of bitumen from the surface; crazing and pitting may eventually occur, allowing access of moisture to the fibre, and deterioration may then be accelerated. The felt also loses flexibility on exposure and so becomes more liable to crack. Walking on old felt that has become blistered or wrinkled may cause cracking and subsequent leakage.

The slope of the roof has an important influence on the durability of the felt. If rain water is not able to drain off completely, but stands on the roof in pools, it hastens the breakdown of the felt and may also cause pimpling or blistering, and loosening of grit from the protective surface dressing. Built-up bitumen felt roofing, therefore, should never be laid on a completely flat roof; a true fall, including allowance for deflection, of not less than 1 : 80 is essential, with provision for complete drainage.

Organic fibre felts can be used on dry roof decks and insulation, but are not recommended as the first layers on concrete or screeds. Felts based on asbestos or glass have better dimensional stability than those based on organic fibres, and so are less liable to cockle where the felt is not fully bonded to the roof.

Choice of felt

Built-up felt roofing will generally consist of three layers of felt bonded with hot bitumen.

If self-finished felt is used, some roofing contractors prefer to use a felt coated with a light dusting of sand instead of the talc normally used to prevent sticking in the roll, because they consider that the sand coating gives a better bond to the roof deck and so helps to avoid blistering. In that event, allowance of 0·35 kg/m² for fine sand and 1·1 kg/m² for coarse sand should be made for the additional weight of the sand in specifying the weight of the roll. Saturated felts, particularly those based on organic fibres, are not recommended for the lower layers of a built-up roof in damp climates, because they tend to hold moisture and may cause blistering.

The type of felt for the top layer depends to some extent on the slope of the roof. On roofs with a slope of not less than 10 degrees, mineral-surfaced felt type 1E can be used, whilst types 2E and 3E are suitable for slopes not less than 5 degrees. On slopes less than 5 degrees the protective mineral chippings should be applied *in situ*, secured with an adhesive of 'cut-back' bitumen (a solution of bitumen in a volatile solvent), blinded with 5 or 10 mm mineral chippings. The chippings should preferably be white or light in colour, so as to keep the felt and the

underlying structure cool in sunny weather and so minimise thermal movements.

Fire protection

Although bitumen felt is a combustible material, the degree of fire protection afforded by a built-up roof depends upon the roof structure as a whole. Requirements relating to the ability of a roof covering to afford protection against the spread of fire into the building or to adjoining buildings are laid down in both the English and the Scottish Building Regulations, and examples are quoted of various felt-covered roofings that would be deemed to satisfy these regulations.

The requirements are two-fold:
 (1) The roof shall resist penetration of fire.
 (2) The roof shall not encourage the spread of flame.

Each requirement is classified from A to D, representing minimum to maximum hazard; thus an 'AA' roof resists penetration of fire for one hour and the flames will not spread. Details of fire resistance for pitched and flat roofs will be found in Table A3 of CP144 and in the Building Regulations.

Roof construction

The various kinds of roof decks on which built-up roofing can be laid are listed in **Table 1** with the appropriate methods of treatment. For example, some that are thermally insulating in themselves require a topping of cement and sand, and the felt is laid on this; with others, the thermal insulation, where necessary, is provided by means of materials such as fibre insulating board, or cork, on which the felt can be laid direct.

In the construction of roof decks, the following three points call for special consideration since they have an important bearing on the performance of the felt covering and the efficiency of the roof as a whole.

Felt roofs must not be laid without falls

Moisture and thermal movement

It was mentioned earlier that tearing of the felt can be caused by movement of the underlying structure, especially drying shrinkage and thermal movements. Since these cannot be eliminated entirely, special precautions are necessary in laying the felt to minimise their effect and these are indicated under **Preparation for fixing**; in addition, everything practicable should be done to ensure that the movements themselves are kept to a minimum. To control the drying shrinkage in screeds or topping the mix should not be richer than 1 : 4 and it is important that the sand should be clean, well graded and fairly coarse. The water content of the mixes should be no more than is necessary for placing, compacting and finishing. The laying of screeds over weak insulating materials supported on roof decks is not recommended. Drying shrinkage of the screed is unrestrained and the screed is liable to crack and curl during drying, and the screed and the insulating material can become waterlogged during the hardening period for the screed.

Thermal movements of a concrete decking or topping are best controlled by a white or light-coloured treatment applied to the felt after it is laid, e.g. a layer of light-coloured grit or stone chippings, or decking slabs; movement joints should also be provided in large areas of concrete decking.

Constructional and rain-water

The average weekly rainfall in the driest part of the United Kingdom is approximately 12 mm; an evaporation rate of 12 mm per week is achieved only in fine summer weather. Porous decking or screed exposed to the weather is more likely therefore to increase its moisture content than it is to dry out.

Once felt has been applied the entrapped water, whatever its origin, can escape only very slowly, and its effects may continue for many months or even years. Apart from any visible effect, moisture retained in the structure reduces the thermal insulation and so encourages condensation, thus defeating the purpose of the insulation. It may, moreover, create a serious nuisance by causing stains, or even appearing as drips, on the ceiling below; also, in sunny weather, the pressure of water vapour may cause the felt to blister. Every practicable step should therefore be taken to reduce the water content. Drainage tubes, inserted in the structural roof and left open until the ceiling is plastered, are of great benefit in reducing the amount of water retained.

A practice known to have given satisfactory results is as follows. The structural deck is drained so that no ponds can form on it. This may be done by inserting drainage tubes at the time of construction; alternatively, the roof may be flooded and the deck punctured at the centre of all ponds. An insulating screed consisting of cement and no-fines lightweight aggregate (e.g. sintered pulverised-fuel ash) is then laid, followed by a thin 1 : 4 cement/sand topping, just sufficient to bring the screed to a fair face. The screed can then be allowed to mature until conditions are suitable for the roofing felt covering to be laid.

Although most of the rain penetrating the screed will drain away, even a drained screed will retain some moisture, and provision must be made for its escape. Small pressure release vents can be provided every 20 m², or when a lot of moisture is trapped brick box ventilators as shown in **Fig. 5** can be used. Various types of proprietary ventilators are available as alternatives to the brick box. One system uses tubes set into the screed and connecting the ventilators. The use of a spot-bonded ventilating felt as the first layer of built-up roofing will also encourage evaporation.

A system of pumping air through porous screeds has been developed in the last few years. Some success has been reported, but it cannot be expected to dry out in a few weeks a screed that may contain 1000 kg of water.

If lightweight concrete were made water-repellent, the absorption of mixing water and rain would be much reduced, and a no-fines screed laid on a drained roof deck would then hold very little water. Trials have shown this practice to be feasible with little or no loss of strength, and British Patent No. 824592 has been taken to cover the process. At present, though, it has not been successful; there has been insufficient demand for the treated aggregate to allow it to be marketed at a realistic cost.

Lightweight concrete slabs, insulating boards, compressed straw panels, wood-wool slabs and other absorptive materials intended to be incorporated in the roof structure should be stored under cover and should be fixed in sections as the roof covering is proceeded with, so that each section may be covered without delay. Wherever possible, tarpaulins or polythene sheets should be used to protect uncovered sections in wet weather.

Condensation

Moisture condensing within the thickness of a roof structure under built-up roofing can, if the condensation is heavy and long-continued, produce the same effects as entrapped construction or rainwater. This is not confined to high humidity buildings such as mills or laundries, it is common in dwellings. Two methods of preventing trouble are possible; to fit a true vapour barrier over the structural roof covered by thermal insulating material and waterproof covering. The vapour barrier should consist of one layer of bitumen felt fixed with hot bitumen using the 'pour and roll' technique, with the insulating material stuck to the vapour barrier with more hot bitumen. The whole job, from vapour barrier to covering, should be completed at one time, with precautions taken to protect raw edges overnight. The other technique is to use a vapour check at ceiling level, combined with ventilation of the roof space. The ventilation of the roof is necessary because it is impossible to make a true vapour barrier at ceiling level, but the ceiling should be as free as possible from gaps and holes, and should have a low vapour permeability. A ceiling could, for example, be constructed of foil-backed plasterboard, or have a polythene membrane immediately above it. The minimum ventilation is given in CP144 as 1000 mm² opening per metre run of eaves on two opposite sides of the building. Air must have a free passage from one side to the other, and if the roof is wider than 12 m, the size of the openings should be doubled. With some types of roof slab, especially those of cellular concrete, these precautions can be relaxed. The slabs give satisfactory service in most situations without a vapour barrier beneath them.

Preparation for fixing

Splitting of the covering can be avoided by careful design. Research into the behaviour of bitumen felt in tension has indicated that it will tear at about 5 per cent extension, so if in the design a limit of 2 per cent is adopted, the risk of tearing would be avoided. Thus, if a crack or joint is likely to open 2 mm, the width of felt left unbonded over the crack or joint should be 100 mm.

Recommended preparatory measures for roofs of different kinds are outlined in **Table 1** and particular attention is directed to those relating to lightweight concrete screeds and woodwool slabs.

Observations on roofing felt laid on these materials demonstrate that the practice of bonding the felt to a blinding coat of slurry often leads to cracking of the felt in places where shrinkage cracks develop in

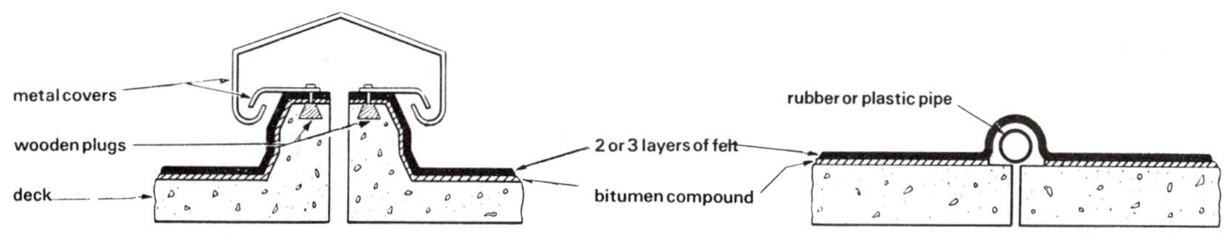

Fig. 1. Twin-kerb expansion joint for major movement

Fig. 2. Expansion joint for moderate movement

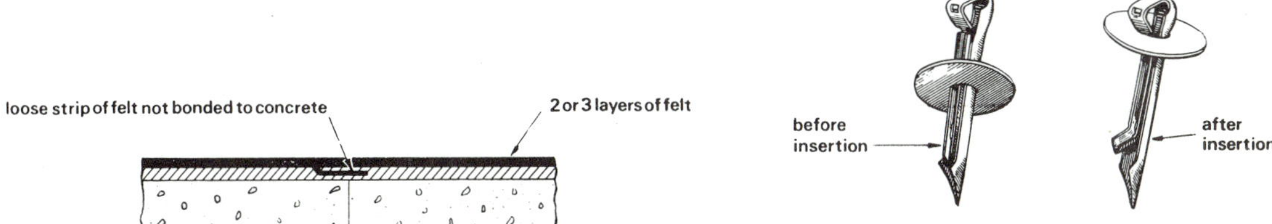

Fig. 3. Flat expansion joint for minor movement

Fig. 4. Expanding nail

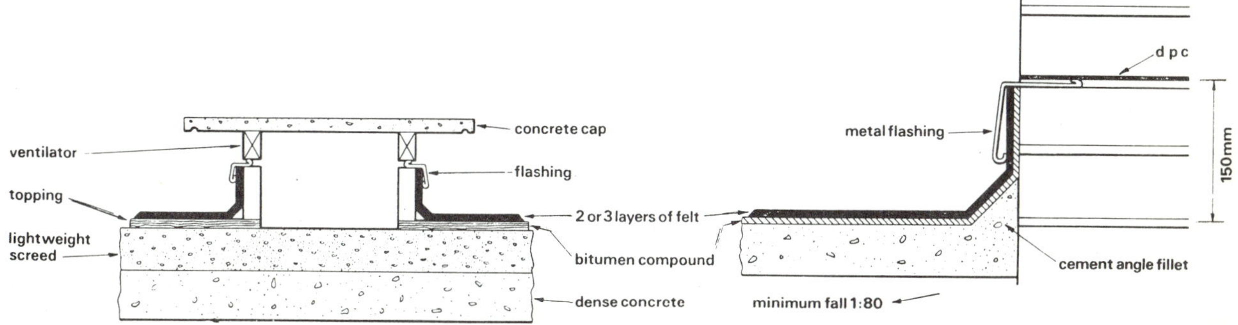

Fig. 5. Brick box ventilator

Fig. 6. Treatment of upstand

the roof structure. Felt laid directly on aerated concrete is sometimes prone to blister.

For these reasons it is recommended that a 12 mm topping of one part of cement to four parts of clean coarse sand should be used, and that this should be laid in alternate bays, each of not more than 10m². For aerated concrete screeds, however, some contractors prefer to prime the surface with a bitumen primer instead of using the topping, and this is permissible.

Since shrinkage cracks often develop along the line separating two areas of roofing of different widths, or at junctions between lengths of roof, or in places where the run of the roof is interrupted by roof-lights, it is further recommended that cuts should be made in the topping and preferably through the full depth of the screed along the lines of such junctions and along lines joining the adjacent corners of roof lights, and that any large areas of topping should be similarly subdivided at junctions between the bays.

Apart from providing a fair surface for adhesion of the felt and facility for correcting any inadequacy in fall, the purpose of the topping is two-fold. It aims to restrain the drying shrinkage of the woodwool slab or lightweight concrete screed and also to encourage any cracking of the topping to occur along the lines of the cuts rather than in random directions across the slab. Subsequently, loose strips of felt should be laid along the lines of the cuts, **Fig. 3,** to prevent adhesion of the overlying felt for a distance of about 100 mm on either side. The aim here is to allow the felt to stretch rather than tear in response to any small movements that may occur in the topping. The method illustrated in **Fig. 2** is better where considerable movement is likely. Expansion joints in the roof structure should preferably be of the twin-kerb type, **Fig. 1.**

Fixing

Methods of fixing are also set out in **Table 1.** The surface should be clean, free from dust or debris, and at least surface-dry at the time of laying. Except for wooden boarded roofs, where the first layer is nailed and should not be bonded to the boards, roofing felt is customarily bonded to the roof with hot bitumen spread at the rate of about 1·5 kg/m². On highly absorptive surfaces, this is preceded by an application of a bitumen primer to reduce frothing of the bitumen and assist adhesion. The primer must be allowed to dry before proceeding.

Table 1 Recommended constructions

Roof decks	Treatment before laying felt	Method of fixing first layer
Aerated concrete screeds, or lightweight aggregate concrete screeds, on dense concrete slabs, precast concrete beams, hollow pots, etc.	12 mm topping of 1 : 4 cement : clean coarse sand, laid in alternate bays not exceeding 10 m²*. Drain tubes through the deck and ventilators may also be required.	Partial bonding. Where cuts are made in the topping (see text), strips of felt to be laid along the cuts.
Asbestos-cement	Fibre insulation board, laid to break joint, bonded with hot bitumen to a vapour barrier	Bonding overall
Dense concrete, shell roofs, etc.	Nailing battens should be provided at the top of steep slopes. Fibre or cork insulation board may, if required, be bonded to the roof over a vapour barrier	Bonding overall, unless cracking is expected
Metal deck	Fibre insulation board, laid to break joint, bonded with hot bitumen to a vapour barrier	Bonding overall
Compressed straw slabs	Tape joints	Bonding overall
Wood boards	If additional insulation is needed, insulation board, laid to break joint, may be fixed	Nailing
Woodwool slabs	12 mm topping of 1 : 4 cement : clean coarse sand, laid in alternate bays not exceeding 10 m²	Partial bonding. Where cuts are made in the topping strips of felt to be laid along the cuts.

*With aerated concrete screeds, some manufacturers recommend priming the surface in preference to a cement : sand topping ; either is permissible.

It is seldom necessary to bond the felt overall, but there should be sufficient bonding to prevent lifting and tearing of the felt in high winds. Where there is a substantial loading of tiles or stone chippings, it may be enough to bond the edges of the strips of felt. For fixing to the cement/sand topping recommended for use over lightweight concrete screeds or woodwool slabs, partial bonding at the edges and in strips or spots elsewhere is preferable to over-all bonding, since this gives the felt more scope for stretching in response to any differential movement in the roof slabs.

The suggestion for partial bonding applies only to the first layer of felt. Partial bonding is essentially a precautionary measure imposed by the impossibility of predicting precisely where cracking may occur. It is not necessary or desirable to use partial bonding where the felt is laid on fibre insulating board, cork or compressed straw slabs.

Table 2 Defects in built-up felt roofs

Defect	Cause	Remedial treatment
Tearing or cracking	1. Differential movement at cracks or joints in the substructure	1. Adopt details shown in Figs. 1, 2 or 3. If defects extensive, re-lay over fibre insulating board
	2. Perforation of unsupported felt	2. Provide a supporting fillet and patch with felt
	3. Perforation of blisters by traffic	3. Cut, bond down and patch over blister
	4. Sharp bends in felt	4. Round off and patch, or use a metal flashing
Pimpling on top layer of felt, eventually breaking open and exposing fibres	Expansion of more volatile fractions of bitumen, or of air or water, in sunny weather. May occur with self-finished felt, particularly where pools of water can stand on it, or with mineral-surfaced felt laid at too low a pitch	In serious cases on felt roofs, dress with cut-back bitumen and grit
Blisters between layers of felt	Expansion, in sunny weather, of entrapped air or water	Cut the blister, re-bond and patch. If the trouble is persistent, a heat-reflecting treatment, or the weight of a heavy dressing, may help
Blisters between felt and deck	Expansion in sunny weather, of entrapped air or water. Particularly liable to occur on surfaces such as lightweight concrete screeds or asbestos cement, if not dealt with as in Table 1	
Cockling and rippling	Moisture movement of either felt or roof, especially where felt is not bonded	Little can be done, except to relay. Site dressing, using a fair weight of grit, may make the defects less conspicuous
Lifting of laps	1. Bad bonding initially 2. Pulling as a result of formation of blisters	1. Re-bond 2. Cut, bond down and patch
Local deterioration associated with ponding	Inadequate fall, or obstruction of outfall	Clear outfalls. Site dressing with cut-back bitumen and coarse grit may limit further damage
Arching of tiles, screeds, etc.	Inadequate allowance for thermal or moisture movement	Provide 25 mm gaps around bays of about 9 m² and fill with a bitumen compound. If necessary, rake out old compound and re-point
General deterioration and embrittlement	Premature deterioration indicates inadequate falls, wrong choice of felt, neglect of any recommended surface dressing, or wrong laying	Re-lay if deterioration is far advanced; if not, site dress with cut-back bitumen and grit
Loss of grit	1. Site dressing in adverse weather conditions 2. Use of hot bitumen or emulsion instead of cut-back bitumen 3. Inadequate fall	Site dress, using cut-back bitumen and fairly coarse grit

Partial bonding may also be achieved by using a proprietary underlay perforated with holes about 12 mm in diameter and gritted on the undersurface. Hot bitumen is spread over this so that the bonding of the felt to the roof is restricted to the areas of the holes in the underlay. The gritted under-surface provides a continuous air space through which vapour can escape at the verges, thus preventing a build-up of pressure beneath the felt.

A form of expanding nail, **Fig. 4,** has been introduced for nailing felt to lightweight concrete of medium strength without a topping, but there has not yet been much experience of its use. Because of the alkaline character of concrete, nails in an aluminium alloy would be unsuitable for this purpose.

Upstands at parapet walls, etc., need careful detailing to exclude rain. The upstand, which should be at least 150 mm high, should preferably be covered with a metal flashing as shown in **Fig. 6.** Where there is a damp-proof course in the wall, the flashing should be made continuous with it. Bends in the felt in these or similar circumstances should be supported, preferably with a wood or mortar fillet.

Where the edges of a roof are finished with a kerbing under the felt, the development of shrinkage cracks in the joints of the kerb may lead to tearing of the felt. Laying a loose strip of felt over each joint, **Fig. 3,** to break the bond with the overlying felt will reduce this risk. A metal trim should be firmly anchored at close centres to a substantial member in the roof deck.

Roof paving

Where a built-up roof has to carry much traffic, a more substantial surface is needed. This may be of tiles bedded in bitumen, or a topping of bitumen macadam not less than 12 mm thick or a mortar screed not less than 25 mm thick. Since there must be enough freedom of movement in tiling or screed to prevent lifting or arching under the combined effects of thermal and moisture movement of the roof structure, tiling or screed should be laid in areas of not more than 10m² and these areas should be separated by 25 mm joints. Tiles should be butt-jointed. Each area of mortar screed should be sub-divided into units not larger than 0·6 m by 0·6 m by

cutting through the full thickness of the screed before it has set and hardened. Subsequently these cuts and the joints between the separate areas should be filled with a bitumen jointing compound.

Treatment of defects

The appearance of dampness, or even of water drips below a built-up roof does not necessarily mean that the roof covering is leaking, for these defects can also be produced, as already explained, by construction water, by rainwater or by heavy condensation within the roof structure. If, therefore, on careful inspection, the felt covering appears to be intact, these possibilities should be considered, remembering that water can remain entrapped for many months after the covering has been laid.

There is no quick way of drying out a roof that, from any cause, has become waterlogged. If water is dripping from points in the ceiling or, as often happens, from electric conduits, holes may be bored near the places affected to assist drainage. Drying can also be assisted by building brick box ventilators, **Fig. 5,** on the roof at appropriate places, or by the other methods already described.

If the conditions within the building are warm and humid, condensation may be suspected; if it is in fact the cause of the dampness, the only remedy, in severe cases, may be to provide more thermal insulation over an effective vapour barrier. In milder cases it may be sufficient, if practicable, to improve the ventilation of the building.

Sources of dampness caused by cracking or deterioration of the felt or failure to cover upstands with flashings can usually be traced by inspection. Where the felt has become torn or cracked but is otherwise in good condition, patching on the lines indicated in **Figs. 2** or **3** will be well worth trying; it can best be undertaken by a specialist contractor. Other defects can be dealt with by the methods given in **Table 2.** If there is reason to think that thermal movement of the roof has been a contributory factor in damaging the felt, it may help to prevent a recurrence if, after repair, the felt is covered with a layer of light-coloured grit to reflect solar heat.

Further reading

BS 747 Roofing felts Part 2 : 1970
BS CP 144 Roof covering Part 3 : 1970 Built-up bitumen felt
Built-up roofing; Felt Roofing Contractors Advisory Board:
   1966

Asphalt and built-up felt roofings: durability

A study of flat roofs covered with mastic asphalt and built-up bitumen felt roofing has disclosed the most common causes of failure and made possible an assessment of these roofings when properly designed and laid. It is suggested that they should be inspected at least annually so that action can be taken to deal with defects early enough to prevent their leading to failure.

Although many complaints are made about leakage and maintenance costs of flat roofs, these still remain popular with designers. A survey of flat roofs on Crown buildings covered with mastic asphalt and built-up bitumen felt was therefore mounted by the Building Research Station at Garston in order to provide more detailed information about the frequency and causes of failure and success. Figures 1 and 2 show the number of examples studied and the age and performance of each up to the date of the survey. The failures recorded could all be attributed to failure to comply with published recommendations for design and installation (*see* 'Further reading'). A study of similar roofings on local authority dwellings in Scotland is also nearing completion. Failure to comply with published recommendations, particularly in regard to edge details, at verges, curbs and parapets was also noted. A further publication incorporating the details of the lessons to be learned from both surveys is planned.

General characteristics

A flat roof is normally a composite structure. Its functional requirements are met partly by the load-bearing deck and insulation, partly by the covering and, where appropriate, by a vapour barrier under the insulation. Only the covering is considered here.

Mastic asphalt Asphalt to BS 1162 tends to develop a grey appearance when weathered whereas the BS 988 material remains dark in colour, though finishing by sand rubbing lightens its appearance; the two are visually indistinguishable when given a solar reflective treatment. In direct sunshine, both types harden gradually and shrink slightly; this causes shallow surface crazing which does not usually develop into cracks. Cracking from other causes, such as the drying shrinkage of a coating of unsuitable paint or the differential movements of paving tiles not laid on a separating layer, may progress through the full thickness of the asphalt.

Asphalt being, in effect, a stiff liquid can tolerate slow movement involving the full thickness of the material, but otherwise it behaves as a brittle solid and is liable to crack under any suddenly applied strain, particularly under impact and in cold weather. Because of its high coefficient of thermal expansion, and because partial bonding to a roof deck of 'wet' construction tends to cause blistering due to concentrated vapour pressure, it is now usual to separate the asphalt from any substrate, except bitumen-bonded lightweight aggregate screeds, by an isolating membrane of sheathing felt. The purpose of this is to avoid blistering or cracking and to bridge discontinuities, but where significant movement may occur, for example at a butt joint in the deck, it is desirable to break the continuity of the asphalt by means of a movement joint. The dead weight of the roofing, including sheathing felt, is about 40 kg/m^2, and this, with proper edge detailing, is usually sufficient to prevent uplift by wind suction in the UK.

Built-up bitumen felt roofing The early felts were based wholly on organic fibres which were very strong when new but dimensionally unstable, and could rot, when moisture eventually penetrated the bitumen coating. The asbestos-based felts are more stable but they may still include up to 20 per cent of organic fibre. Bitumen felt based on glass fibre is stable and rot-proof but the glass fibre is in the form of a bonded tissue, not a woven fabric, and does not provide very high strength, so that the felt needs careful handling. The merits of each type can be exploited by correct design, often using two or more types in the same system. None of these three standard types can be stretched, before splitting, more than about 5 per cent (say 2 per cent for design purposes) and if fully bonded to a substrate which subsequently cracks they will split. On concrete and screeded surfaces, it is now usual to fix the first layer by only partial bonding to a pattern that allows the escape of water vapour, during drying, and reduces the risk of splitting. The felt can be fully bonded to dry insulation materials; on a wooden deck, the first layer is fixed by nailing.

If exposed to the weather, a bitumen surface is gradually attacked by solar radiation, both ultra-violet and heat-producing, and by atmospheric oxidation; the effects are increasing embrittlement and crazing, or superficial pimpling, leading to exposure of the fibres and progressive attack of the bitumen underlying them. Some form of surface protection is therefore necessary. Unless a special finish for solar reflection or foot traffic is required, the roofing should be given a surface dressing of bitumen compound and mineral chippings; in the Scottish study it was found that stone chippings smaller than 25–32 mm grade are liable to be scoured by the winds to expose the bitumen surface below. By its added weight this dressing reduces the risk of lift by the wind; it also contributes to the fire grading of the roof. To reflect solar heat, white chippings may be used. The weight of built-up felt roofing varies with the nature of the surface dressing and the risk of uplift by wind suction may limit the proportion in area that can be safely left unbonded.

For more detailed recommendations for built-up felt roofs, see Digest No 8 (1970 edition).

Performance

Not all of the sites were visited and though a great deal of information was obtained by technical questionnaires some of the detail requested, particularly as to falls, was not provided. Official records of minor repairs (under £100) are almost non-existent and some reliance had to be placed on the memories or personal notes of the maintenance personnel. Published guidance has undergone some changes, for example in the revision of Codes of Practice, the effects of which could not in all cases be isolated.

About one-third of the roofs studied from Garston had 'failed', this being defined as a roof, or its weathering detailing, that has permitted water to penetrate at some time. The roofs shown in Figs 1 and 2 to be 'defective' exhibit visible imperfections such as blisters, cockling and ponding which could lead to failure though this has not yet occurred.

The incidence of failure seems to vary between the two types of roofing mainly in that the probability of failure remains practically constant throughout the life of an asphalt roof but increases steadily during the life of a built-up felt roof.

Asphalt roofing, properly designed and laid, should prove capable of lasting 50–60 years; the natural ageing of bitumen felt is likely to limit its life to about 20 years.

Mastic asphalt Of 123 coverings, 34 were classified as having failed: 20 by splitting, cracking or tearing under strain, as by differential movement of a deck from which the covering was not properly isolated; 14 due to disruption of associated weathering, skirtings or flashings, at parapet walls, roof lights or around rainwater outlets. After repair, none of these failed coverings had needed to be replaced.

Built-up bitumen felt Of 200 coverings, 77 had failed. Most of these were due to the same causes that damaged the asphalt roofs, inability to accommodate movements in the roof decks (37 roofs) or disruption of weathering (26 roofs). A few failures attributable to entrapped moisture or condensation had also resulted from avoidable inadequacies in design or workmanship. A single instance of wind damage could be ascribed either to inadequate fixing or to a gale of abnormal severity. Five of the roof coverings had been replaced after 20–30 years of service (see Fig 2) because of deterioration by natural ageing.

Findings common to both types

Use and location of building No correlation was found between the performance of the roof covering and the function or structural type or form of the building, or with the exposure and topography of the site.

Falls There was no evidence that the provision of falls had any effect on the life of the coverings. But a truly flat surface is unlikely to be achieved in practice, and ponding is likely to occur unless a fall sufficient to drain depressions is provided. Ponding is less harmful to the covering than was at one time supposed but if a leakage occurs in a ponded area water will enter the building in greater quantity than through an effectively drained surface.

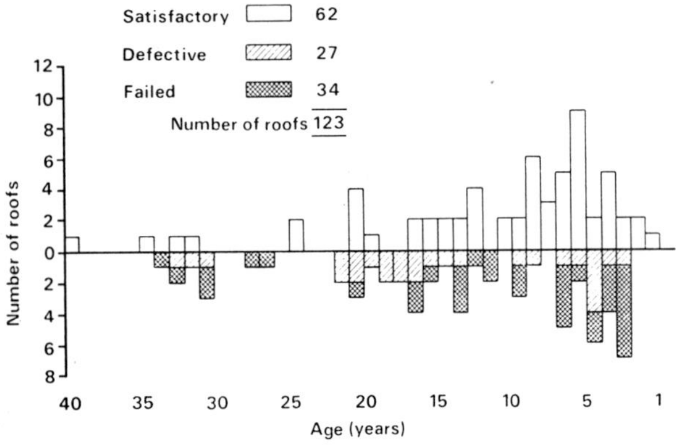

Fig 1 Mastic asphalt coverings

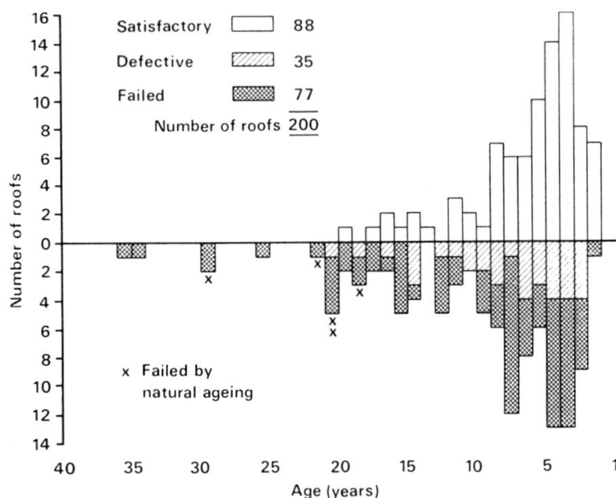

Fig 2 Built-up bitumen felt coverings

Visual appraisal of defects

Roofings should be inspected at least annually by a maintenance surveyor to ensure that rainwater outlets are kept clear and to decide on any action necessary to prevent defects from becoming failures. The photographs (3–9) may assist in identifying defects.

Mastic asphalt

Ponding: Slight localised deflections below the general plane of the drainage falls can result in ponding. In asphalt roofing, this can cause crazing which may be only shallow and in the condition illustrated (**3**) does not warrant the cost of remedial measures on technical grounds alone though it might do so on grounds of appearance.

Cracking: Differential movement between precast concrete roofing units can cause cracking (**4**). This might have been avoided had the perimeter of the asphalt not been bonded directly to the concrete beyond the edges of the sheathing felt, which stops short as can be seen from the change in level of the surface. Rain has not yet penetrated but it could do so by capillary action. Repair is desirable, but not urgent.

Blistering: (**5**) shows a combination of blistering and sagging in an asphalt skirting. The most likely sequence of events was that, during hot weather, vapour pressure from water leaking into the parapet wall produced the large blister and in so doing either fractured the tucked-in portion above the blister or pulled it out of its chase. The detached portion of the asphalt when warmed by the sun would tend increasingly to sag under its own weight and so produce the fold seen below the blister. If not arrested the sagging will progress and rainwater will enter the building. The damaged area should be cut away and made good without delay.

Built-up bitumen felt covering

Splitting: Tearing (**6**) occurred over an edge or verge curb, the capping of which is of plywood; the plywood has delaminated as a result of entrapped moisture and as it distorted it imposed sufficient tensile forces to split the felt. Adherence to recommended details would avoid this defect. Metal edge trims were found also to split the felt at or about the butt joints in the trims.

Cracking: A portion of replacement felt at the northeast corner of a roof (**7**) shows cracking of the type sometimes called 'crocodiling'. The horizontal area of felt is protected by mineral aggregate dressing but on the fillet at the base of the upstand to the parapet wall the felt suffers maximum exposure to the sun. The combination of solar heating, ultra-violet radiation and atmospheric oxidation has crazed the coating of bitumen, applied generally to the area to stick down the mineral aggregate and then cracked the top layer of self-finished felt. Further progress of the

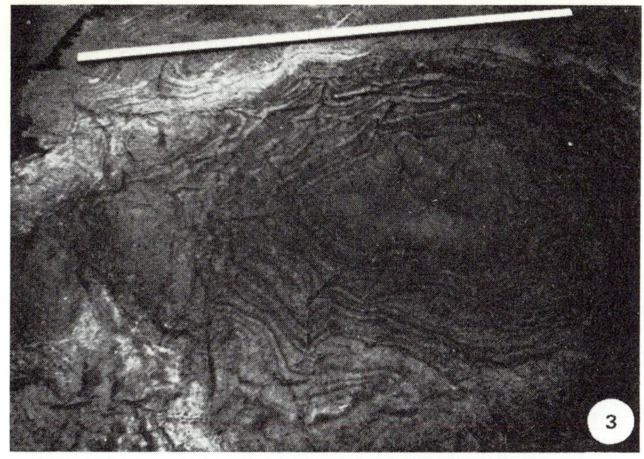

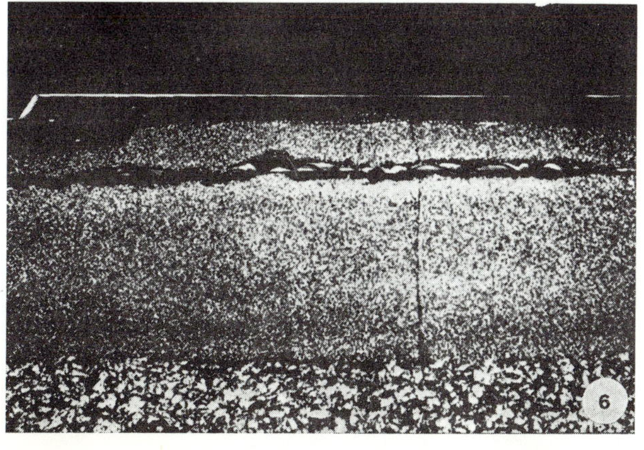

Further reading

Mastic asphalt:

    BS 988 Mastic asphalt for roofing (limestone aggregate)
    BS 1162 Mastic asphalt for roofing (natural rock asphalt)
    BS CP 144 Roof coverings Part 4:1970 Mastic asphalt
      (metric units)
    Application of mastic asphalt; Mastic Asphalt Advisory
      Council:1967

Built-up bitumen felt:

    BS 747 Roofing felts Part 2:1970
    BS CP 144 Roof covering Part 3:1970 Built-up bitumen felt
    Built-up roofing; Felt Roofing Contractors Advisory Board:
      1966

cracking into the underlying layers of felt can be arrested by a reflective treatment, and as the roofing is laid on insulation of vegetable origin this should be done immediately.

Blistering: The brick indicates the size of the blisters (**8**)—up to about 1 m long and 75 mm high. The top layer of the three-layer roofing is mineral-finished. Cutting open a blister revealed that the bottom layer of felt is adhering firmly to the base and the middle layer is firmly bonded to the top; the blisters have developed between the middle and bottom layers. This could result from insufficient pressure when rolling the middle layer into the hot bitumen bonding compound, or from moisture entrapped between the two layers (either rain or dew). As none of the blisters shows signs of bursting, and the bottom layer of felt will still provide cover if they do, repair can be delayed. Meatime, the alternating expansion and contraction of the blisters resulting from thermal variations, particularly in sunny weather, can be reduced by the application of a lime/tallow wash, taking care not to tread on the blisters during application.

Ruckling, rippling, cockling: Undulations (**9**) to which these descriptions have variously been applied can result from inadequate pressure on the roll or insufficient or badly distributed bitumen compound. It is particularly important to avoid these faults in the intentional partial bonding of the first layer. But this did not apply to the present case in which the bottom layer of felt was fully bonded to a fibreboard underlay. Asbestos-based felt type 2A was specified for the bottom and middle layers, but type 1C, which can contain a considerable proportion of vegetable fibre, for the top layer. The vegetable fibre in the top layer of felt could, therefore, have been exposed to the ingress of moisture and the resulting dimensional changes would then aggravate the effects of vapour pressure corresponding to variations in ambient temperature. However, the regular, parallel pattern of the undulations suggested that most, if not all, of the blame stemmed from not allowing the top layer of felt to flatten before fixing.

Except in one case where an undulation is associated with partial lifting of a lapped bond in the felt, which needs to be re-sealed, no immediate repairs are considered necessary so long as the undulations are not likely to be trodden on by maintenance staff servicing plant on the roof.

Roof drainage: part 1

This digest is published in two parts: Part 1, Sizes for eaves gutters and down-pipes, based on experimental data; Part 2, Sizes for valley and parapet gutters, outlets or receivers, and rainwater pipes, based on experimental data and theoretical studies.

The two parts together replace Digest 107 which is now withdrawn.

Design rate of rainfall

Roof drainage calculations can usually be based on a rate of rainfall of 75 mm/h. Short storms of higher intensity — 150 mm/h or more — do occur and should be assumed for situations where overflowing cannot be tolerated. Regional differences are more significant in relation to total rainfall than to intensities and can usually be ignored for the present purpose.

As an indication of the frequency of these intensities: 75 mm/h may occur for 5 minutes once in 4 years, or for 20 minutes once in 50 years.

150 mm/h may occur for 3 minutes once in 50 years, or for 4 minutes once in 100 years.

Sizes of eaves gutters and downpipes

Flow load

The rate of run-off from a roof is the product of the design rate of rainfall (usually 75 mm/h) and the effective roof area. (1 mm of rainfall on an area of 1 m² is 1 litre of water).

Information on the strength of the wind during periods of intense rainfall is sparse, but it is suggested in CP 308 that the angle of descent of wind-driven rain should be taken as one unit horizontal for every two units of descent (*see* Fig 1). This means that the effective roof area should be taken as the plan area plus half the elevation area — in Fig 1, (b + c/2) × length.

Prepared at Building Research Station, Garston, Watford WD2 7JR
Technical enquiries arising from this Digest should be directed to Building Research Advisory Service at the above address.

Eaves gutters

Each eaves gutter should be designed for the wind direction that will give the maximum rate of flow in the gutter. In practice, it may be convenient to calculate the flow load per metre run of eaves.

A square angle within 4 m of an outlet will reduce the flow capacity of a gutter. This can be allowed for by multiplying the rate of run-off by one of the following factors, instead of adjusting the values for flow capacity of the gutter:

angle within 2 m of the outlet:
— sharp-cornered × 1.2
— round-cornered × 1.1

angle within 2-4 m of the outlet:
— sharp-cornered × 1.1
— round-cornered × 1.05

Table 1 gives flow capacities for level gutters, of half-round, segmental and ogee section, with outlet at one end. The flow capacities of gutters of other profiles and sizes can be calculated from the formula given in Part 2.

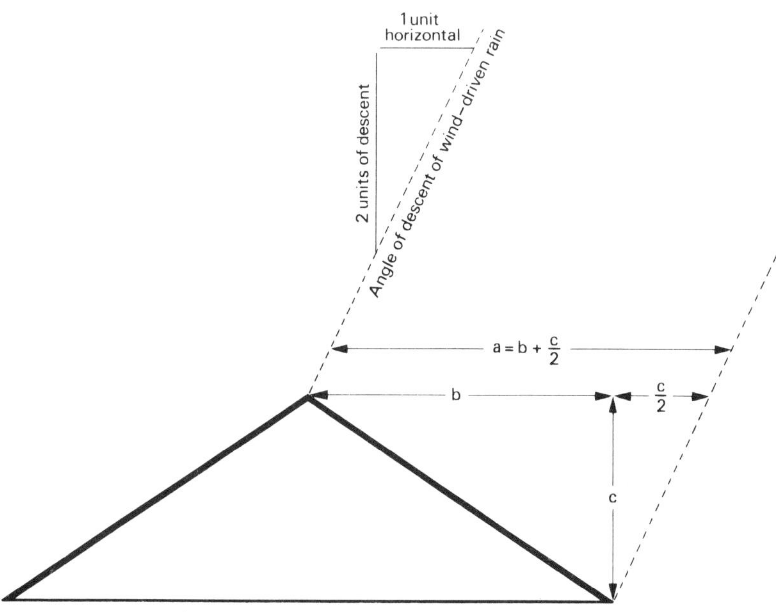

Fig 1 Effective roof area

Table 1 Flow capacities (litres/second) of level gutters with outlet at one end

Nominal gutter size mm	True half-round gutter (1, 2, 5)	Nominal half-round (segmental) gutter (3, 4, 5)	Ogee gutter (2)	(3, 4)
75	0.4	0.3	—	—
100	0.8	0.7	0.9	0.5
115	1.1	0.8	1.4	0.7
125	1.5	1.1	1.7	0.8
150	2.3	1.8	2.6	—

1 Asbestos-cement to BS 569: 1973
2 Pressed steel to BS 1091: 1963
3 Aluminium to BS 2997: 1958
4 Cast iron to BS 460: 1964
5 Unplasticised pvc to BS 4576: Pt 1: 1970

The effective area of roof (in m²) drained by level eaves gutters with outlet at one end, at a rate of rainfall of 75 mm/h, can be found by multiplying the flow capacities given in Table 1 by 48. For other rates of rainfall, the areas will be inversely proportional.

Downpipe sizes appropriate to stated gutter sizes are listed in Table 2. Figures are given both for sharp and round-cornered outlets (see Fig 2). The table allows also for the position of the outlet in relation to the end of the gutter.

Table 2 Minimum downpipe sizes (nominal dia in mm) for various gutter sizes

HR gutter size	Sharp or round-cornered outlet	Outlet at end of gutter	Outlet not at end of gutter
75	sc	50	50
	rc	50	50
100	sc	63	63
	rc	50	50
115	sc	63	75
	rc	50	63
125	sc	75	89
	rc	63	75
150	sc	89	100
	rc	75	100

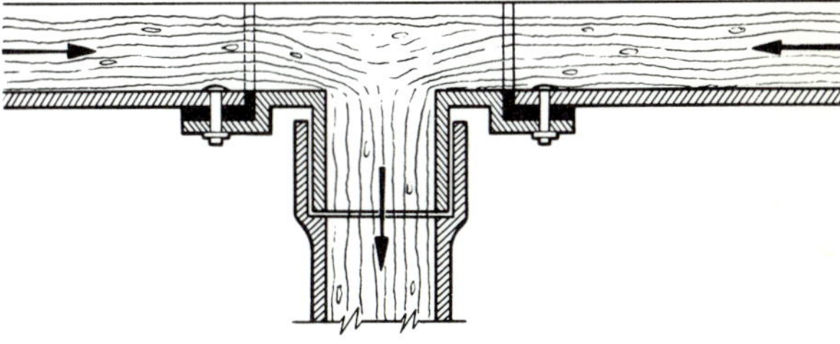

(a) sharp-cornered outlet

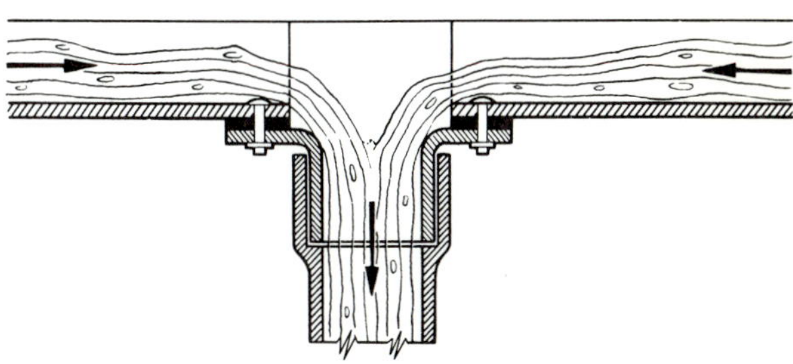

(b) round-cornered outlet

Fig 2 Outlets receiving flow from two directions

Example 1 *Calculate the sizes of eaves gutter and downpipes required to drain a roof 40 m long × 10 m (ridge to eaves) 30° pitch.*

First calculate the effective roof area per metre run of eaves. By calculation, or measurement from drawings, plan length ridge-to-eaves is 8.65 m; the height is 5 m.

Plan length + half the height = $8.65 + \dfrac{5.0}{2} = 11.15$ m.

The effective roof area is thus 11.15 m² per m of eaves.

With rainfall at 75 mm/h (1.25 1/min m²) the rate of run off will be
$$11.15 \times 1.25 = 13.94 \text{ 1/min}$$
$$= 0.23 \text{ 1/s}$$

From Table 1, a 125 mm (nominal) true half-round gutter could cope with the flow from a length of eaves of $\dfrac{1.5}{0.23} = 6.5$ m

Alternatively, a 150 mm (nominal) true half-round gutter could cope with the flow from a length of eaves of $\dfrac{2.3}{0.23} = 10$ m

Outlets would be needed at not more than double these distances apart, viz 13 m or 20 m for the two sizes of gutter, respectively.

Two possible solutions are shown in Fig 3: the roof could be drained (a) by a 125 mm true half-round gutter and three 75 mm diameter downpipes, or (b) by a 150 mm true half-round gutter and two 100 mm diameter downpipes. The downpipe sizes are read from Table 2 and assume the use of round-cornered outlets.

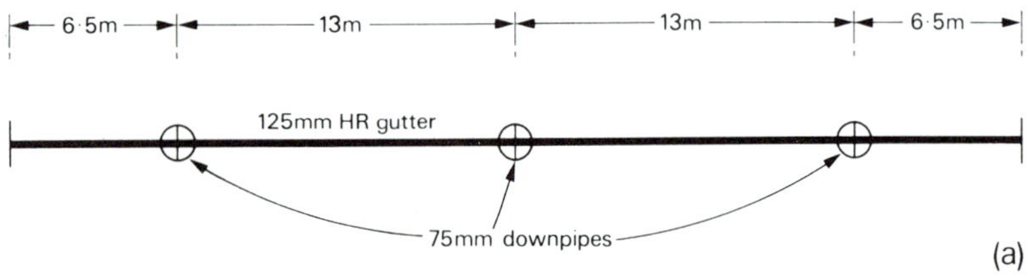

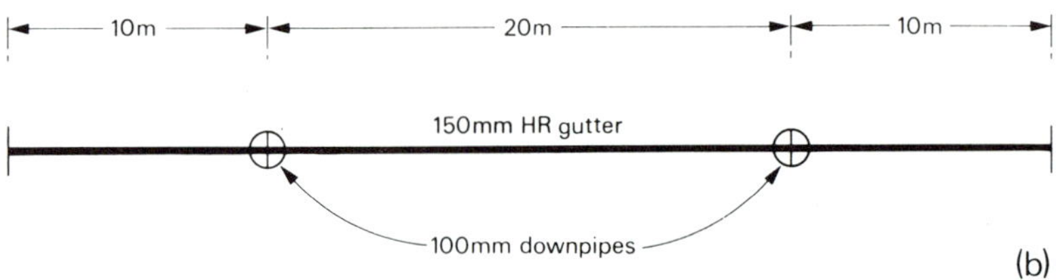

Fig 3

Roof drainage: part 2

Sizes of valley and parapet gutters, outlets or receivers, and rainwater pipes

For valley gutters, the *plan* area of the double roof should be taken as the effective area to be drained.

The sole of a valley or parapet gutter should be at least 300 mm wide; a convenient profile is obtained by sloping the sides to the same pitch as the roof, up to a height which gives sufficient flow capacity, and then turning a vertical upstand as the freeboard. The depth of the freeboard should be two-fifths of the gutter depth, up to a maximum of 75 mm (see Fig 4).

A gutter that cannot be allowed to overflow should be provided with two or more outlets; a weir overflow at the end of the gutter may also be needed.

Flow capacity for free discharge

Gutters should normally be provided with outlets designed to accept the flow from the gutter without increasing the depth of flow in the gutter ('free discharge') except where the minimum practicable gutter size provides a large surplus capacity (see page 3).

The depth of flow in a straight level gutter discharging freely decreases from the maximum at the far end (the high end) to the outlet. It may be assumed that the depth of flow at the outlet (H_0 in Fig 4) is half the depth at the far end. When the gutter is flowing at its maximum capacity, therefore, the depth at the outlet is half the gutter depth excluding freeboard (D). A fall in the gutter will increase the rate of flow but no figure can be quoted for this; it should be regarded as an increase in the safety margin against overflow.

In a gutter discharging freely, the flow capacity (Q) is approximately

$$\sqrt{\frac{A^3}{B}} \times 10^{-4} \text{ litres/second}$$

where A = area of flow at outlet (mm²) and
B = width of water surface at outlet (mm)

The chart in Fig 5 shows the plan area of roof to be drained for values of B and A. Thus for any rectangular or trapezoidal gutter section and any depth of flow, values of B and A can be calculated or measured from drawings of gutter sections and the equivalent roof area can be obtained; alternatively, the depth of flow at the gutter outlet (required in calculating the size of rainwater pipe) can be determined for a given roof by trial and error.

The charts, Figs 5–7, are based on a rate of rainfall of 75 mm/h. For any other rate of rainfall, the measured plan area must be multiplied by $r/75$ (where r is the design rate of rainfall in mm/h) to obtain the appropriate gutter or rainwater pipe size from the charts. Conversely, the plan area of roof that can be drained by a given gutter obtained from the charts must be multiplied by $75/r$.

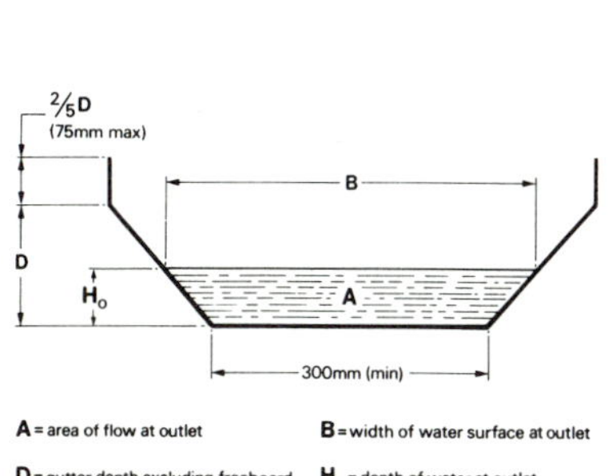

Fig 4 Dimensions at gutter outlet

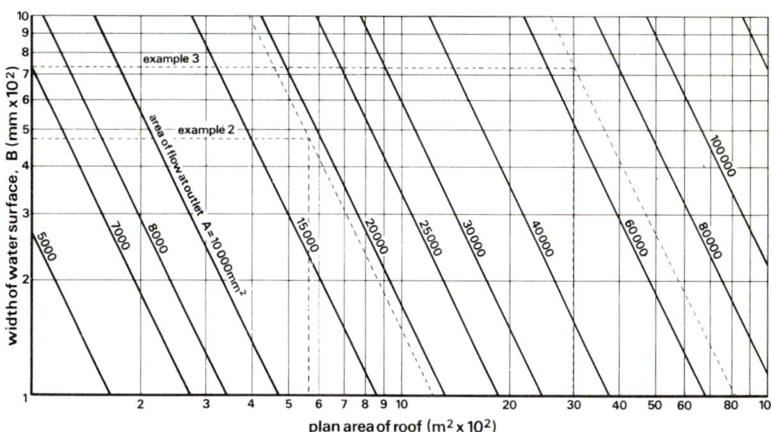

Fig 5 Relationship of roof area and flow dimensions

Prepared at Building Research Station, Garston, Watford WD2 7JR
Technical enquiries arising from this Digest should be directed to Building Research Advisory Service at the above address.

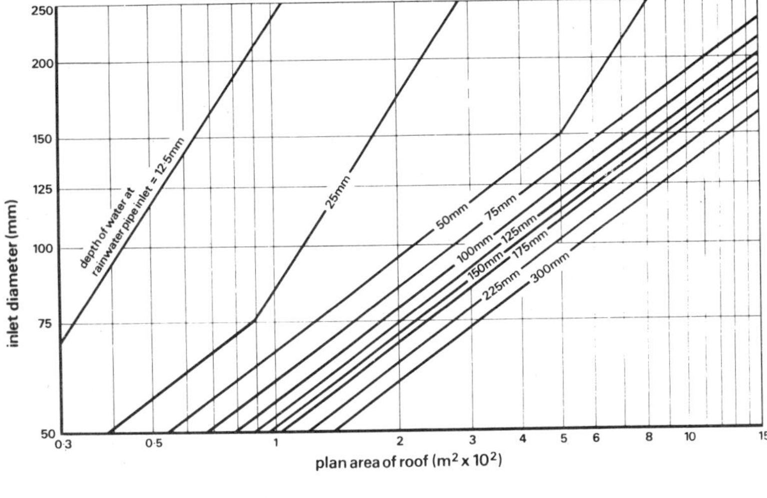

Fig 6   Roof area to be served by rainwater pipe inlet (no swirl)

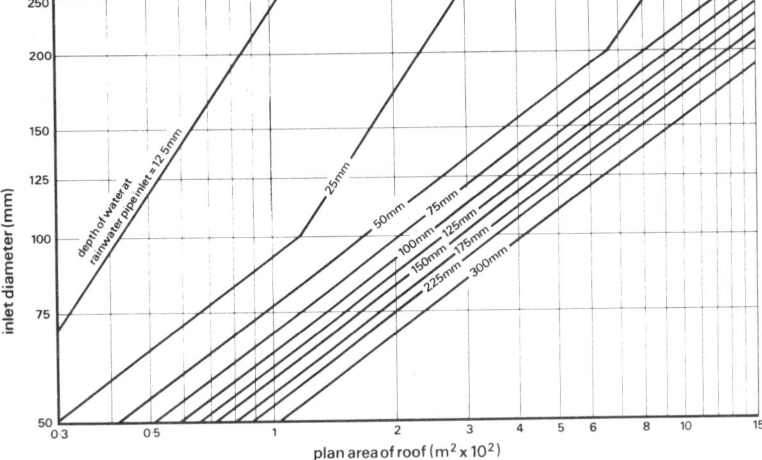

Fig 7   Roof area to be served by rainwater pipe inlet (with swirl)

Box-type receiver

Wherever possible, gutters should discharge into a box-type receiver, the depth of which can be chosen so as to permit the use of a rainwater pipe of convenient size.

The receiver should be at least as wide as the maximum gutter width and it should be long enough to prevent the flow from the gutter over-shooting the box. The top of the box should be level with the top of the gutter except where the box is external to the building, when the outer edge is kept lower to provide an emergency overflow. A suitable shape for an external box-type receiver is shown in Fig 8, in which the recommended depth of the edge is $H_d + 2/3 H_d$, where H_d is the depth of water at the inlet to the rainwater pipe required to give the necessary discharge capacity (obtained from Figs 6 or 7 according to whether or not there is swirl, see below). The length L_b of the box should be $2\sqrt{F.H_0}$, where F is the fall between the water surfaces (not less than $H_0 + 2/3 H_d$).

Where a box is situated part way along a length of roof, a separate value for L_b should be calculated for each gutter length draining into it and the two lengths added to give the total length of the box.

Rainwater pipes

Where the gutter discharges directly into a rainwater pipe, the rainwater pipe inlet (or gutter outlet) should be designed to allow free discharge from the gutter. The discharge capacity of the rainwater pipe inlet depends on the diameter of the inlet, the depth of water (head) over the inlet and on whether the water swirls about the inlet.

The charts (Figs 6 and 7) give the roof area that can be drained by an inlet of given diameter, at various heads, for inlets with and without swirling, respectively.

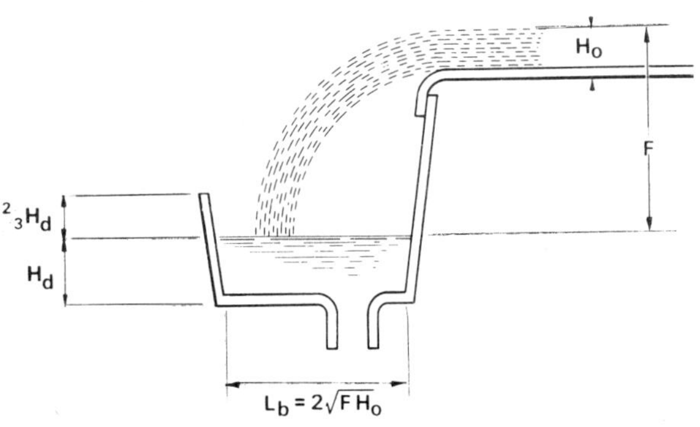

Fig 8   Dimensions of box-type receiver

Swirl can be neglected if the centre line of the inlet is at a distance less than its diameter from the nearest vertical side of the gutter or box. Swirl can also be suppressed by placing a vertical guide vane along the axis of the gutter, over the rainwater pipe inlet. In the absence of swirl, the inlet acts as a weir for values of the ratio H_i/D up to 1/3 and the discharge capacity of the inlet (Q_i) in litres/second is

$$\frac{D}{5000}\sqrt{H_i^3}$$

where D = diameter of inlet (mm) and

H_i = head of flow over the inlet (mm).

Where H_i/D is greater than 1/3, the inlet acts as an orifice and

$$Q_i = \frac{D^2}{15\,000}\sqrt{H_i}$$

Fig 6 has been prepared from the above expressions. *

Swirl will occur when the centre of the inlet is at a distance greater than its diameter from the nearest vertical side of the gutter or box (unless suppressed by a vane). The inlet then acts as a weir with the same discharge rate as in the absence of swirl for values of H_i/D up to 1/4; above this, the inlet acts as an orifice but swirling will reduce its discharge capacity to

$$\frac{D^2}{20\,000}\sqrt{H_i}.$$

The chart in Fig 7 is applicable to this situation. *

When sizing the inlet to a rainwater pipe leading directly from the gutter, the head H_d should be taken as the depth of flow at the gutter outlet (H_0 in Fig 4). If, however, the gutter discharges into a box-type receiver to which the rainwater pipe connects, the head is the depth of water in the box (H_d in Fig 8).

The diameter of the rainwater pipe can be reduced to two-thirds of the effective inlet diameter provided that the transition is gradual over a length of not less than the diameter of the inlet.

If the inlet is covered by a grating, the effective discharge capacity of the inlet must be calculated from the unobstructed area of the grating.

Gutters with large surplus capacity

The minimum practicable gutter size may be considerably greater than the hydraulic considerations demand. The excess capacity can then be utilised to reduce the required size of rainwater pipe by permitting the depth of flow in the gutter to rise above that for free discharge. This applies only where the rainwater pipe leads directly from the gutter.

The procedure involves some trial and error to discover the smallest size of rainwater pipe inlet

that will not cause the gutter to overflow, and is as follows:

Select a convenient size of rainwater pipe inlet and, for the roof area to be drained, read the head H_d (ie the depth of water at the rainwater pipe inlet) from Figs 6 or 7, as appropriate.

From H_d, calculate the width of the water surface B at the gutter outlet, and the area of flow A.

Using these values of B and A, obtain from Fig 5 the equivalent plan area of roof for free discharge.

Calculate $R = \dfrac{\text{plan area of roof to be drained}}{\text{equivalent plan area for free discharge}}$

Using R, from Fig 9 read $X = \dfrac{\text{max. depth in gutter}}{\text{depth at gutter outlet}}$

Calculate $X \times H_d$. If this is greater than the gutter depth (excluding freeboard), the procedure must be repeated using a larger diameter rainwater pipe inlet. If, however, this depth is less than the gutter depth, a smaller inlet can be tried until the smallest which will not result in overflowing of the gutter is arrived at.

Example 2

The plan area of a roof to be drained by a valley gutter is 25 m × 20 m = 500 m². The roof pitch on either side of the gutter is 30°. The gutter is to be trapezoidal in section, with sides sloping at 30° and a sole width of 300 mm. What depth of gutter and what size of rainwater pipe are required?

Gutter

The required depth is found by trial and error for assumed values of the depth, H_0, at the gutter outlet.

Try H_0 = 25 mm:

then, for the given gutter section,

$$B = 386 \text{ mm}; \qquad A = 8580 \text{ mm}^2$$

From Fig 5, the corresponding roof area is under 200 m² which is substantially less than the area to be drained.

Therefore, try H_0 = 75 mm:

then, $B = 560 \text{ mm}; A = 32\,250 \text{ mm}^2$.

The corresponding roof area (from Fig 5) is over 1000 m², which is on the safe side, but uneconomical.

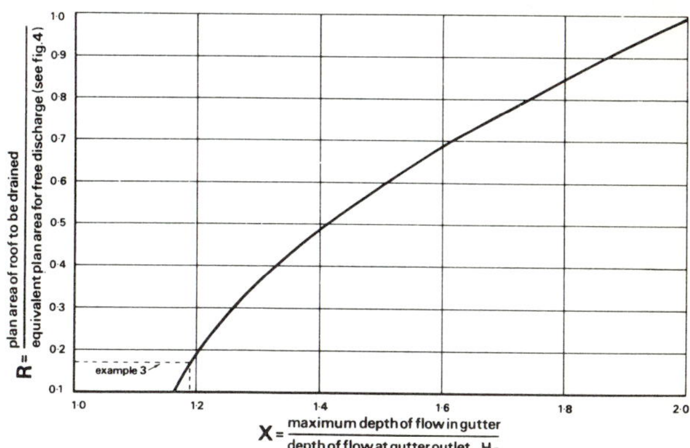

Fig 9 Relationship between depth of flow at outlet and maximum depth of flow for gutter with restricted outlet

---

Therefore, try $H_0 = 50$ mm:

then, $B = 473$ mm; $A = 19\,300$ mm².

and, from Fig 5, the corresponding roof area is 560 m² (approx) which is the approximate plan area of the roof to be drained. (The roof area obtained from the assumed depth of flow will rarely equal exactly the area of roof for which the gutter is required but the exact depth of flow can be obtained by interpolation). For free discharge, the maximum depth of flow in the gutter is twice the depth at the outlet; therefore the required gutter depth is 2×50 mm $= 100$ mm. To this should be added a freeboard of 40 mm.

Rainwater pipe

Where the rainwater pipe is connected directly to the gutter, the depth of flow at the gutter outlet found above, ie $H_0 = 50$ mm, is used in calculating the diameter of the rainwater pipe inlet (gutter outlet). As marked on Fig 7, with $H_0 = H_d = 50$ mm, and a plan area of 500 m², the required diameter of the rainwater pipe *inlet* is 180 mm. The *pipe* diameter can, however, be reduced to 120 mm; as explained on p. 135, a reduction to not less than two-thirds of the inlet diameter is admissible.

If the gutter discharges into a box-type receiver, it may be possible to use a smaller rainwater pipe inlet by increasing the head over this inlet.

Try a 150 mm rainwater pipe inlet:

from Fig 7, the required head $H_d = 90$ mm, and the difference F between levels of the water surfaces in the gutter and the box should not be less than

$$H_0 + 2/3H_d$$

that is, $F \geqslant 50 + (2/3 \times 90)$ mm $= 110$ mm.

The length of the box should not be less than

$$2\sqrt{F.H_0} = 2\sqrt{(110 \times 50)} = 148\cdot3 \text{ mm}$$

and the bottom of the box should not be less than

$$(H_d + 2/3H_d) = 150 \text{ mm}$$

below the bottom of the gutter.

Try a 125 mm rainwater pipe inlet:

from Fig 7, the required head $H_d = 175$ mm.

$$F \geqslant 2/3H_d + H_0 = (2/3 \times 175) + 50 = 170 \text{ mm}$$
$$L_b \geqslant 2\sqrt{(F.H_0)} = 2\sqrt{(170 \times 50)} = 185 \text{ mm}$$
$$H_d + 2/3H_d = 175 + (2/3 \times 175) = 291 \text{ mm}$$

that is, the use of the smaller diameter inlet requires that the bottom of the box should be at least 291 mm below the bottom of the gutter.

Example 3

A gutter of the same shape as in Example 2 is 200 mm deep, excluding freeboard.

(i) *What is the maximum area of roof that can be drained to it?*

(ii) *If the gutter is to drain a roof 25 m × 20 m = 500 m², what is the smallest rainwater pipe into which it can discharge directly?*

(i) The maximum capacity of the gutter is obtained when it discharges freely. The depth at the outlet is then half the gutter depth, that is,

$$H_0 = \frac{200}{2} = 100 \text{ mm};$$
$$B = 646\cdot6 \text{ mm};$$
$$A = 47\,330 \text{ mm}².$$

From Fig 5 the maximum roof area which this gutter can drain is 2000 m².

(ii) If the roof area to be drained is only 500 m², the maximum depth of flow in the gutter for free discharge is only 100 mm (*see Example 2*). The excess capacity can be utilised by letting the head at the outlet exceed that for free discharge by reducing the diameter of the rainwater pipe inlet.

Try a 135 mm rainwater pipe inlet:

from Fig 7 the required $H_d = 125$ mm and therefore

$$B = 732 \text{ mm}; \quad A = 64\,500 \text{ mm}².$$

From Fig 5 the equivalent plan area of roof for free discharge is 3000 m².

Therefore,

$$R = \frac{500}{3000} = 0\cdot15$$

and using Fig 9 as indicated,

$$X = 1\cdot19$$

Therefore the maximum depth in the gutter

$$= 125 \times 1\cdot19 = 150 \text{ mm}$$

which is on the safe side. Therefore, try a 120 mm rainwater pipe inlet. From Fig 7 the required H_d is 225 mm, which is greater than the gutter depth. The minimum size of rainwater pipe inlet is therefore 135 mm.

The rainwater pipe diameter can then be reduced to two-thirds of its inlet diameter, ie to 90 mm, as already explained.

5 Joints

Principles of joint design

This digest brings together the design considerations concerning fit that are common to most kinds of joint, whatever their primary function. For this purpose the term 'fit' includes not merely the requirements for matching size—whether a component is too big or too small for the space provided—but also the closely related capability of the completed assembly of components and joints to function satisfactorily.

Only recently have attempts been made to identify the factors that are fundamental to the process of fitting and jointing and to relate them in a way that enables designers to take systematic account of them. Knowledge is still far from complete: the values that are assumed to be appropriate for various kinds of inaccuracy in building are as yet based on a limited collection of data, and the functional requirements and capabilities of particular joints are still in many respects unknown in precise terms. However, a framework within which designers can make use of existing information—and of more comprehensive information as it becomes available—can now be put forward as an aid to joint design.

The essential is for designers to reconcile the dimensions required by the joint so that it can function satisfactorily and the dimensions that are imposed by the unavoidable inaccuracies of building. When this reconciliation has been achieved in practice then the parts concerned can truly be said to fit.

Deviations

Discrepancies between the intended and the actual sizes and positions of building components arise from two sources: man-made inaccuracies of manufacture and assembly, and unavoidable dimensional changes resulting from the inherent properties of the materials.

It has been common to describe man-made inaccuracies as 'tolerances', often without regard to whether they were in fact tolerable from the design point of view and usually assuming (wrongly) that they represented absolute limits. Man-made inaccuracies are now called *induced deviations* and acceptable limits for them, known as *permissible deviations,* can be set in the knowledge that there will usually be an unavoidable proportion of dimensions falling outside the limits.

Inaccuracies which are associated with inherent properties—such as thermal and moisture movements —are known as *inherent deviations.*

The dimensional needs of a joint

For a joint to perform as intended, its finished width must lie within certain limits. For example, if a gunned mastic is to be used, limits are imposed below which gun application is not practicable and above which the material may slump; a lower limit is dictated also by the least width of mastic that can accommodate the expected movement, while the upper limit may be more or less fixed by material costs. Upper and lower limits may similarly be fixed by limits of lip-seal pressure for gaskets (Fig 1a), by depth of engagement of a baffle in its grooves or by the amount of overlap in a lap joint. Thus the acceptable limits of width stem from decisions about the functions that the joint must perform and then about the materials and techniques of construction used in the joint to enable it to perform those functions.

Joint functions

The functions that any particular joint may be required to perform depend primarily on its environment, its location and on the functions and properties of the joined components. Table 1 is a comprehensive list of joint functions from which the designer may select those that apply. As an aid to selection it is useful to consider which functions of the joined components must be maintained at the joint and then any additional functional requirements that arise from the presence of the joint itself. The functions required of the joint form a basis for the selection of jointing materials, the shape of the joint and the corresponding profiles for the edges of the components.

A convenient way to approach this stage of joint design is to take each functional requirement in turn and produce a schematic design solution identifying the characteristics which will be needed in the jointing product and any restriction on its location within the joint. For example, the functional requirement 'to resist the passage of water vapour' will call for a vapour-resisting layer and this must be sited on the warm side of any thermal insulation in the joint. Often there will be alternative solutions—in this example to employ thermal insulation which itself

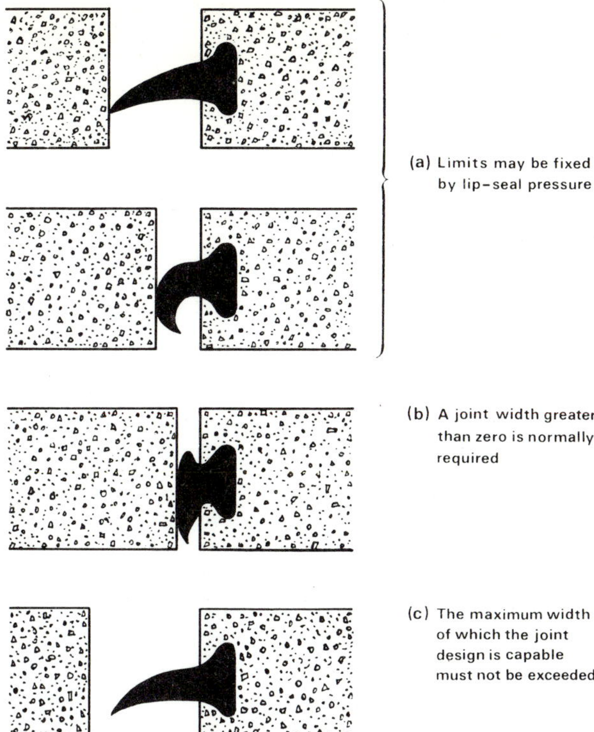

(a) Limits may be fixed by lip–seal pressure

(b) A joint width greater than zero is normally required

(c) The maximum width of which the joint design is capable must not be exceeded

Fig 1

has adequate vapour resistivity. When schematic solutions for all the relevant functions have been devised they can be examined to see how they can best be combined into one joint design, or how far the required characteristics are embodied in an existing design. Again there may be alternative combinations of functions.

At this stage the designer can select specific jointing products having, as far as possible, the required properties. Some jointing products may be subjected to strain by changes in joint width; an estimate of the effects of inherent deviations on changes in joint width in service will provide a basis both for selecting a jointing product and for deciding the minimum joint width needed to avoid excessive strain.

The designer now has a joint design featuring all the functions required of it except one: the function of accommodating induced deviations. It is at this point that the extent to which consideration is given to the question of fit very largely determines the success of the joint. Failure of the joint usually undermines substantially the performance of the joined components and often leads to costly remedial work or recurring maintenance expenditure. Careful consideration of the effects of induced deviations is likely to be well repaid.

Induced deviations

Suppose that components from two batches are to be used end to end; one batch is of components 100 mm long with a manufacturing permissible deviation of ±7 mm and the other is of components 200 mm long with a permissible deviation of ±10 mm. If it is assumed that the worst will happen, then an assembly of two components may have a combined length of 317 mm, or 283 mm. But this is an extremely remote possibility and to apply this assumption to all the component combinations is to assume not only that the worst will happen but that it will happen every time.

An expression such as '100 mm long with a manufacturing permissible deviation of ±7 mm' means that a finished length of 100 mm is aimed at, but because no process is perfect some of the components will be longer than this and others shorter. A distribution curve (Fig 2) of the sizes actually produced will show that the majority are near to the intended size and that those near to the outside limits are relatively few.

The spread of the curve is an indication of the accuracy of the process. Figures 2 and 3 could be distribution curves for the same component, manufactured accurately (Fig 2) and less accurately (Fig 3). A measure of the spread is called *standard deviation* and when calculated for a particular distribution this can be used to identify the points on the distribution curve that may be chosen as *permissible deviations*. Experience based on a very large number of distributions has shown that the shapes of such curves are essentially similar. In Fig 4 the calculated value for standard deviation has been used to divide a distribution curve into separate parts and shows the chances of deviating more than one, two or three standard deviations from the mean. These proportions can be applied to all distributions of induced deviations.

Occasionally, very large errors occur. The manufacturer of a 100 ±7 mm component does not mean that he will never manufacture a component longer than 107 mm or shorter than 93 mm. In order to tell his customers exactly what he does intend he could make known the standard deviation for his product; in the absence of this, an assumption must be made. It may be reasonable to assume, for example, that the quoted accuracy of ±7 mm represents three standard deviations either side of the mean and thus that only about one in three hundred of the components will fall outside the stated range.

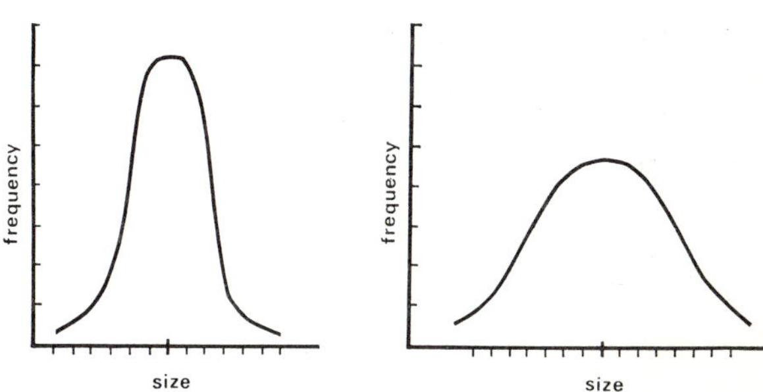

Figs 2 and 3

Fig 4

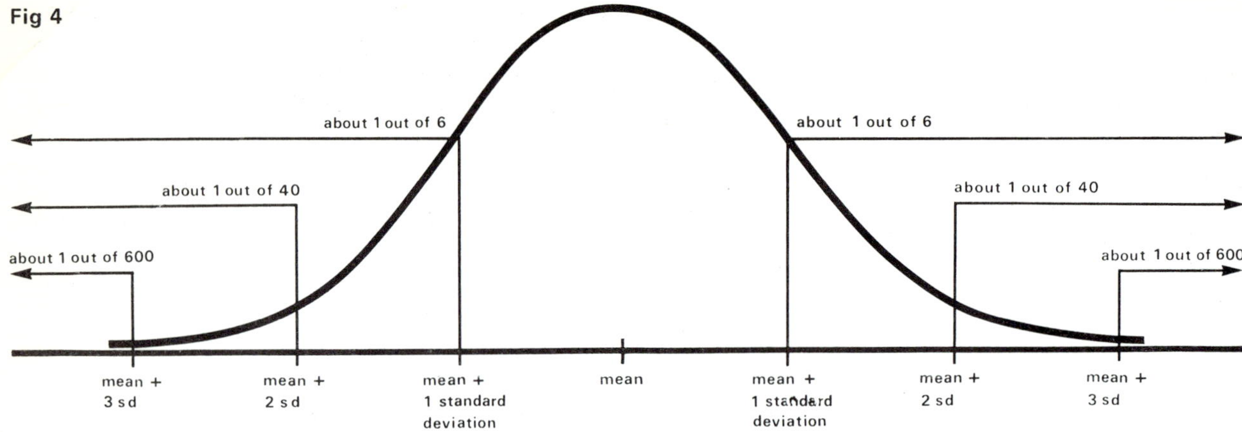

A standard statistical method for summing any number of random variables reflects the chance combinations that occur in practice. The total effect is obtained by taking the square root of the sum of the squares of the separate deviations. Thus, where one component has a permissible deviation of ±7 mm and another a permissible deviation of ±10 mm, the total deviation is $\pm\sqrt{7^2+10^2}=\pm12$ mm, approximately. This is significantly less than ±17 mm, which would have been assumed on the basis that the worst combinations would consistently occur. Since the components of the combination have one chance in 600 of falling outside their permissible deviations, it is illogical to design a combination which has a very much smaller chance; a deviation of more than 17 mm will occur less than once in 100,000, whilst the calculated deviation ±12 mm has the same chance of occurring as have the permissible deviations of the components.

Not only does this method of calculation reflect the real consequences of random combinations more accurately; it enables the designer to produce joints that are not unnecessarily large and that are much more likely to be within the dimensional capabilities of the joint design finally chosen.

Reconciling joint and component dimensions
Induced deviations in building comprise three main types of inaccuracy: marking setting-out lines,* positioning of components and manufacture of components. The extent to which any one of these affects the dimensions of the joint depends on the nature of the assembly concerned. If, for example, the assembly is of components that are small or light enough for their positional errors to be readily corrected then no allowance for their positional deviations need be made. If they are required to fit within a space of which the deviations in size are known then the total allowance, Y, for induced deviations can be obtained from

$$Y = \sqrt{m_1^2+m_2^2+m_3^2\ldots\ldots+s^2}$$

where m_1, m_2, etc, are the manufacturing deviations ($\pm m$) for the components and s is the deviation

($\pm s$) in the size of the space. Since all the constituent deviations were plus and minus values so also is the total Y.

The positive value of Y is the amount by which the whole assembly of components must be reduced in size in order to make it small enough to enter the space, but the total clearance within the assembly will vary by $\pm Y$, that is, over a range of $2Y$, and a share of this range will be borne by each joint. If the joints are equalised in width and if there are x joints each will vary in width by $2Y/x$, and since the deduction made was only enough just to secure entry of the components into the space it follows that each joint width may range from 0 to $2Y/x$. Because joints normally require minimum widths greater than 0 (Fig 1b), the component sizes must be further reduced by the amount needed to provide the required minimum at each joint.

Before making this further reduction, it is necessary to check that the selected joint design is able to function adequately when its width may be any value between the required minimum and the minimum plus the range—in this example $2Y/x$—produced by induced deviations. Finally, inherent deviations must be estimated and the amount by which the components will expand must be further deducted from their size. The full range of expansion and contraction that may occur at each joint must be added to the earlier calculated maximum joint clearance to check that the maximum width of which the joint design is capable is not then exceeded (Fig 1c).

The designer's options
The designer may change the values of any variable within the relationship between fit and joint function provided that this relationship is maintained. He may, for example, decide to assume smaller values for deviations than those likely to exist and accept that a correspondingly higher proportion of misfits will occur; or he may require deviations to be kept within those lower values, where this is practicable, in order to use a particular type of joint. In this case the design should incorporate a specification of the values of permissible deviation for the relevant parts of the construction.

---

* *see* Digest 114 'Accuracy in setting-out'

Other options are available. Some or all of the dimensional problems can be avoided by using *in-situ* construction to absorb inaccuracies, by employing 'make-up' pieces, or by concentrating deviations at a special joint. Lapped joints generally accommodate deviations very readily; the process of reconciling joint dimension can be used to calculate the required amount of lap. Deviations affecting the thickness, as opposed to the overlap, of the lap joint zone may also be important in terms of fit and satisfactory function.

Good joint design

Every joint probably has a paramount function, often reflected in its description—'weatherproof joint', 'structural joint'. But to design a joint as though the paramount function is the sole function is likely to produce a joint that is less effective than it might have been. The check-list of functions should help to identify other functions that should be considered. Without clear identification of functions there is little if any basis for determining the characteristics required of the joint and which jointing products and profiles have those characteristics.

In the past, when joint failure has not stemmed from wrong choice of jointing product, it has stemmed from the occurrence in practice of joint dimensions that were not foreseen by the designer and that exceeded the capabilities of the joint design employed. Reconciliation at the design stage of the dimensions needed by the joint and those dictated by the unavoidable inaccuracies of building is essential to successful joint design.

A systematic approach to joint functions and dimensions should help the designer to give joints the attention they deserve as important and often vital parts of the construction.

Table 1 Check list of joint functions

A Control of environment To resist : A1 Passage of water A2 Condensation A3 Passage of water vapour A4 Passage of air A5 Passage of heat A6 Passage of light A7 Passage of sound A8 Generation of sound A9 Passage of electro-magnetic radiation A10 Passage of insects and vermin A11 Passage of plant leaves, roots, seeds and pollen A12 Passage of odours A13 Generation of odours	**G Durability** G1 To have specified minimum life. G2 To resist damage by extremes of temperature. G3 To resist damage by freezing of water. G4 To resist damage by light. G5 To resist damage by polluted air. G6 To resist damage by airborne or structure-borne vibrations, shock-waves or high-intensity sound. G7 To resist damage by water, water vapour or aqueous solutions or suspensions. G8 To resist damage by electro-magnetic radiation. G9 To resist damage by plants and micro-organisms. G10 To resist unauthorised dismantling or damage by man.

B   Loadbearing capacity
To resist stress in one or more directions :

(a) Compression	(d) Shear
(b) Tension	(e) Torsion
(c) Bending	(f) Stresses due to impact

C   Safety
C1   To resist passage of fire
C2   To resist pressure due to explosion

D   Maintenance
D1   To permit partial or complete dismantling and reassembly

E   Accommodation of deviations
E1   To accommodate variations in the sizes of the joint at assembly due to deviations in the sizes and positions of the joined components.
E2   To accommodate continuing changes in the sizes of the joint due to thermal, moisture and structural movements, vibration and creep.

F   Fixing of components
F1   To support joined components in one or more directions.
F2   To resist differential deflection of joined components.
F3   To provide fixing of components to structure.

H   Ambient conditions
H1   To perform required functions over a specified range of joint clearance variation.
H2   To perform required functions over a specified range of air pressure differentials.
H3   To perform required functions over a specified range of atmospheric humidity.
H4   To perform required functions over a specified range of temperatures.
H5   To exclude if performance would be impaired :

(a) water	(e) insects
(b) ice	(f) plants
(c) snow	(g) polluted air
(d) micro-organisms	(h) solid matter

J   Appearance
J1   To have acceptable appearance.
J2   To avoid discoloration due to algae, moulds or efflorescence.
J3   To avoid promotion of plant growth.
J4   To avoid pattern staining.

K   Economics
K1   To have economic first cost.
K2   To have economic maintenance cost.
K3   To have economic depreciation.

Joints between concrete wall panels: open drained joints

This digest deals briefly with some of the main factors, principally the threat of rain penetration, which determine the performance standards required of joints between precast concrete wall panels, and discusses in detail the design of one type of joint—the open drained joint—which when properly designed and executed has proved as successful as any in accommodating these factors.

Industrialised building, in particular the increasing use of storey-height wall panels, has placed severe demands on the performance required of joints. The components are large and their surfaces are relatively impermeable to water. The larger the components, the more difficult it is to make them accurately, to place them accurately, and to provide for the larger movements—structural, thermal, and moisture—associated with them. Their impermeability has serious implications for joint design because rainwater striking the surfaces of components can be discharged over the joints.

Traditionally, joints could be left to take care of themselves; even now, the notion of designing for what are really only gaps formed by the juxtaposition of components is not always grasped. It must be appreciated that such designs cannot simply be taken off the shelf and expected to perform equally satisfactorily in a variety of situations. Performance is too easily influenced by factors already mentioned; in no other aspect of building is the need for integration of the design and construction processes so urgent.

Joints at present used for buildings of precast wall panel construction are of two main types: those relying exclusively on a seal provided by sealants and/or gaskets—the filled joint, and those in which the geometry of the joint, in conjunction with an air-tight seal, is deployed in a two-stage defence against the weather—the open drained joint.

Filled joints

These rely mainly on mortars, sealants and gaskets to fill the joints, often at their faces (Fig 1). Too often the joints are given a low priority in design and assembly and the filling is required to take up any deficiencies which become apparent in the finished building. Mortars are liable to shrink, they have poor adhesion and cannot accommodate subsequent movement; their use should be limited to small components. Sealants have a greater range of uses but must be chosen with care in relation to joint width and movement; fracture or failure to adhere to the sides of the joint will allow air and water infiltration. These risks may indicate that the demands made on sealants are excessive and that an alternative solution to the problem of weather-tightness must be found. Gaskets can be shaped to suit the internal profile of a joint, though they are sometimes designed

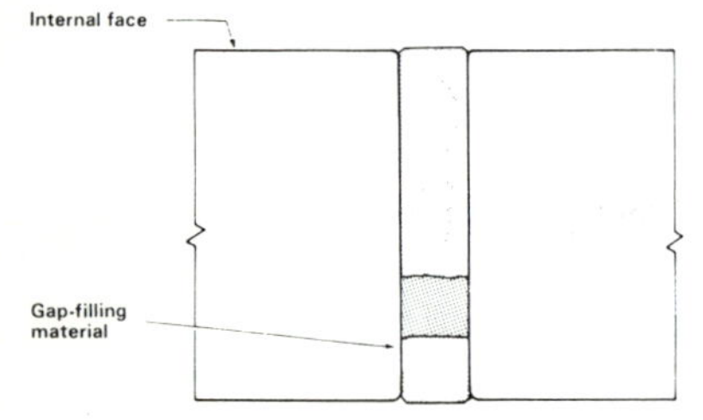

Fig 1 Vertical joint filled at face

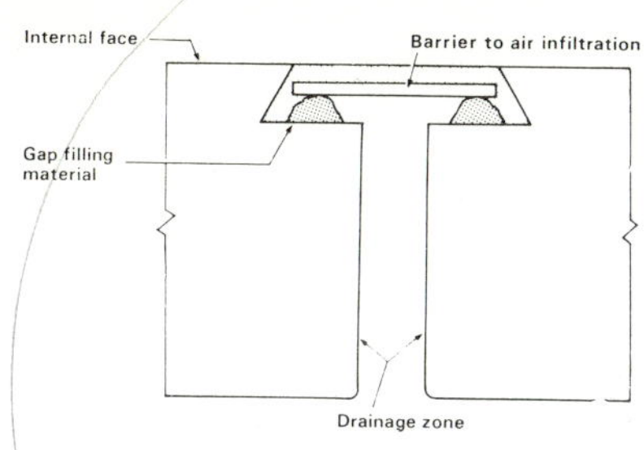

Fig 2 Vertical joint sealed at back

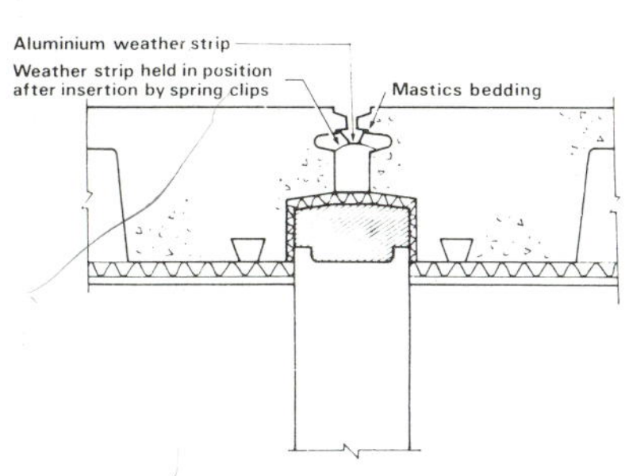

Fig 3 Vertical drained joint, developed from the original design (Reema)

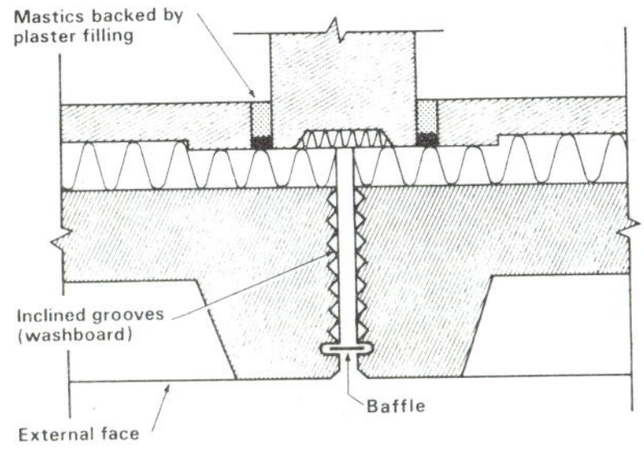

Fig 4 Vertical drained joint (Malmstrom)

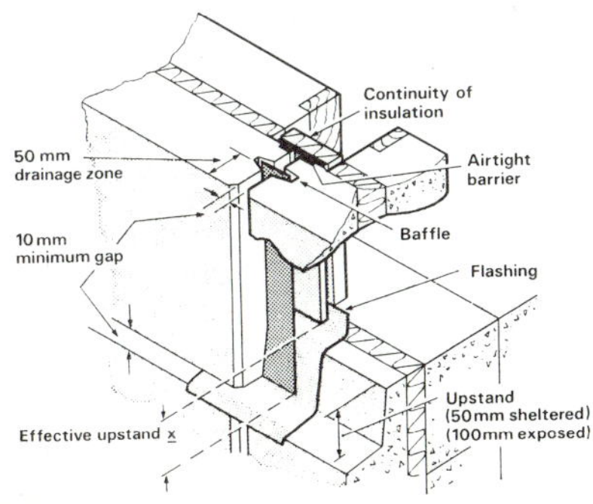

Fig 5 Drained joint between facade panels

with greater emphasis on ease of application than on weather-proofing; concrete surfaces are often not smooth enough for gaskets to seal effectively. Whatever the type, gaskets share with mastics fundamental limitations when relied upon exclusively for weather protection. These limitations originate from uncertainties of accuracy in manufacture and assembly, and often place unreasonable demands on the properties of available sealants.

Design principles of the open drained joint

The open drained joint aims at overcoming the weaknesses noted above. It has no jointing material at the face but relies instead on (i) an air-tight barrier at the back of the joint to prevent air flow through it, and (ii) the geometry of the joint to trap most of the rain in an outer zone (Fig 2). An intermediate baffle usually divides the vertical joint into two zones: the outer, from which most of the water entering the joint is drained away, and the inner, which drains away what little water is able to penetrate past the baffle. The joint is thus effectively sealed only at the back, where the seal is least exposed to weathering action.

The air-tight and water-tight barrier at the back has a vital role; without it much of the water would be blown past the baffle, with consequent risk of rain penetration into the building. Even so, in conditions of severe exposure, some water penetrates to the air barrier and must therefore be deflected by it.

The idea of open drained joints is not new. Examples can be found in early types of patent glazing and, in the 1950s, applied to concrete cladding panels by Reema and in jointing systems of Scandinavian origin. Examples of these designs are shown in Figs 3 and 4.

Design details

One type of vertical joint embodying these important design features (and its intersection with a horizontal joint) is shown in Fig 5. Such a jointing system is likely to perform satisfactorily even in conditions of severe exposure. Features and dimensions important to the successful performance of the joint are as follows:

1   An air barrier at the back of both horizontal and vertical joints prevents air infiltration.

2   Most of the water entering a vertical joint drains within a zone 50 mm deep measured from the face of the joint.

3   A baffle at a depth of 50 mm, i.e. placed at the back edge of the drainage zone in a vertical joint, arrests wind-driven rain.

4   Horizontal joints are protected by a 50 mm upstand on sheltered sites and a 100 mm upstand on exposed sites. This upstand must be carried effectively across the intersection with the vertical joints.

The flashing at the intersection of vertical and horizontal joints must ensure that rainwater shed by the baffle, plus any that has penetrated to the air barrier, is brought to the front of the joint at each intersection. This precaution is necessary to overcome the effects of any misalignment of vertical joints, and to ensure continuity of the upstand in the horizontal joints. The effective upstand overlap at the joint intersection is indicated as x and is measured from the foot of the baffle. This is why the positioning of the baffle is important; the lower ends of baffles, if dragged downwards into position during assembly, may subsequently creep back and so reduce the overlap. This should be guarded against. Similarly, the top ends of the baffles should be carried well up under the flashings.

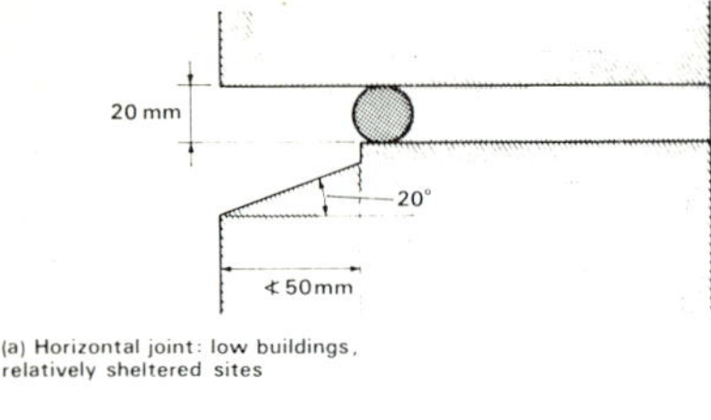

(a) Horizontal joint: low buildings, relatively sheltered sites

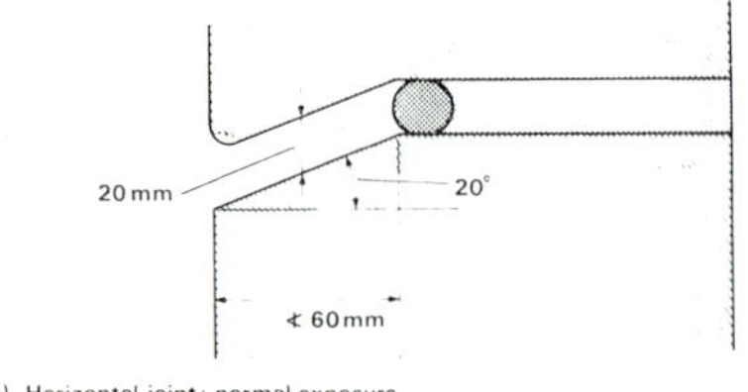

(b) Horizontal joint: normal exposure

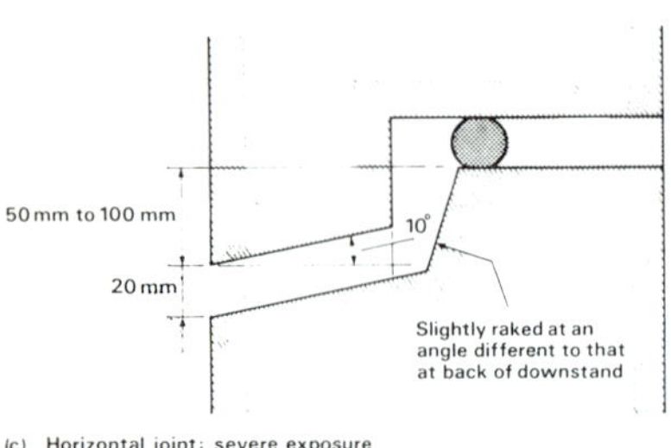

(c) Horizontal joint: severe exposure

Fig 6 Horizontal joints

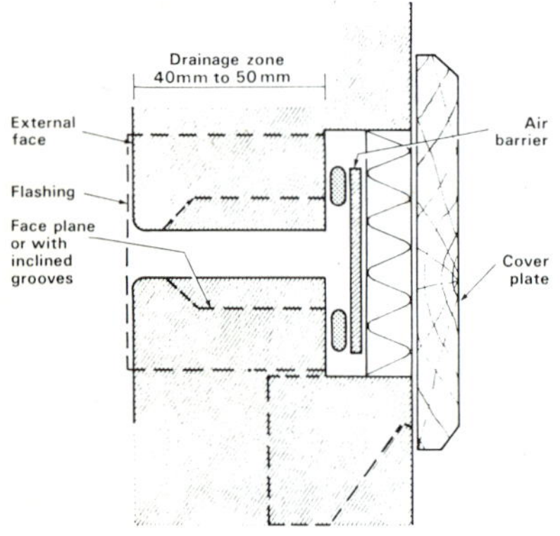

Fig 7 Vertical drained joint without baffle, sealed from the back

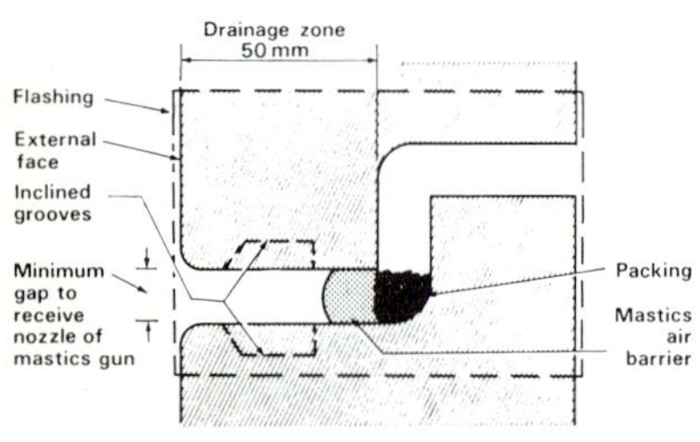

Fig 8 Vertical joint without baffle, sealed from the front

Experimental background to recommendations

An experimental rig set up in 1960 to explore the performance of joints in conditions of natural exposure allowed measurements to be made of the amounts of rainwater penetrating the joints. With subsequent experience in Scandinavia and the UK, the results provide the basis of the recommendations given in this digest.

Horizontal joints

Measured amounts of water entering horizontal joints showed clearly the value of upstands and sloping cills in reducing water penetration. The effect of sealing the backs of horizontal joints showed that an effective air barrier is the best defence against water penetration. Given perfect sealing, the results suggested that a joint with an unfilled section in front of the gap-filling material probably provides adequate weather protection if the lower surface of the joint is slightly weathered, as in Fig 6a. With less than perfect sealing, slight weathering of both upper and lower surfaces of the joint is advisable Fig. 6b, while with severe exposure, an overlap is needed to protect the joint Fig 6c. Experiment showed that little is gained by providing an upstand with an effective height greater than 50 mm, unless the filling is likely to be ineffective or the exposure severe. In these circumstances up to 100 mm may be necessary. These dimensions take no account of tolerances; these must be allowed for by additional overlap. In general, the safest recommendation is that given in Fig 5: to design for a 50 to 100 mm upstand as appropriate for sheltered or exposed situations. If, for the purposes of interchangeability, an upstand cannot be formed in the components, much the same effect can be obtained with suitably designed flashings, provided an air seal is maintained at each side of the flashing, but weather-tightness should not be sacrificed for interchangeability.

Vertical joints

The value of an effective air seal at the back of the joint was demonstrated; the amount of water entering sealed and unsealed joints approached a ratio of 1 : 3. Increasing the width of a joint had relatively little effect on the catch since by far the largest proportion of water entering a joint did so from the surfaces of adjacent panels. Consequently vertical features on the face of the panel which reduce the side flow of water give considerable protection to the joint. The evidence was that vertical grooves do not act as drainage channels, but that they do provide a sharp edge in the face of a joint and this probably curtails the sideways flow of water.

Open joints with inclined (washboard) grooves in their faces do not prevent rainwater penetrating to the back of the joint but help to drain it towards the front. Thus washboard grooves offer sufficient additional protection to suggest that intermediate baffles could probably be omitted from joints in all but high or very exposed buildings—but only if careful workmanship by operatives fixing the air barrier can be guaranteed. Loosely inserted baffles of type shown in Fig 5, and hollow neoprene tubes, provide some additional protection against water penetration.

Although tolerances are less critical with the open drained joint than with other kinds of joint, difficulties may arise when adjustment of position is difficult. Recent work has shown that it is possible to make the selection of a suitable width of joint in a less arbitrary manner than previously—*see* BRS Current Paper CP 5/71.

The drainage zone

In some instances the drainage zone will provide adequate protection to the air barrier at the back of the joint, without the aid of a baffle. This is especially so if the zone has vertical edges to prevent the

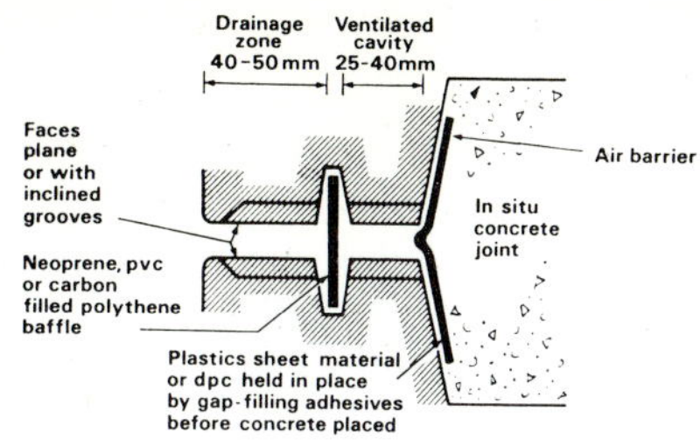

Fig 9 Vertical drained joint with baffle and ventilated cavity

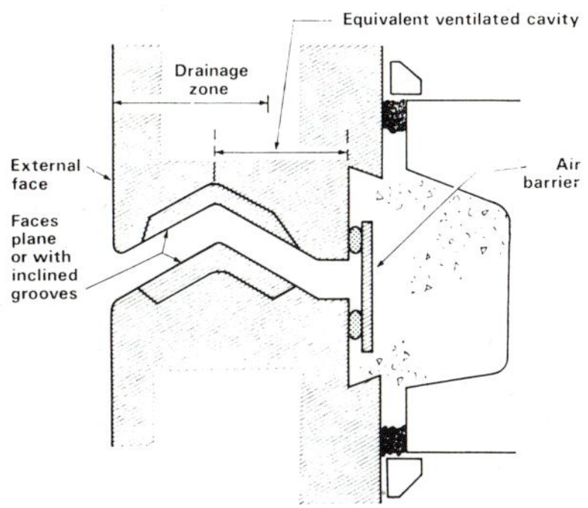

Fig 10 Vertical drained joint with a ventilated cavity equivalent

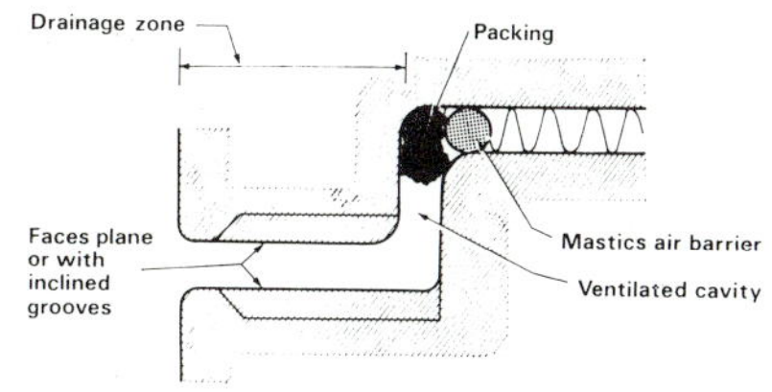

Fig 11 Vertical drained joint with ventilated cavity

sideways flow of water, or has inclined grooves to lead water to the front of the joint. Joints of this kind offer no problems in construction when the air barrier can be sealed from the back (Fig 7), but when panels are hung from a frame, joints are best sealed from outside the building, and a slightly lapped joint (Fig 8) will provide a stop against which the gap-sealing material can be gunned into position.

The baffle and ventilated cavity

The joint shown in Fig 9 gives almost certain protection to the air barrier because the small quantities of water penetrating the baffle will tend to drain in the cavity rather than impinge on the barrier, especially if the faces of the joint are provided with inclined (washboard) grooves. The simple strip baffle shown in Fig 9 does not need to be a tight fit in the vertical grooves and is tolerant of variations in joint width. The effect that deviations in size and position of the panels will have on joint width should be calculated or estimated from previous experience to determine the required depth of baffle grooves and width of baffle strip. It may be necessary, where joint widths are likely to vary greatly, to provide several widths of baffle from which a suitable one can be selected for each joint.

Alternatively, the equivalent of a baffle and a ventilated cavity could be provided by a labyrinth, for example, by a joggled drainage zone (Fig 10) or by returning the open section of the joint in the plane of the panel (Fig 11).

The air barrier

When proper provision is made for the drainage zone, the baffle and the ventilated cavity, the air barrier is relieved of most of the waterload which it would otherwise receive. Ideally sealants forming the air barrier should be in rolling shear (Fig 7) rather than in tension (Figs 8 and 11). When joints in low buildings are backed by *in-situ* concrete, very simple air barriers—perhaps of dpc felt temporarily held in place with gap-filling bituminous adhesive—have been found to work satisfactorily (Fig 9). Care is needed to overcome difficulties in sticking the air seal across adjacent faces when these are not in line.

Workmanship

The design of drained joints reduces the risk of water penetration; it does not, however, guarantee that a building will remain weatherproof if workmanship is unsatisfactory. Operatives engaged in jointing must therefore be given proper training with the principles of drained joint design fully explained. Finally, a systematic inspection procedure should be instituted so that the intentions of the designer are seen to be met.

Further reading

The relationship between component size and joint dimensions, Bonshor R B and Harrison W H. Reprinted from Building. BRS Current Paper CP 5/71.

Tolerances and fits for building: the calculation of work sizes and joint clearances for building components. Draft for Development to be published shortly by British Standards Institution.

PD 6440 Accuracy in building, Pt. 1 1969 Imperial, Pt. 2 1969 Metric, British Standards Institution.

6 General

Condensation

Principles involved in condensation; conditions producing condensation—atmospheric conditions and artificial influences. The behaviour of absorbent materials and surfaces. An explanation of interstitial condensation. Designing to avoid condensation—characteristics of the building fabric—characteristics of the environment. Estimating condensation risk—a worked example. Lightweight sheeted roofs.

Principles involved in condensation

The amount of water vapour that air can contain is limited and when this limit is reached the air is said to be saturated. The saturation point varies with temperature—the higher the temperature of the air, the greater the weight of water vapour it can contain. Water vapour is a gas, and in a mixture of gases, such as when present in the air, it contributes to the total vapour pressure exerted by the mixture. The ratio of the vapour pressure of any mixture of water vapour and air to the vapour pressure of a saturated mixture at the same temperature is the *relative humidity* (RH), which is expressed as a percentage. Alternatively, relative humidity can be regarded as the amount of water vapour in the air expressed as a percentage of the amount that would saturate it at the same temperature.

In conditions of, for example, 20°C and 80% RH, all the moisture can be held in the air. If more water vapour is introduced into the air and the temperature remains constant, the relative humidity will increase; saturation point (100% RH) may be reached and thereafter any further vapour will be deposited as condensation. If on the other hand the amount of water vapour remains constant but the temperature falls, because the colder air can support less moisture, the RH will rise until at about 15°C it is 100% and any further cooling will cause water to condense. This is the *dew-point* of that air which at 20°C had an RH of 80%.

Conditions producing condensation

It has shown that changes in temperature or in moisture content can cause condensation to occur. These changes can occur naturally—by changes in atmospheric conditions, or artificially—by living habits or industrial processes.

Atmospheric conditions When warm damp weather follows a period of cold, the fabric of a heavy structure which has not been fully heated will not warm up immediately but may remain comparatively cold for several hours or, if the walls are very thick, for a day or more. When the warm, moist, incoming air comes into contact with cold wall surfaces which are below its dew-point, water will condense upon them, but as the walls warm up and eventually exceed the dew-point, condensation ceases and the condensed moisture evaporates. A building of light construction will warm more rapidly and is less likely to suffer condensation from this cause.

A solid floor, with a non-insulating finish, has a surface that is slow to warm, and if there is a rise in temperature and humidity of the air above, it may suffer condensation for several hours. In general, the bigger the heat capacity of the structure, the longer will condensation persist on its surface in adverse conditions.

Artificial influences The humidity inside an occupied building is usually higher than outside. People themselves and many of their activities increase the amount of moisture in the air. Sedentary persons breathe out more than a litre of water, as vapour, in twenty-four hours; physical exertion may raise this to four times the rate. Moisture vapour is released by cooking, by clothes-washing and drying and by the combustion of oil or gas; a litre of oil burnt produces in vapour form the equivalent of about a litre of water and if burnt in a flueless appliance this vapour is emitted into the air within the building.

Many industrial processes require high humidities and temperatures and some release large quantities of steam. The risk of condensation is great when the

RH and temperature of the air remain above 60% and 20°C for long periods.

Condensation, particularly in dwellings, does not necessarily occur in the room where the water vapour is produced. A kitchen or bathroom in which vapour is produced may be warm enough to remain free from condensation except perhaps on cold, single-glazed windows, cold-water pipes and other cold surfaces. But if this water vapour is allowed to diffuse through the dwelling into cold parts such as the stair-well and unheated bedrooms, condensation will occur on the cold surfaces of those rooms, which may be remote from the source of the moisture. Soft furnishings, including bedding, and clothing may become damp because of this, especially as some of these materials are slightly hygroscopic.

Removal of the moisture-laden air from the building, from a point near to the source of the moisture, will greatly reduce the likelihood of condensation.

Water absorbed during construction During its early life, a building may be prone to condensation from the evaporation of water which entered the structure during construction, either as mixing water for the concrete, mortar and plaster or by exposure to the weather before the roof was completed. Much of this moisture evaporates into the internal air in the building and then condenses in the colder regions, usually at night. The amount of water can be as much as 4000 kg, the drying period may be as long as a year.

Some benefit can be gained by leaving all internal doors open and the upper storey windows ajar to facilitate the drying-out process; incoming occupants should be warned of the drying-out period.

Absorbent surfaces and materials

Temporary or intermittent condensation which is clearly visible on a non-absorbent surface may pass unnoticed on an absorbent surface or material. Condensed water can be absorbed and held until conditions change and allow it to dry out, but condensation can only be accommodated in this way if the periods of condensation are short enough and drying periods long enough to avoid complete saturation of the absorbent material. This is the principle on which anti-condensation paint works.

Fig 1 Temperature conditions in wall which may lead to interstitial condensation

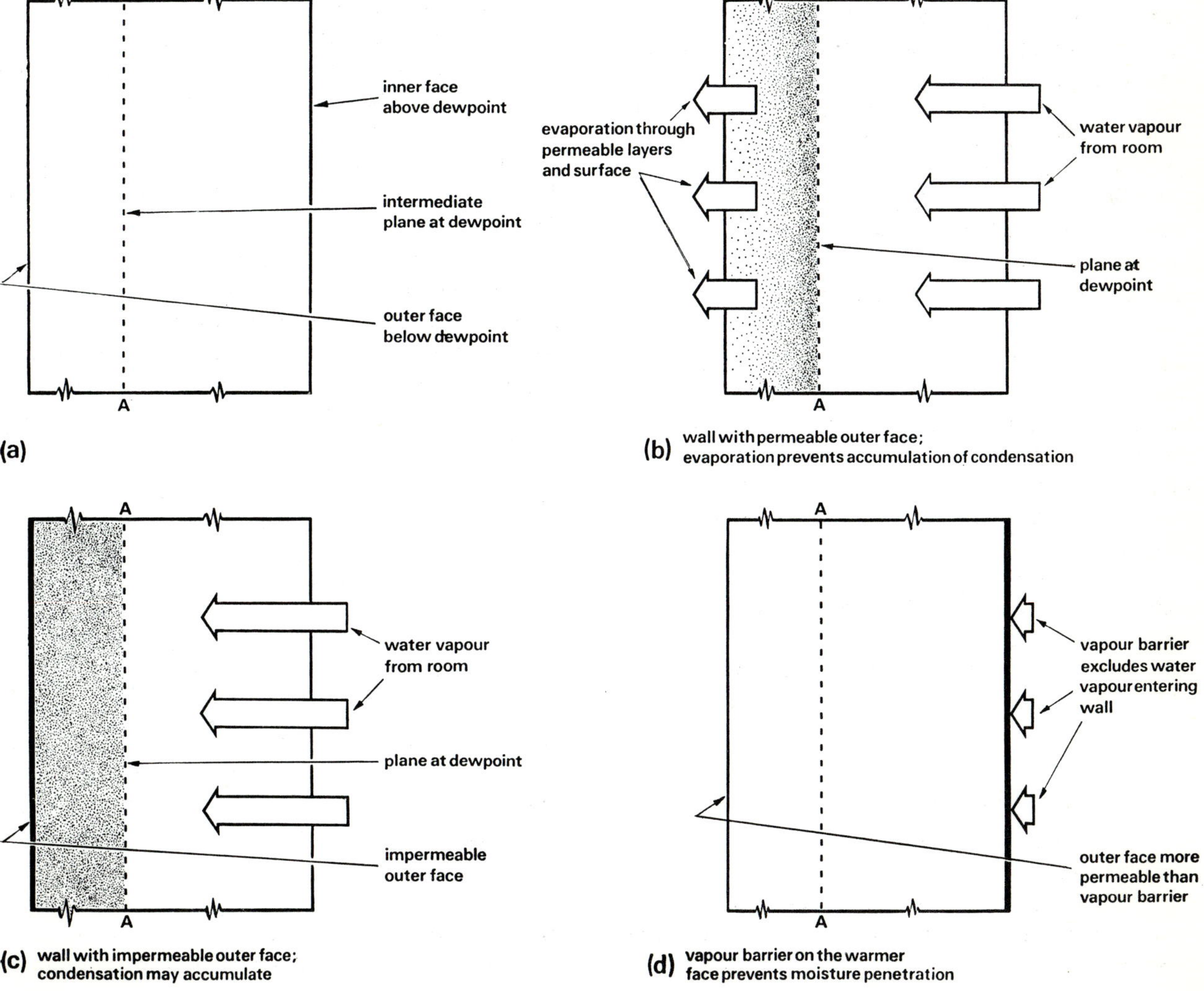

(a)

(b) wall with permeable outer face; evaporation prevents accumulation of condensation

(c) wall with impermeable outer face; condensation may accumulate

(d) vapour barrier on the warmer face prevents moisture penetration

Interstitial condensation

When the material of a wall, roof or similar building element is permeable to water vapour—and this applies to nearly all building materials—a dew-point temperature is associated with each point within the material. In the same way that a temperature gradient exists through a structure, depending on the thermal properties of the component materials, so a dew-point gradient depending on their water vapour diffusion properties exists also. If at any point the actual temperature is below the dew-point, then condensation will occur at that point within the material or structure. For example, the temperature through the thickness of a wall may vary from the inner face being above dew-point to the outer face being below dew-point; at some intermediate position the temperature will then be equal to the dew-point and condensation will begin at this plane (Fig 1a).

The exact processes taking place are complex and are further complicated by the fact that the condensed water changes both the thermal and vapour-transmission properties of porous materials, but the simple concept above is adequate for determining situations in which the risk of condensation trouble is unacceptably high.

If the outer portion of the wall is permeable to moisture, or if ventilation is provided behind impermeable wall or roof claddings, condensation will not be troublesome because the moisture can evaporate gradually to the outside air (Fig 1b).

If the outer surface is impermeable, the condensed moisture tends to accumulate in the wall and may ultimately saturate the material (Fig 1c). The situation will be most severe when the humidity of the indoor air is high.

A vapour barrier on the inner face of the wall (on the potentially warm side of any layer of insulating material) will prevent the passage of water vapour into the wall but only if it is undamaged and continuous. If the outer face of the wall is more permeable than the inner vapour barrier, any moisture contained in the wall can escape to the outside air (Fig 1d). If, however, the outer face of the wall has an impermeable cladding, or if the cladding is of organic material that would suffer in prolonged damp conditions, a ventilated cavity should be formed between the cladding and the wall so that any moisture evaporating from the wall surface is removed.

In some forms of construction, for example, behind timber cladding or tile hanging, a moisture barrier may be needed near to the cold outer face of the wall, to exclude wind-driven rain or snow: a 'breather' type of membrane, i.e. one that will bar the passage of liquids but will transmit vapour, is then required for this position. It should be used in conjunction with a vapour barrier on the warm side of the construction so that less vapour can enter the wall from the warm side than can escape by the breather membrane and there will be no accumulation of condensed moisture on the inner face of the breather membrane.

Designing to avoid condensation

The chart (Fig 2) shows the interdependence of relative humidity, on the left-hand vertical scale, dry-bulb temperature, on the horizontal scale, and the concentration of moisture in the gaseous mixture on the right-hand vertical scale, expressed for this purpose as mixing ratio, the mass of water vapour per unit mass of dry air.

An example of this relationship is shown by the reference points A–D, marked on the chart, which represent the following conditions:

At A, the outdoor dry-bulb temperature is $0°C$, the mixing ratio of the air is 3·4 g/kg of dry air, which gives an RH of 90%.

If this air is warmed to $20°C$ and the mixing ratio remains the same at 3·4 g/kg, the RH will become 23% (point B).

If excess moisture amounting to 7 g/kg is now introduced as a result of activities within the building, at the same temperature the RH will increase to 70% (point C).

D shows the dew-point temperature of the resulting air/moisture mixture, from which it follows that condensation will not occur if the adjoining parts of the building fabric to which the air has access are kept above $15°C$.

Two additional scales, not so far mentioned, have rather different uses. The scale of partial pressure of water vapour is directly related to that of mixing ratio, and it may be used, as explained later, to estimate the rate of diffusion of water vapour through the structural fabric. The scale of wet-bulb temperature denotes instrumental readings that are commonly taken with a sling psychrometer. Simultaneous wet-bulb and dry-bulb readings, when transferred to the charts, define the physical properties of an air sample at the conjunction of the respective oblique and vertical temperature lines.

Characteristics of the exposed building fabric

In all buildings heated and occupied during the winter, it may be assumed that the air temperature and water vapour pressure indoors will be in excess of those outside. As a result, heat and water vapour will attempt to flow outward through the fabric in an effort to restore the balance. Their progress is determined by the precise construction through which they must pass, and the result is a characteristic distribution of temperature and vapour pressure throughout the exposed structure.

The relationships governing this distribution are

depicted in Figs 3 and 4. The gradients shown are determined by the total differences of temperature and water vapour pressure across the structure and by the succession of resistances to flow that must be overcome. Table 1 gives a selection of typical values of thermal and water vapour resistance. The estimating procedure described later makes direct use of vapour *resistivity*, for which values may be obtained from the table, but other sources will be found to present figures for vapour permeability, or *diffusivity*, which is the reciprocal of the resistivity.

In Fig 3 it will be seen that the boundary surfaces of structures offer resistance to heat flow. Unless the structure is independently heated, the internal surface is at a temperature lower than that of the indoor air, and it will consequently cool the layer of air in contact with it. How far the local air temperature is

depressed depends on what proportion of the structure's total thermal resistance is contributed by the internal surface. Figure 5 shows the appropriate depressions for a range of temperature differences and comparative levels of structural insulation. It is

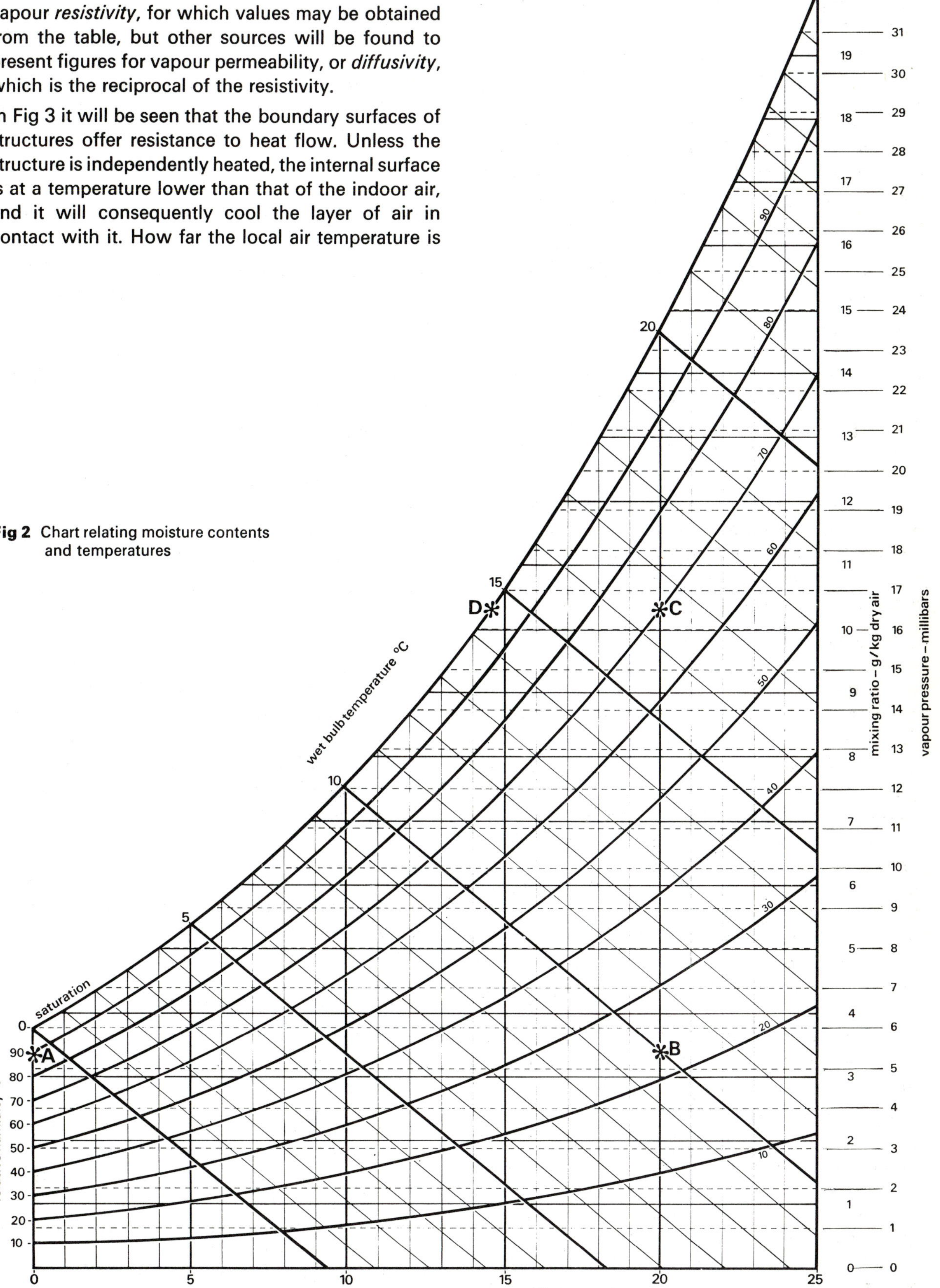

Fig 2 Chart relating moisture contents and temperatures

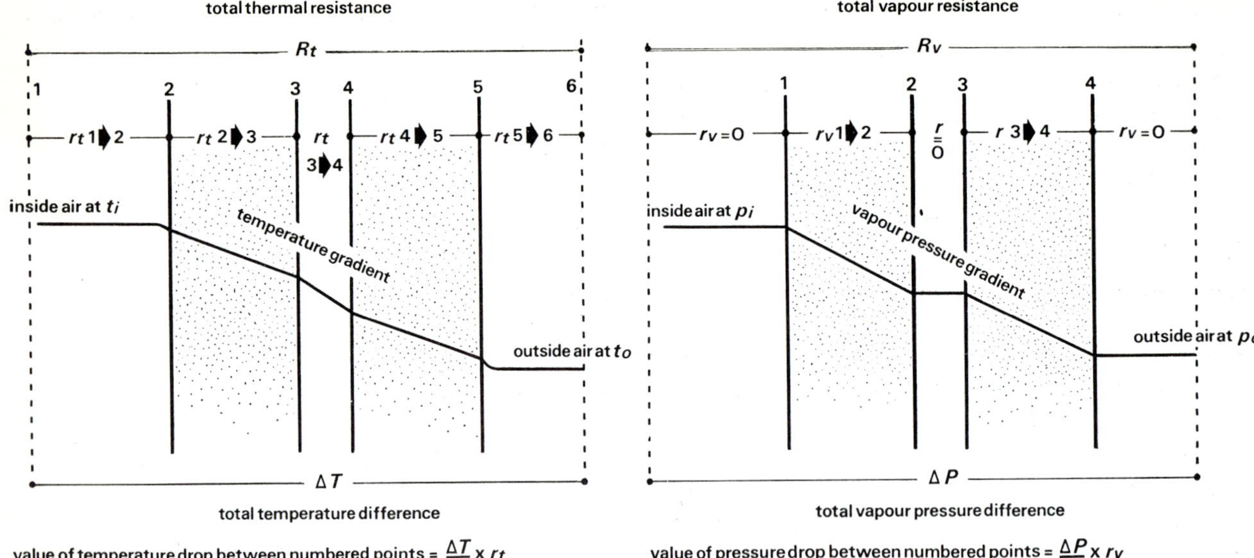

value of temperature drop between numbered points = $\dfrac{\Delta T}{R_t} \times r_t$

Fig 3 Temperature gradient through a structure

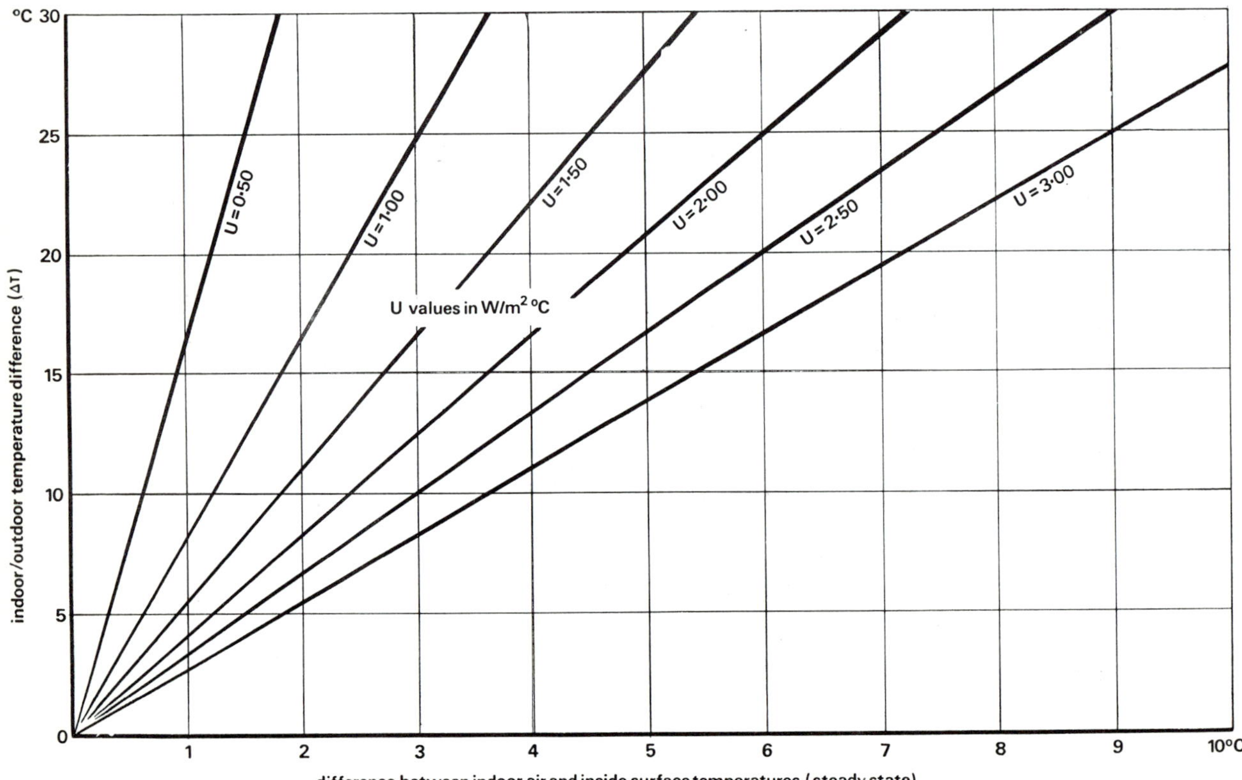

value of pressure drop between numbered points = $\dfrac{\Delta P}{R_v} \times r_v$

Fig 4 Vapour pressure gradient through a structure

applicable only to steady state conditions. In buildings of high thermal capacity construction, heated intermittently, the depression of temperature at the internal surface will often be greater, thus increasing the likelihood of condensation. Relating Fig 5 with the earlier example of outdoor air at 0°C, indoor air at 20°C and dew-point at 15°C, the maximum allowable drop of 5°C in the inside surface temperature will be seen to require a structural U-value not greater than 2·1 W/m² °C.

In the gradient of Fig 4, no vapour pressure drop occurs at the internal surface. Furthermore, it is practically possible to maintain the total indoor/outdoor pressure difference across the internal surface if the latter can be made impervious to water vapour. In this event the sole provision against condensation is that the surface should remain above the internal dew-point temperature. Not all surfaces can be so chosen to stop the passage of moisture vapour, and the more permeable they are, the greater is the chance that, at some internal part of the structure, a sufficiently low temperature will be

Fig 5

reached to cool the vapour below its dew-point. The respective temperature and pressure gradients, when charted, can be used to identify the point where condensation might occur in given circumstances, as demonstrated in the worked example later in this digest.

Characteristics of the environment

Except when a warm front brings moist air to follow a cold dry spell, the risk of condensation through natural ventilation in a heated building is slight. It is almost wholly dependent on the amount of excess moisture introduced by the occupants and their activities.

At normal ventilation rates, the gain by the air of body moisture from persons not engaged in physical exertion is roughly 45 g/person in one hour. This results in the indoor atmosphere having an excess moisture content over outdoor air of some 1·7 g of water vapour per kg of dry air. Provided the ventilation rates were properly controlled, this would be a suitable design assumption for shops, offices, classrooms, public meeting-places and dry industrial premises. For dwellings, taking account of the moisture produced by cooking and bathing and the likelihood of restricted ventilation in cold weather, a safer design value for moisture excess might be 34 g/kg. Catering establishments and industrial workshops requiring humid atmospheres or using wet processes may well contribute 68 g/kg or more to the internal air. In naturally ventilated premises such design values may be added to the assumed mixing ratio of the outdoor air.

Table 1. Typical values of heat and vapour resistance

	Thermal resistance (r_t) m² °C/W
Surfaces	
Wall surface—inside	0·12
—outside	0·05
Roof (or ceiling) surface—	
inside	0·11
outside	0·04
Internal airspace	0·18

	Vapour resistance (r_v) MN s/g
Membranes	
Average gloss paint film	7·5–40
Polythene sheet (0·06 mm)	110–120
Aluminium foil	4000

	Thermal resistivity m °C/W	Vapour resistivity* MN s/g m
Materials		
Brickwork	0·7–1·4	25–100
Concrete	0·7	30–100
Rendering	0·8	100
Plaster	2	60
Timber	7	45–75
Plywood	7	1500–6000
Fibre building board	15–19	15–60
Hardboard	7	450–750
Plasterboard	6	45–60
Compressed strawboard	10–12	45–75
Wood–wool slab	9	15–40
Expanded polystyrene	30	100–600
Foamed urea-formaldehyde	26	20–30
Foamed polyurethane (open or closed cell)	40–50	30–1000
Expanded ebonite	34	11,000–60,000

*Resistivity—1/diffusivity

Fig 6

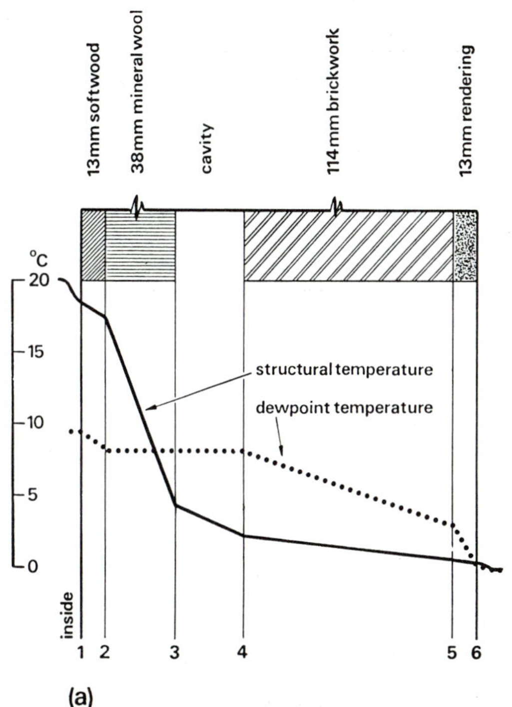

(a)

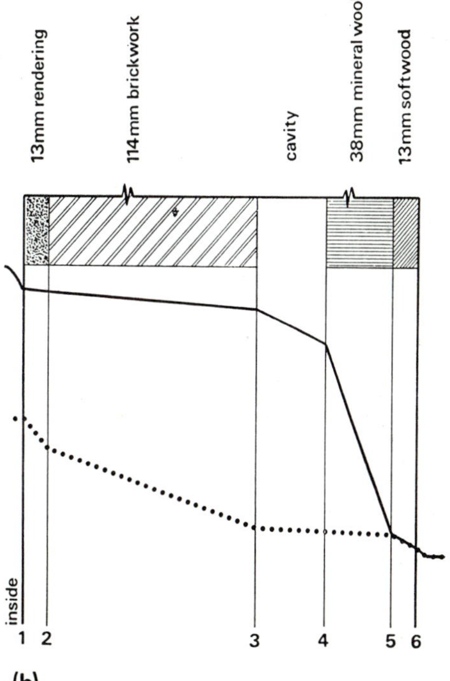

(b)

Estimating condensation risk

The principles used in this digest to predict the likelihood of condensation, and to design so as to avoid it, may be applied to wall, floor or roof constructions, but lightweight sheeted roofs present special problems, see page 159.

Figs 6a–6d show schematically some variations of a wall construction. The four illustrations are drawn with the thicknesses of the components to scale, the boundaries of the materials are numbered in sequence and the temperature difference across the structure is set up on an adjacent vertical scale. Using Fig 6a as an example, the design assumptions and the estimating procedures are then as follows:

1. Indoor/outdoor air temperature difference, ΔT, assumed to be:
$$20-0 = 20°C$$

2. Thermal resistance
(value from Table 1 × thickness of component):

Inside air to point 1		0·12
1–2	13 mm softwood 6·93×0·013	= 0·09
2–3	38 mm mineral wool 27·72×0·038	= 1·05
3–4	cavity	0·18
4–5	114 mm brickwork 0·9×0·114	= 0·10
5–6	13 mm rendering 0·83×0·013	= 0·01
6—outside air		0·05

$$R_t = 1·60$$

3. Temperature drop between points:
$$= \frac{\Delta T}{R_t} \times r_t = \frac{20}{1·6} \times r_t = 12·5 \times r_t \ (°C \text{ point to point})$$

Inside air	=	20°C
Inside air—point 1	= 12·5×0·12 =	1·5
	drop to	18·5
1–2	= 12·5×0·09 =	1·12
	drop to =	17·38
2–3	= 12·5×1·05 =	13·12
	drop to	4·26
3–4	= 12·5×0·18 =	2·25
	drop to	2·01
4–5	= 12·5×0·10 =	1·25
	drop to	0·76
5–6	= 12·5×0·01 =	0·12
	drop to	0·64
6—outside air	= 12·5×0·05 ×	0·63
		0·01*
Outside air		0°C

*Error due to approximation

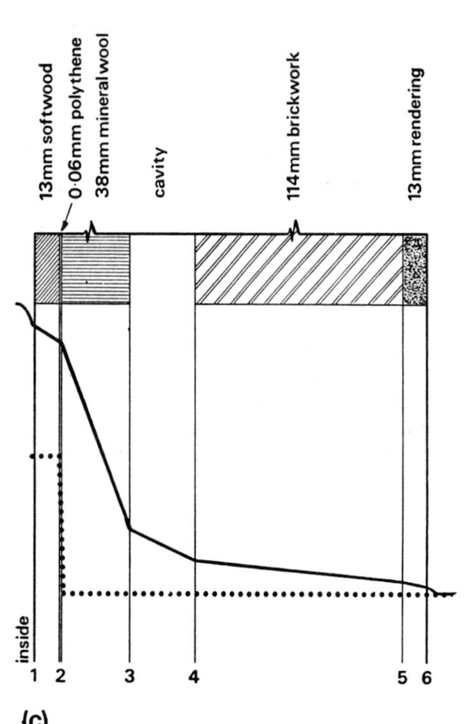

(c)

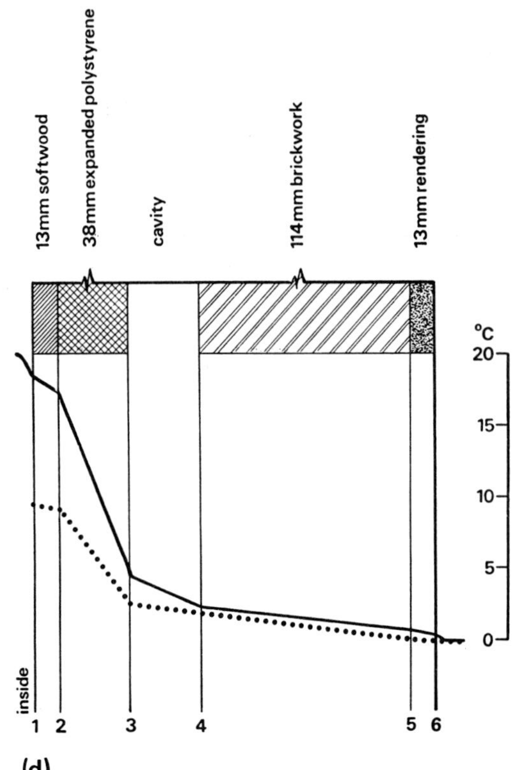

(d)

4. The profile of the temperature drop is plotted against the vertical scale.

5. From Fig 2, using the design assumption that the outside air is at 0°C and is saturated at a mixing ratio of 3·8 g/kg, the outdoor vapour pressure can be read (on the right-hand scale) as 6 mb; if a moisture vapour excess of 3·4 g/kg is contributed by activities indoors, inspection of the two right-hand scales of Fig 2 shows that the total moisture content of 7·2 g/kg gives an indoor vapour pressure of 11·4 mb. The indoor/outdoor vapour pressure difference, ΔP, is therefore $11·4 - 6·0 = 5·4$ mb.

6. Vapour resistance
(value from Table 1 × thickness of component):

$$
\begin{aligned}
\text{point } 1\text{–}2 &= 60 \times 0·013 &&= 0·78 \\
2\text{–}3 \ (\text{virtually nil}) &&&= 0·0 \\
3\text{–}4 \ (\text{nil}) &&&= 0·0 \\
4\text{–}5 &= 25 \times 0·114 &&= 2·85 \\
5\text{–}6 &= 100 \times 0·013 &&= 1·3 \\
&&& \overline{} \\
&& R^v &= 4·93
\end{aligned}
$$

7. Pressure drop between points:

$$
\frac{\Delta P}{R_v} \times r_v = \frac{5·4}{4·93} \times r_v = 1·10 r_v \ \text{(mb point to point)}
$$

			Corresponding dew-point temp. from Fig 2
Indoor vapour pressure	11·4		9·4°C
point 1–2 1·10×0·78 =	0·86		
	drop to	10·54	8·0°C
2–3	= 0·0		
		10·54	8·0°C
3–4	= 0·0		
		10·54	8·0C
4–5 1·10×2·85 =	3·13		
	drop to	7·41	3·0°C
5–6 1·10×1·30	1·43		
	drop to	5·98	0·0°C

8. Profile of dew-point temperature: Reference to Fig 2 will show the respective dew-point temperatures for the vapour pressures at points 1, 2, 5 and 6 from which the dew-point profile may be constructed.

9. Estimation of condensation risk: At any point where the computed temperature is lower than the computed dew-point temperature, condensation can occur in the conditions assumed. In the worked example, liquid may form in a position where, clearly, it can reduce the effectiveness of insulation and it is likely also to put the nearby timber at risk of rot. As an illustration of the effect that structural detailing may have, Fig. 6b shows the construction reversed and free from risk in the same surrounding conditions. Slight modifications, shown in 6c and 6d, are sufficient, however, to limit the potential risk by using materials that modify the vapour pressure gradient.

Some calculated risk may be accepted, as for example when condensed moisture can do no damage to the weather side of a structure and can be prevented from soaking back. In a great deal of masonry construction it is probable that condensate is soaked up harmlessly until it has an opportunity to distil outwards or evaporate from the surfaces during favourable periods.

Where there is an estimated likelihood of intra-structural condensation the greatest risks arise if vegetable-based materials are exposed to the consequent dampness, particularly where the weather side of the construction is highly resistant to vapour flow. Then, ventilated cavities may offer some help and their value increases as the vapour resistance of the warm-side materials is improved.

Lightweight sheeted roofs

The lowest temperature to which a roof cladding falls occurs in winter, on calm frosty nights when the sky is unclouded. The cooling effect of the low outside temperature is then intensified by the roof radiating heat to the sky, which in these conditions has an effective temperature of about −45°C for radiation from the earth.

With roof cladding of low thermal capacity, e.g. single-skin metal or asbestos-cement sheeting, the change in temperature of the inner surface as a result of a sudden drop in the effective outdoor temperature takes place rapidly, and the cladding may be about 5°C below the outdoor temperature for several hours. The use of vapour barriers, ventilation and insulation to combat the high risk of condensation is discussed in National Building Studies Research Paper 23, A. W. Pratt. *Condensation in sheeted roofs*. HMSO 1958 (*now out of print*).

Further reading
Digest 108, Standardised U-values.
BRS Current Papers Design Series 24, E. F. Ball. Condensation in large panel construction. BRS 1964.
Condensation in dwellings, Part 1: A design guide, £1.00; Part 2: Remedial measures, £1.25. HMSO.

Security

Crime in this country is approaching an annual total of one and a half million indictable offences. Most of it is committed against property and a very large part of this is by entry into buildings—residential, commercial and industrial.

In the construction industry, security considerations arise at almost every stage of a project:

Site selection and layout

Sketch plan

Detailed design

Workmanship and supervision

Protection of materials on site and work during construction

Payment of wages on site

For every firm involved, there is the further consideration of security of its own premises.

This digest has, therefore, been prepared by the Building Research Station from material contained in reports of specialist sub-committees of the Home Office Standing Committee on Crime Prevention and with the advice and assistance of the Home Office Crime Prevention Centre, the Metropolitan Police and the Hertfordshire County Constabulary, all of whose help is gratefully acknowledged.

The construction industry should hand over to its clients buildings that are secure in design, in internal layout and in their fixtures and fittings, so that with the operation of sensible security routines by the users, many crimes will be prevented. The discipline of security routines is not appropriate matter for this digest, but just as the provision of an efficient lock is of little value if the key is left under a convenient stone, so is the converse true, that security routines can only be as effective as the standard of the structure and its equipment.

The first parts of this digest are applicable to the design and construction of buildings of all types; later sections make recommendations specific to dwellings, schools and shops.

Consultation with police

Police forces have a corps of trained and experienced crime prevention officers whose function is to advise on security measures. In order to obtain the most benefit from their advice, they should be consulted from the outset of the project. They should see plans, bills of quantities and detailed specifications as these become available and should be invited to help in the final selection of ironmongery and other equipment.

Site selection and layout

The incidence of crime, which can vary even between parts of a county, is known in detail only to the police; the prospective developer of a site should seek their guidance on the extent of the hazard in the locality of a particular site. Knowledge of this will help to guide him in the standard of security that will be required, and, in extreme cases, might affect his decision to use the site.

In preparing a layout of a site, it should be kept in mind that clear visibility is an accepted deterrent. Points of access to buildings, or parts of a building that accommodate the most vulnerable contents should not be shielded by other buildings, or by other parts of the same building or by perimeter walls and solid gates. Provision should be made for good lighting of exteriors throughout the hours of darkness and for unimpeded visibility by police or security patrols.

Even a good perimeter fence will not deter a determined intruder but it will hinder trespass and restrict the unlawful entry of vehicles and the easy transfer of goods. A yard enclosed by a solid wall and with solid gates permits a thief who has succeeded in entering to operate undisturbed. Chainlink fencing (*see* BS 1722 parts 1 and 10) topped by barbed wire provides a good standard of protection; gates should have locking bars bolted through, with ends burred over, and secured by a close-shackle padlock. It should not be possible to lift gates off their hinges or to press out hinge pins. The number of entrances should be restricted.

If considerations other than security require the provision of a solid wall, grille gates should be provided in positions offering a good view of the enclosed area.

Car parks should be lit during the hours that they are likely to be in use.

Sketch plan stage

Although much improvement in security standards could often be made with only small additional expenditure, the provision of a very high standard of security to protect high risk goods, equipment, money, jewels, etc, is necessarily costly. The cost can be minimised by looking at this problem at the sketch plan stage with a view to forming only a part of the building into a secure cell—a strongroom or a security store—which can be properly located in the centre of the building, avoiding external walls, roof lights, etc. The earlier this is considered, the more effective can it be made with the least outlay.

Detailed design

Throughout the detailed design stage, security should be regarded as one of the functional requirements of the building; it should be put on equal terms with requirements such as heating, lighting, thermal insulation and sound insulation.

It should be applied as a criterion not only to the selection of ironmongery but even to the choice of materials and components for cladding the exterior of the building, its walls and roof.

For example, a window, normally regarded as a means of admitting light or providing ventilation, should be viewed also as a means of admitting a criminal if it is not properly protected.

Dwellings

From the statistics available, it appears that windows are more vulnerable than doors, but the number of break-ins through doors is very high. The number of attacks at ground floor level is about ten times the combined total of attacks at basement and first floor levels.

Windows

There is scope for further developments in the design of domestic types of window and fastening but of the types currently available metal frames appear to offer slightly better security than wood. Some manufacturers offer a built-in ratchet cockspur lock for casement windows and a transom security lock. One metal window has a lockable stay preventing its being opened sufficiently to enable a person to enter.

Internal glazing will prevent the removal of putty while it is still soft or of outside glazing beads and the whole pane of glass from the outside of the building; some forms of double-glazing will also be effective against entry in this way.

If panes of glass are to be small enough to prevent entry they must be not more than $0 \cdot 05 \, \text{m}^2$, but if this minimum is to be exceeded, then the larger the better.

Doors

All external doors should be equally secure, although, of course, if a dwelling is left empty the 'last door' cannot be bolted from the inside. The second door is better situated where it may be overlooked by neighbours. The ideal construction is of solid wood, but this ideal has to be balanced against cost and appearance. Door frames, particularly back-doors, are often very flimsy, offering no protection against the jemmy. The door should be constructed so that panels cannot be destroyed, and if it is not solid it should be ledged and braced and filled with a strong material. Hardboard which can be gouged away, should not be used in exterior door construction. Care should be used in the fitting of doors to prevent large gaps, and the frame should overlap the door to prevent access to the tongue of the lock.

It is just not possible to design a burglar-proof door—the compromise is an economic one between the simple lock (in a survey, it was found that 700 dwellings on one housing estate could be entered with only 36 keys), and the lock that is so secure that it is easier to drill it out of the door. The determined burglar may drill it out, but the casual burglar or house-breaker will be deterred.

Although the only means of admitting light into a hall is sometimes through glass panels in the door, each piece of glass represents a weakness. If glass must be fitted in a position where its breakage will give access to the lock or the latch, it should be reinforced.

Adjacent to an external door, glass bricks are preferable to window glass but window glass in strips up to 125 mm wide will offer less facility to burglars than larger panes of glass.

Hinges should be of adequate length and positioned so that the pin is inside and the screws are concealed when the door is closed. Letter boxes must conform to GPO Regulations, but they should be as small as is permissible and positioned away from the lock. A juvenile can reach an arm through a BS 2911 letter plate to the shoulder, and operate a lock up to about 400 mm away—over 50 per cent of people arrested for house-breaking are juveniles.

All exterior doors should ideally be fitted with a BS 3621:1963 mortice lock, but doors less than 45 mm thick are weakened by such locks and should be fitted with an automatically deadlocking rim lock. Back or side doors should be reinforced with securely fitted shoot bolts, top and bottom. On some doors, such as casement doors to a garden, a keyhole need be provided only on the inside of the door so that the lock cannot be picked.

Access and shelter

Particularly on local authority flats, continuous balconies are often built around the exterior of the building. Once a thief has gained an entrance to one flat he then has fairly easy access to all the other flats on the same floor by the balcony. From a security point of view, therefore, balconies should be restricted to individual dwellings.

If any form of shelter is provided for a door, it must not shield the door from view or provide cover for a thief to work on the door. This applies equally to screen walls, porches, gossip screens, etc.

Careful attention should be paid to the siting of external pipework; it can provide a convenient ladder to upstairs windows. Thieves will even climb plastic rainwater pipes; it may appear insecure, but it is a risk they are sometimes prepared to take. If pipework could provide access, painting with non-setting paint will be a deterrent.

Garages

Lock-up garages should be provided as normal practice; the most secure type of door is the up-and-over, fitted to a metal frame. If the garage has windows, they should be glazed with obscured glass.

In the absence of garages on housing estates, hard standings should be provided and they should be illuminated during the hours of darkness. In addition, tool sheds or out-houses should be provided in which tools can be stored securely not merely to prevent their loss but also to prevent their use in the commission of crime.

Meters

Coin-operated gas and electricity meters constitute a considerable security risk; if they have to be installed, they should be accessible only from inside the dwelling.

Street lighting

Adequate street lighting should be provided, and it is of value to fit exterior lights to dwellings as a deterrent against break-ins at doors.

Fencing

Fencing around and between gardens will, if it is of adequate quality, hamper a thief's progress to and from a house. It is suggested that chainlink fencing not less than 1·2 m high should be provided.

Schools

Schools comprise a particularly vulnerable type of premises, many having large numbers of accessible windows and glazed doors; they are known to be unoccupied for long periods and their internal layout is well known to the local populace.

Windows

About three-quarters of the number of illegal entries are made through windows. All ground-floor windows and upper-floor windows that are easily accessible, for example from flat roofs, should be fitted with good quality internal fastenings, if necessary with window locks.

Doors

All glass in external doors should be either of the toughened variety or reinforced with small mesh wire. Doors should be fitted with locks at least of the standard of BS 3621 : 1963. Double doors should have two flush bolts fitted to the edge of the first closing leaf and should be at least 45 mm thick to prevent their being weakened by such bolts.

Storeroom

The biggest losses are in the field of valuable equipment. Appropriate classrooms, laboratories and workshops should possess a substantially built interior storeroom in which valuable equipment can be stored. If there are windows to the storeroom they should be fitted internally with window bars, preferably vertical since these are more difficult to work on than horizontal bars. The bars should be at least 20 mm diameter or section, square section bars being stronger than round rods, at not more than 125 mm centres and the ends should be built-in to a minimum depth of 75 mm. Bars more than 600 mm long should be provided with cross ties. When designing for the provision of bars, the requirements for means of escape in case of fire must also be borne in mind.

Properly designed provision should be made for the secure storage of poisons, dangerous chemicals, guns and ammunition, as appropriate.

Safes

In spite of the claims of most head teachers that money is not kept on school premises overnight, quite large sums are stolen. It is suggested that every school should be provided with a small safe (see page 166).

Alarm system

Consideration should be given to the incorporation of an alarm system into the automatic system for class bells with which many large modern schools are equipped.

Shops

Windows

In the design of new shops, it is desirable to avoid windows giving access to the back of the premises but where these are unavoidable, or in existing buildings, they should be of glass brick or be provided with window bars as described above, or have expanded metal grilles.

Doors

All doors other than that used for final exit should be fitted with bolts, top and bottom, in addition to a secure mortice lock. The lock should be at about the centre height of the door, but if this is not possible there should be two locks, one at the top and one at the bottom of the door. The final exit door should be at the front of the shop.

Extruded aluminium sections may allow locks to be easily forced out of their aluminium housing unless the lock housing is reinforced by a length of steel angle.

Roofs

Precautions should be taken to ensure that access to the roof space cannot be gained by way of the roof space or flat roof of an adjoining building.

Flat wooden roofs, covered with bitumen felt or similarly easily cut roof covering, are very vulnerable to attack as are skylights and plastics domes. Skylights of any type are best avoided but if they must be used they should be protected by window bars.

Alarm system

The expense of an intruder alarm system cannot always be justified and in no case is an alarm a substitute for physical security. But if careful attention is given to other physical protective measures, security will be much improved by a properly installed system, well maintained by a reputable company.

Shops with high value goods, such as jewellers and furriers, may require a sophisticated alarm system; where the cost of a comprehensive system cannot be justified, the alarm system could be restricted to one secure room. An installation can be arranged to give direct warning to the police of an unlawful entry or can give an alarm signal elsewhere, for example to a caretaker's residence where a telephone is available.

Many new shops are built as part of a large development scheme; block wiring for burglar alarm installations can then be included in the entire shopping precinct; this is likely to encourage the occupiers to have an alarm installed.

Safes

Cash kept on premises overnight should be kept to a minimum but it must be assumed that cash and other valuables will be retained overnight both in schools and shops, and possibly for much longer periods in some dwellings.

For small overnight cash holdings, say up to fifty pounds, a small floor safe or a wall safe of one or two brick size may be adequate but the value of property to be kept in any safe should be decided only after reference to the safe manufacturer and the insurer. Any safe weighing less than one tonne and sited on a ground floor should be secured against bodily removal or manipulation. Anchorage of safes sited above or below ground level will depend upon the facilities for their manipulation or removal, for example, lifts, ramps, hoists, etc, but as a guide any safe weighing less than 600 kg should be secured to the fabric of the building. In shops in particular, a safe should be located in such a position that it can readily be observed from outside the building and it should be illuminated at night by at least two lamps, as a safeguard against the failure of a single lamp.

Security on building sites

Although thefts from building sites present a serious problem, many builders still seem not to be security conscious.

Special attention should be given to the provision of secure compounds for storage and to adequate protection of sites by lighting and by alarms, which could be battery-operated if mains electricity is not available.

It would be helpful if building equipment, machinery and materials were made identifiable in case of theft. More care should be taken in the phasing and delivery of equipment and fittings; these are often delivered well in advance of requirements and then left out in the open. Deliveries should be phased so that there is a minimum quantity of material left on sites over the week-end.

There should be a security compound, preferably with chainlink fencing of a minimum height of 2·5 m. Additional security could be provided by the use of existing and completed buildings. There should be supervision where the value of equipment, etc, warrants this, provided either by watchmen or by a security organisation. Attractive waste, such as offcuts of wood and copper piping, should be cleared.

Cash payment of wages on site

Consideration should be given to the employment of a firm specialising in wage packeting and carrying.

When it is necessary to move cash on foot, this should never be undertaken by the elderly or infirm or by any new employee whose references have not been checked. There is often an advantage to be gained by carrying small amounts of cash on the person rather than in a bag or case; alternatively, notes only may be carried on the person and coin in a bag or case. Where an appreciable amount of cash is to be carried, more than one person should be employed. Times and routes should be varied whenever possible but where the choice lies between a busy and a quiet route, the busy one should be used. The person carrying the cash should walk facing oncoming traffic so as to reduce the risk of a surprise attack from a vehicle from behind. Regular or inflexible procedure should be avoided.

Movements of cash by vehicle, and the vehicle used, should also be varied. At least one escort should be provided in addition to the driver (whose primary duty is driving the vehicle), the number of men to be determined by the amount to be carried. The most dangerous times are when arriving at or leaving the bank or business. At these times the escort should ensure that there is no apparent danger before the persons carrying the money leave the security of the vehicle or premises. The money should be conveyed in a properly installed car safe or locked in a boot. The decision as to the route to be followed should be made by the responsible person immediately prior to departure. The route should be varied as much as practicable and busy roads used in preference to side roads.

Secrecy of cash movements is essential. The knowledge of such movements must be restricted to the smallest number of people necessary for its safe handling.

Counting the cost

The cost of introducing basic security measures into new dwellings was investigated for one industrialised construction system. The schedule of door and window locks and fastenings actually provided totalled £6 19*s*. An alternative scheme meeting the standards thought by the crime prevention officer to be desirable totalled £16 7*s*. The difference, expressed as a percentage of the total cost of the dwelling, is small, but crime prevention cannot be assessed only in terms of cost effectiveness, which takes no account of the cost to the community of police time, the cost of taking an offender to court and of punishment or reformation of the offender. Moreover there is a moral duty upon every citizen to do all that he can to prevent the spread of crime. There are many people who are not basically criminals, but who may be too weak to resist the temptation if goods are left where they can be removed undetected. If they find that their first offence escapes detection only too easily, they may well be tempted further and further down the road to crime.

Concrete pumping

Pumping concrete can be easier, quicker and therefore cheaper than placing by more conventional methods. The pumps, usually lorry-mounted, need be hired only for the period when concrete is to be placed and discharge points can be reached that are otherwise difficult of access. The now common practice of specifying concrete by its required strength at 28 days is valid for concrete that is to be pumped. If the maximum benefits are to be obtained, the site organisation must be planned from the outset with pumping in mind and the supply of concrete must be adequate in timing as well as in quality.

Since the introduction of concrete pumps to the UK in the 1930s, their main use has been in placing large volumes of concrete in civil engineering projects. Builders have tended to regard pumping as uneconomic, and some who did try encountered problems, not having had sufficient time to experiment with mix design.

Although the basic principles of pump design have not changed, higher efficiency and greater reliability have been achieved. As a result, the use of pumps on building sites is now increasing, most of the pumps being owned and operated by specialist pumping firms.

Why use pumps?

Pumping concrete can be easier, quicker and therefore cheaper than placing by conventional methods. Very roughly, the volume of concrete moved in one hour is: by wheelbarrow 2 to 3 cubic metres, by skip and hoist 15 cubic metres, by chute or conveyor 30 to 40 cubic metres; a pump can deliver up to 100 cubic metres if required.

The conventional method of moving concrete on large sites is to use a tower crane with skips. The crane may have to be hired for the duration of the building operation but cannot be used in bad weather or high winds. Pumps, which are usually lorry-mounted, need to be hired only when concrete has to be moved and since access from above is not required, the roof can be constructed before concrete is laid, allowing the placing of concrete in bad weather.

Complex design can be produced with less complex shuttering than by conventional methods. Well-compacted concrete with a good finish can be obtained without vibrators, although their use does help to reduce the pressure on formwork and aid the flow of concrete through reinforcement.

A pump can give a lower unit placing cost but the savings from a reduction in overall building time will often exceed the direct savings in handling costs.

The modifications to mix design necessary to make concrete pumpable often give better durability and shrinkage characteristics than a mix designed purely for compressive strength.

Plant and equipment

Mechanical pumps

Mechanically driven pumps are the simplest types, but they are relatively heavy with large-diameter cumbersome pipes and are best suited for placing large volumes of concrete over long distances and where only limited flexibility is required at the discharge end. The twin-cylinder models can operate either individually into separate pipelines or in tandem into a single pipeline to produce a more continuous flow. Outputs claimed vary from 14 to 24 m³/h per cylinder with a range of 500 m horizontally and 50 m vertically through a 180 mm diameter pipeline.

Fig 1 Flexibility at the delivery end.

Hydraulic pumps

The majority of pumps in use are twin-cylinder, hydraulically driven. They are lighter than mechanically powered pumps and can be lorry-mounted; in conjunction with small bore pipelines (75 or 100 mm) they have been developed into highly mobile units. Hydraulic power has advantages of smooth operation and infinitely variable control from zero to maximum. Using 100 mm pipeline, concrete has been pumped vertically up to 85 m and outputs of up to 100 m³/h are claimed.

Pipes

Rigid seamless steel pipes in 3 m lengths are usually used, the diameter being dependent upon working conditions. Bends of 22½, 45 and 90 deg are available and are of large radius (up to 1 m) to minimise frictional resistance and heat. Pipe diameters should be in accordance with pump manufacturers' instructions, but in general the smaller diameters (75 and 100 mm) are used for greater vertical distances and the larger diameters (up to 180 mm) for greater horizontal distances and for concrete with high frictional resistance. For pumping large aggregate concrete, the pipe diameter should be three to four times the maximum aggregate size. Where flexibility in the pipeline is required, for example on booms and at the delivery end, rubber pipes are normally used.

Crew

A pump normally requires a two-man crew, one to operate the pump and one at the delivery end. Both men are needed for cleaning the equipment and also for erecting and dismantling pipelines (unless a folding boom is used).

Concrete specification

The commonest method of specifying concrete for placing by any means is to state the minimum requirement for crushing strength at 28 days—and for concrete that is to be pumped, it is a sensible method.

Not all concrete is pumpable. When handled by other methods concrete remains at or near atmospheric pressure but in a pump the pressure exerted on the concrete may be as high as 5 MN/m². High pressure can cause bleeding and segregation of the mix resulting in a blocked pipeline. The concrete mix, therefore, has to be designed not only to meet the specification but also to avoid bleeding and segregation during pumping. A further requirement is that the frictional resistance of the mix must not be so great that the pump is unable to push the concrete along the pipe.

In the past the design of pumpable mixes presented difficulty and blockages were frequent but, with the increasing application of pumps, the experience of users in the UK and research at the Station have shown how to design pumpable concrete. The subject is dealt with in detail in BRS Current Papers 19/69 and 29/69 and in 'Guide to Concrete Pumping'. Briefly, the aim is to ensure that the concrete has the following properties in addition to those specified: cohesiveness (the ability to retain its homogenous nature under pressure), plasticity (the ability to change shape as it passes through valves, tapered pipes and bends without the necessity for excessive pressure), and self-lubrication (a limited amount of water should escape to the walls of the pipe to act as a lubricant for the solids). Important factors in attaining these properties are the grading, particle shape and proportions of the aggregates and the total quantity of fines (sand and cement). If the mix has not been designed for pumping in the first instance, the building contractor should check that it is pumpable well in advance of pumping operations, so that there is time to modify the mix, if necessary, before the event. The design or modification of mixes for pumping requires an expertise which, if not available within the contractor's organisation, can be found in experienced specialist concrete pumping firms and in some ready-mixed concrete companies.

Whilst large quantities of dense concrete are now pumped every day, the pumping of lightweight concrete still presents difficulty (see BRS Current Paper 29/69 and 'Guide to Concrete Pumping').

Site organisation and costing

A concrete pump is an expensive machine designed for a single purpose—unlike a crane which can handle a variety of materials. It follows, therefore, that its utilisation must be high if it is to be operated economically. For this reason and because their operation requires experience the majority of concrete pumps are operated by specialist companies on a hire or contract basis. Some large contractors, however, particularly those that undertake civil engineering work, operate their own pumps.

The speed at which concreting operations can take place depends on a number of factors, such as the type of work, its location, the supply of concrete, the speed of compaction and finishing and site conditions. Cost comparisons of methods of handling concrete based only on the potential outputs of machines are misleading.

The total time and labour content of a concreting operation on a particular site can only be estimated by taking account of all the factors for several different methods of handling concrete and converting the figures to costs. No two sites are exactly alike, each must be considered separately.

If a specialist pumping sub-contractor is to be employed, he can assist in estimating time and labour content as well as costs. Having completed this exercise at the planning stage, the main contractor will not only know the relative speeds and costs of concreting operations but will also be able to determine the effect that the speed of concreting will have on the overall building time, and hence on the overall costs. The savings due to a reduction in building time are often greater than direct savings in handling costs; a more expensive method of handling can reduce the overall cost because of its speed and a consequent saving in building time.

Planning

The methods by which concreting operations are to be carried out need to be considered in the early stages of planning a contract. The effects of different methods of handling concrete, for example by hoist, crane or pump, can be estimated in terms of overall cost and building time by drawing up production programmes. A decision to use a concrete pump may affect the selection of other plant: for example, a crane may still be needed to handle other materials but its lifting capacity can be less because it is not required to handle concrete. On a large job it is possible that the use of a concrete pump may mean that one crane can be used instead of two. The production programme will indicate the pump's utilisation, so that a decision can be taken whether to purchase the equipment or to employ a specialist firm. In either case pumping operations require detailed planning by the main contractor, in collaboration, if its services are to be used, with the specialist pumping company to determine the type of pump, concrete mix, concrete supply, sequence of operations, locations for pump that provide access for

truck mixers and space for washing down, pipeline layouts, or coverage with the boom and size of placing gang and number of vibrators.

Concrete supply

Whether mixed on site or supplied ready-mixed the concrete should fulfil the following requirements; it must be designed to be readily pumpable, be consistent throughout each mix and consistent from mix to mix and the supply must be regular and at the desired rate.

The concrete supplier should be notified at the earliest opportunity that a concrete mix is required for pumping because, with certain aggregates, it takes time to design a pumpable mix and to obtain the approval of the designers of the structure concerned.

If ready-mixed concrete is used it is essential that those in charge of site operations are able to make contact with the suppliers by telephone. Instructions to delay supply can only be implemented on loads that have not been batched. If sufficient notice has not been given several loads may be on the site or in

Fig 2 Typical boom range of a mobile pumping unit.

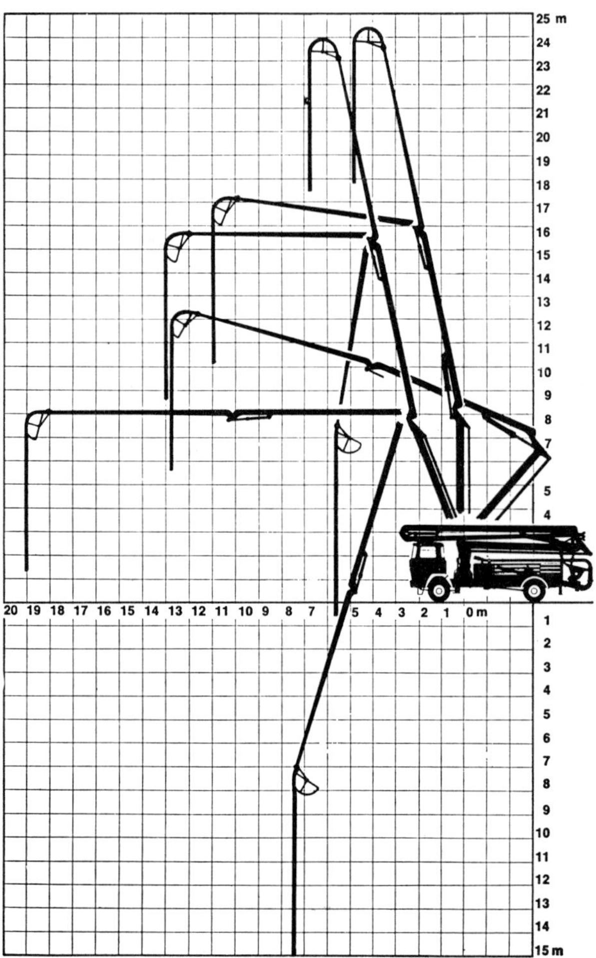

transit. The supplier should provide a contact with whom alterations in the design of the mix can be discussed, should some undefined factor affect pumpability; and the supplier should be prepared to make alterations in the proportion of the mix at short notice.

Applications

When the hydraulically operated mobile pumps were first introduced by specialist pumping companies, builders tended only to make use of the pumps for two reasons: because the work was inaccessible or awkward to mechanise by any other means, or because the contract was behind schedule and a very rapid method of concrete placing could help to restore the situation. Now that builders, pump operators and concrete designers have gained more experience, concrete pumping is recognised not only as a very fast method of concreting but also as a method that proves to be economical in many circumstances.

Pumps are used for a variety of concreting operations in building but the main applications are:

Foundations and basements
Pumps are frequently used for work at or below ground level when there are extensive or mass foundations or basements to be concreted. The ability to place concrete rapidly means that the number of stop-ends is kept to a minimum and that a large volume of work can be completed in a day or when, for example, a brief thaw occurs during a prolonged cold spell.

Other advantages of pumping in this type of work are that the pipeline can be laid over ground that is inaccessible or impassable to vehicles and that concrete can be placed without the need for long chutes which can cause segregation.

Oversite slabs
Pumps can distribute concrete laterally, even for thin slabs, and alternate bays can be concreted without the problem of access through a complex of recently completed bays.

Suspended floors and beams
Pumped concrete is widely used for concreting suspended floors of rib and hollow block as well as solid slab construction in both low and high rise building. In low rise buildings of brick construction with concrete floors considerable savings in time and labour are obtained by pumping instead of using a platform hoist.

Beams as small as 0·04 m² in section have been successfully filled; with a boom the delivery hose can easily be moved along the beam during pumping.

Walls
Concrete can be distributed rapidly along the length of a wall so that the level of concrete rises at a steady rate. When the thickness of the wall and the spaces between the reinforcement allow, the delivery hose can be inserted so that the danger of segregation caused by the free fall of concrete is eliminated. Better compaction and finish can be achieved by keeping the end of the hose below the level of the concrete.

Columns
The delivery hose can be immersed, as for walls, to avoid segregation and to obtain good compaction, and experiments show that tall columns can be concreted in continuous pumping operations.

Tunnelling
Pumps can be used to convey concrete down shafts and along tunnels, the pipeline occupying very much less space than is required for other handling methods. Tunnels can be lined by injecting compressed air in short bursts into the pipeline near the formwork so as to project slugs of concrete into confined spaces.

Further reading:
Guide to Concrete Pumping   HMSO 1971

Drying out buildings

The aims in drying out a building are to bring it to a fit state to receive internal finishes and decoration and to avoid damage or disfigurement of work already done. This digest describes some ways of drying out and also methods of testing the condition of walls, floors and joinery. Although economic considerations will usually call for early completion and, hence, rapid drying, excessively fast drying will cause troubles.

A great deal of water is used in the construction of a building, mainly as mixing water for concrete, plaster and mortar. More is all too often introduced with materials taken from stacks; for example, of bricks, blocks and timber, that have not been protected from rain. Further rain is likely to fall on partially completed work, before the building is roofed in or otherwise made weathertight. Although in favourable conditions the drying process begins as soon as the wet work is placed, there may remain some thousands of litres of water to be dried after completion of the structure.

The heat taken to evaporate five litres of water, even if the temperature of the building never rises above 15°C, is at least 3·6 kWh. So a lot of heat and a lot of time is required for drying out.

Certainly the practical problems presented by drying-out set a premium on protection from the weather both of materials on site and of partially completed work.

The main object in drying out a building is to bring it to a fit state to receive internal finishes and decoration. This should avoid troubles such as expansion of woodwork, particularly floors lifting in compression, loss of adhesion of floor finishes, and paint failures on plaster or wood surfaces. A damp building might also menace the health of its occupants and lead to the development of unsightly mould growths.

Unfortunately, excessively rapid drying can produce its own crop of failures of other kinds so that a balance must be struck between speed and caution, (*see* 'Cautions' on page 174).

The drying out of porous materials, such as brick and concrete, takes place in three distinct phases:

First, all the free water evaporates from the surface; this occurs quite quickly.

Next, the water from the larger pores is lost. This takes longer, partly because there is more water within the material than on the surface but also because it takes time for the water vapour to work its way through the labyrinth of pores to the surface of the material.

Finally, water is lost from the fine pores or cells. This is a very slow process that will continue possibly for several years. Absolute dryness is never attained in actual buildings. Brickwork protected from rain might eventually dry to about one per cent moisture content, concrete about three per cent; brickwork and concrete exposed to rain, about five per cent (all by volume). The equilibrium moisture content of wood varies widely according to its position within the building and the standard of heating: it could be as low as six per cent for wood flooring on a heated sub-floor, or as high as 20 per cent, or more, in some roof timbers, expressed as a percentage of the dry weight.

Drying out is concerned mainly with the second phase.

Two important points to remember are that evaporation inevitably cools the material that loses the water and that the rate of evaporation reduces as the temperature falls.

In a laboratory it is possible to evaporate water so quickly from a brick that the water still contained in it freezes and evaporation comes to a virtual halt until the brick warms up again.

It is not uncommon on walking into a finished building to find that it is noticeably cooler than the air outside This is the result of the materials drying out, and is, incidentally, the principle on which the old-fashioned earthenware butter-cooler worked.

How to dry out the building

The following alternatives are available:

Use natural ventilation by keeping windows open

Natural ventilation will remove the moisture-laden air from a building replacing it with drier air (except in unfavourable weather conditions) from outside which, in turn, will pick up water vapour and remove it from the building. If the windows are not kept open, water will still evaporate from the damp materials but will accumulate until it condenses not only on windows but also on walls and other surfaces. This often leads to the development of unsightly mould growths, which will incur extra cost in cleaning and, in some instances, redecoration.

Use heaters and keep the windows open

Heat not only increases the rate of evaporation from surfaces but also increases the quantity of water vapour that the air will support. Thus, with the same degree of ventilation, more water vapour will be carried to the outside when heaters are used than when they are not. If the windows are closed in an attempt to reduce the cost by conserving heat, this will build up moisture and the effects will be even more severe than if an unheated room is shut up. Flueless heaters burning gas or oil, a combustion product of which is water, should not be used for drying.

Use dehumidifiers and keep the windows closed

Dehumidifiers (*see* p. 174), by removing water from the air, induce the evaporation of water into the air from the surfaces of the construction materials, but as a consequence, the materials cool. The refrigeration types of dehumidifier are very inefficient at low temperatures and therefore it is advisable to provide heaters as well; these should be of the 'dry' type.

Equilibrium vapour pressures within a building can easily be achieved by leaving internal doors open; moving the dehumidifiers from room to room is not necessary. Windows must be kept closed or the equipment will endeavour to dry the outside air as well.

Dehumidifiers remove water more quickly from thin timber sections than from thick masonry walls, and some distortion of the lighter units can therefore occur.

How to tell when a building is dry

Timber

The moisture content of timber can be tested with an electrical moisture meter (*see* page 174) to give immediate results that are accurate enough for practical purposes over the range of moisture contents likely to be encountered in normal usage.

Digest 106 *Painting woodwork* suggests that timber should be as near to its eventual moisture content as possible when painted, ie 10–12 per cent for interior joinery.

Floor screeds

Floor screeds can be tested with a hygrometer, as described on p. 175 When the hygrometer readings fall to within the range 75 to 80 per cent, the sub-floor can be assumed to be dry enough for flooring to be laid.

An electrical moisture meter, as described for use on timber, can also be used but it should be noted that although a high reading indicates a wet screed, a low one does not necessarily indicate more than surface dryness: below the surface layer, the screed may still be wet. A more reliable indication can be obtained by taking a reading of the screed surface layer, covering the area with impervious sheeting about one metre square for 24 hours and then taking a second reading on the same spot. It is likely that this will be significantly higher than the earlier reading and, if so, it is the reading on which to decide subsequent action. No actual values can be recommended here because readings for this purpose will usually be taken on an arbitrary scale of numbers, the interpretation of which is explained in the instrument manufacturer's instructions. The presence of contaminating salts in the screed, either as deliberate additions of, say, calcium chloride or by accident, may give misleading results. The use of anhydrous copper sulphate (which turns blue in the presence of moisture) or of anhydrous calcium chloride (which liquifies) is not recommended.

Walls

As with concrete screeds, the surface condition of plaster or brickwork is not a reliable guide to the moisture content of the material behind the surface. It may sometimes be possible to estimate moisture content and suitability for painting from a knowledge of the weather conditions since the wall was built, but it is safer to measure the moisture content. The principal methods available are:

Electrical moisture meters as described for use with timber may be used with probes forced into the wall. To avoid misleading results, it is better to cover an area of wall, not less than 300 mm × 300 mm, with a piece of glass, metal or polythene sheet, for several hours (preferably at least overnight) before taking a reading in the centre of the area covered.

Coloured indicator papers. These are fixed to the surface and change colour according to the amount of moisture present. Again, conclusions may be misleading if based on a small area of surface and overnight covering, as for electrical moisture meters, is advised.

Hygrometers. The form of instrument described for use on floors (*see* p. 175) is equally suitable for use on wall surfaces, using a suitable form of prop : overnight readings are advised.

Digest 197 *Painting walls : 1* includes recommendations for treatments appropriate to various moisture contents.

Cautions

It is not possible to tell that a material is dry enough to paint or otherwise finish just by looking at it or touching it. Nor is the length of time since building any reliable guide, although it is true to say 'the longer the better'. A traditional rule 'of thumb has been 'one month per inch of thickness to be dried in *good* drying conditions'. Even so, experience shows that this is often not enough. Up to the end of the 1930s, new houses were often left for three months before decorating. But if it is necessary to decorate earlier, use a porous, inexpensive finish such as emulsion paint and accept that early redecoration might be needed.

Because the digest is about drying out, it has discussed methods of speeding up the process. Drying out should not, however, be accelerated too much by the use of heaters or dehumidifiers, otherwise excessive cracking of screeds and/or plaster finishes may occur. If woodwork is already fitted, checking, splitting and opening of joints may be encouraged by too rapid drying. After the building is occupied, heating should not immediately be run at its maximum level to complete drying as this also could result in excessive shrinkage and distortion of wood.

Notes on some of the equipment referred to in the digest

Dehumidifiers

There are two types of dehumidifier, one uses a desiccant to extract the moisture, the other condenses it by refrigeration coils. They can extract as much as 6 litres of water per hour, depending on the moisture content of the structure and the air. Their effectiveness falls at low temperatures, because cold air carries less moisture from the structure than warm air, and some heating is therefore recommended.

Chemical desiccant dehumidifiers rely on the removal of moisture from the air by the hygroscopic properties of various chemicals. They are of two types : (a) in which the chemical dissolves in the water which it absorbs and the solution is discarded, and (b) in which a paper impregnated with a hygroscopic chemical holds water until the impregnated paper is subsequently reactivated by heat and re-used.

They are thus cheap in capital and operating costs and they work at low temperatures.

Refrigeration dehumidifiers act by drawing a stream of moist air over a series of artificially cooled coils. Water vapour in the air is cooled to its dewpoint and condenses on the coils, whence it is collected and removed. They recover the latent heat of evaporation in the vapour and so deliver slightly warmed dry air back into the room. Some types have a system of automatic defrosting.

Moisture meters for wood

Resistance type meters are the most popular. They measure the electrical resistance of the wood, which varies with changes in moisture content, and are accurate to about ± 2 per cent over a range of about 7–25 per cent moisture content. This meets the normal requirements for wood in buildings.

Measurements are taken by inserting a pair of probes into the wood and reading the electrical resistance between them on a scale calibrated to show the equivalent moisture content. The electrical path measured is only the distance between the probes, usually about 25 mm and to a depth of 6 mm, though some have longer probes that can be driven more deeply. It is therefore important to try to obtain representative values by taking readings at several points on each piece, avoiding any areas of local wetting or visible irregularities such as pitch pockets, knots, stains, etc. Chemically treated wood or wood contaminated, for example, by sea-water, is unlikely to give reliable readings.

The meters are calibrated for a stated temperature and species of timber. Corrections must be made for other temperatures and species, though for periodical measurement of the progress of drying it might be possible to ignore the correction for species.

The instructions of the instrument manufacturer should always be followed : these normally include correction tables for temperature and species.

Dielectric-type meters are based on the dielectric constant of wood. Their use is not restricted to any range of moisture contents but they are generally less accurate than resistance-type meters. Instead of correcting for species a more complicated and careful calibration has to be made for the specific gravity of the wood, on which its dielectric properties partly depend.

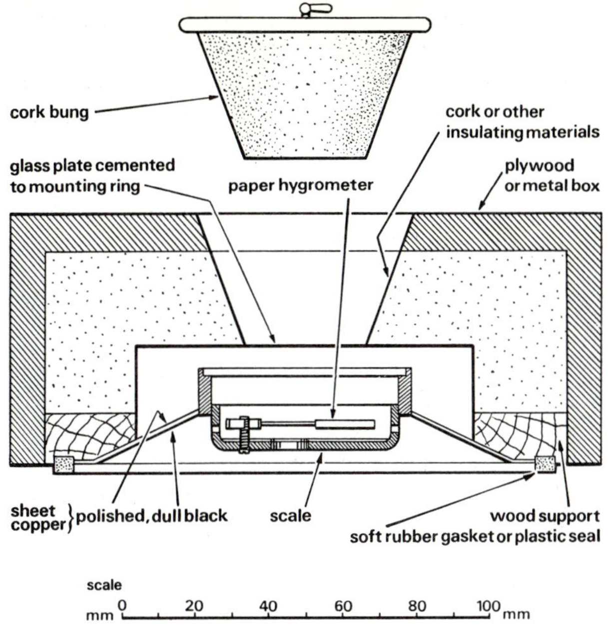

Fig. 1   Apparatus for measuring dampness

Hygrometer

A simple form of instrument measures the relative humidity of a pocket of air in equilibrium with the sub-floor (Fig 1). It consists of a paper hygrometer in a vapour-tight mounting housed in a well insulated box. Provision is made for sealing the edges of the instrument against the surface to be tested—Plasticine is a convenient material for this purpose—and for reading the hygrometer scale whilst the instrument is in position on the floor.

After the instrument has been sealed firmly to the floor surface, a period of not less than four hours should be allowed to elapse for the entrapped air to reach moisture equilibrium with the concrete base before the reading is taken. A longer period is preferable and can sometimes be obtained conveniently by placing the instrument in position overnight and taking the reading in the morning. If readings are to be taken at several points with only one instrument available, subsequent positions may be covered by impervious mats (bitumen felt, polythene sheeting, etc) about one metre square, laid down when the instrument is placed in its first position, to speed up the later readings.

If there is difficulty in obtaining the instrument in the form illustrated, it can be made locally using a paper hygrometer which is obtainable from suppliers of laboratory apparatus; the shape and dimensions of the housing are not critical but the principles of thermal insulation and vapour barrier (the copper sheet) should be followed.

7 Dwellings

Sound insulation: basic principles

The use of novel designs and the recent amendment of the Building Regulations 1965 to include a performance standard in the deemed-to-satisfy provisions demand that increasing attention be paid to sound insulation between dwellings. This digest outlines the basic physical principles involved and complements other digests which contain more specific recommendations.

Types of sound transmission

Insulation is required against sound generated in two different ways. On the one hand, a source such as a radio may produce sound waves in air which in their turn produce vibrations in a party wall or floor (Fig 1a). On the other hand, a wall or floor separating two dwellings may be excited by the direct impact of a solid object, eg footsteps (Fig 1b). The two kinds of insulation have to be measured separately. Only party floors are required to meet a prescribed standard of impact insulation, but both party walls and party floors are required to provide a standard amount of insulation against airborne sound. This digest is concerned mainly with airborne sound.

Measurements to test the airborne sound insulation between two rooms are concerned with the total, amount of energy transmitted, by whatever means, but, to understand what occurs, a further distinction

must be made. It is a familiar fact that a radio or television in a ground floor room of one house can often be heard in the bedroom of a neighbouring house. In this instance vibrations set up in the party wall at ground floor level are able to travel up the party wall and then radiate into the upper floor rooms. This is known as *flanking transmission*. It can still occur when the rooms adjoin, either horizontally or vertically, and often provides a serious addition to sound passing by the obvious path through the common wall or floor.

Specification of airborne sound insulation

In standard tests (BS 2750) the insulation is measured in each of sixteen one-third-octave bands, the centre frequencies of which range from 100 to 3,150 Hz. Measurements to find whether a construction can be deemed to satisfy the Regulations by way of the performance standard must be carried out in actual buildings and the results are expressed in terms of *normalised level difference*; this is the difference in decibels (dB) between the energy levels in the rooms corrected to allow for a standard amount of absorption representative of normal furnished conditions. Laboratory measurements are usually expressed in terms of the *sound reduction index,* which is the difference in decibel measure between the amount of energy flowing towards the wall in the source room and the total amount of energy entering the receiving room. The numerical value of the term which converts from normalised level difference to sound reduction index is usually small and either quantity can be used in considering principles; for this purpose it is therefore convenient to use the general term 'insulation'.

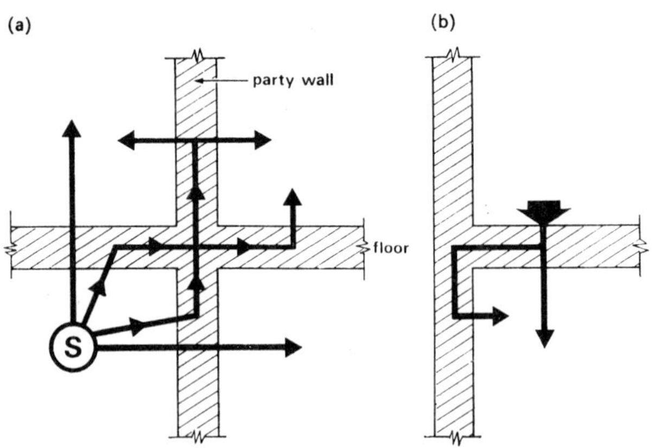

Fig 1 Vertical sections illustrating paths of sound transmission
(a) airborne
(b) impact

Prepared at Building Research Station, Garston, Watford WD2 7JR
Technical enquiries arising from this Digest should be directed to the Building Research Advisory Service at the above address.

The required values of normalised level difference range from 36 dB (for party floors) or 40 dB (for party walls) at 100 Hz to 56 dB at 1,600 Hz and above. The value of a quantity in decibels is equal to ten times the common logarithm of its value, relative to an arbitrary standard, in ordinary measure (joule/m^3 for energy level or watt/m^2 for energy flow). In the usual situation where the conversion term between sound reduction index and normalised level difference is small, this means that the fraction of the energy falling on a party wall which can be allowed to pass through it is only 1 part in 10,000 at 100 Hz and 2·5 parts in a million, at the highest frequencies. A common misconception is that the problem of sound insulation is one of absorbing the sound energy which falls on a wall. In fact, to approach the high levels of absorption that would be required all walls would have to be made like those of the anechoic chambers used in acoustic laboratories and this is not practicable for dwellings. Thus the problem of sound insulation is almost entirely one of reflecting energy back into the source room; the role of absorption is limited to supplementing reflection at high frequencies in some types of wall or floor.

'Mass Law'

Most of the principles involved in direct transmission are revealed by considering a single-leaf wall. The most widely known term in relation to sound insulation is the *mass law*. A strict mathematical derivation is outside the scope of this digest but Fig 2 shows in broad physical terms how the so-called 'mass law' arises. It seems obvious that the amplitude of sound waves radiated into the receiving room must depend only on the amplitude of vibrations in the wall, and not on any other properties of the wall; the amplitude of wall vibrations depends on the amplitude of the oscillatory part of the pressure in the source room which acts on the wall. According to Newton's law of motion,

$$\text{force} = \text{mass} \times \text{acceleration}$$

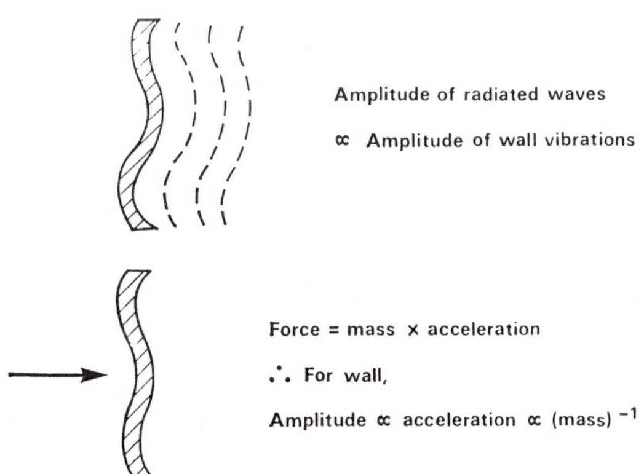

Fig 2 Physical ideas behind the 'mass law'

and, in an oscillatory motion, amplitude and acceleration are proportional to one another. Thus the amplitude of wall vibrations is inversely proportional to the mass of the wall. It follows that the amplitude of sound waves radiated into the receiving room is inversely proportional to the mass of the wall. Transmission is considered in terms of energy. Since this is proportional to the square of velocity, and thus to the square of amplitude, the transmitted energy is inversely proportional to the square of the mass of the wall. This means that by doubling the mass of the wall transmission is reduced to a quarter. In decibel measure, insulation is increased by

$$10 \log_{10} 4 = 6 \text{ dB}$$

The argument can be modified to deduce that a doubling in frequency should also produce a 6 dB increase in insulation. Broadly, these two statements about the effects of changes in mass and frequency constitute the 'mass law'.

When due account is taken of other properties of a wall, theoretical treatment becomes very complicated. However, whatever refinements are introduced, it is nearly always found that, as between single-leaf walls of a similar type, a doubling in mass produces an increase of about 6 dB in insulation, and that, except for the rapid change which is discussed shortly, there is an increase of the same order when the frequency is doubled.

'Coincidence'

There are two important features of the transmission process to which it was unnecessary to draw attention while sketching the basis of the mass law.

First, insulation is always measured for a 'diffuse' incident field which may be regarded as composed of plane waves approaching the wall from all possible directions. The more oblique the incident wave, the better it is transmitted; if the wall were large enough, waves approaching at a glancing angle would always be perfectly transmitted.

As part of the transmission process, a plane wave approaching at an oblique angle produces a forced motion in the wall (Fig 3) which has a greater wavelength (the 'trace' wavelength) than that of the incident waves in air. Because walls possess stiffness as well as mass, it is possible for free bending waves to propagate in a wall. The amplitude of wall vibrations is greatly magnified when the wavelength of the free bending waves is near to that of the wave impressed on the wall by the incident sound wave. This in its turn greatly increases transmission. The effect is known as *coincidence*.

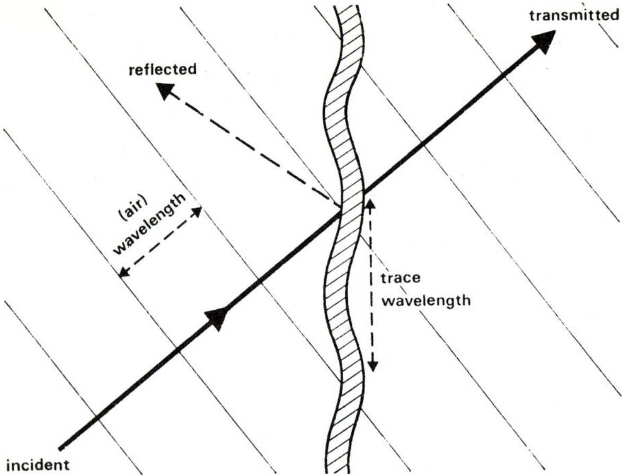

Fig 3 Forced waves in a wall have a longer wavelength than incident sound waves

The velocity of free bending waves increases with frequency. Since wavelength = velocity/frequency, the wavelength of free bending waves is less than that of sound waves in air at the lowest frequencies and greater than that of waves in air at the highest frequencies. It has been seen that the trace wavelength is greater than that of waves in air. Consequently coincidence can only occur when the wavelength of free bending waves is greater than that of waves in air; the frequency where these two wavelengths are equal is known as the *critical frequency*. Above this frequency transmission is dominated by coincidence. Near the critical frequency, coincidence occurs for glancing waves. Thus in this region two effects which lead to high transmission act together, and there is a large decrease in insulation.

The result is that, apart from some modification at low frequencies due to the relatively small area of practical walls, all single-leaf walls have a curve of insulation against frequency approximating to that shown in Fig 4. No origin is shown on either scale, since both the general level of insulation and the position of the critical frequency depend on the particular wall considered; it may be necessary to take measurements above or below the standard range of 100 to 3,150 Hz to detect the critical frequency dip. As 'coincidence' depends on a kind of resonance, the depth of the dip depends on the amount of damping present. For common amounts of damping, insulation at the critical frequency is often as much as 10–15 dB below the level suggested by the trend at lower frequencies. Insulation usually remains 5–10 dB below the low frequency trend for an octave above the critical frequency.

With sufficient mass, adequate insulation can be obtained even at the critical frequency. However, the depth of the dip suggests that it is desirable that a wall should have a critical frequency either below 100 Hz or above 3,150 Hz.

For reasons touched on later it is not too difficult to cope with a critical frequency dip at frequencies somewhat below 3,150 Hz. However, the general principle that critical frequencies between just over 100 Hz and 1,000 Hz should be avoided whenever possible is one of the most basic in sound insulation.

The critical frequency is given by

$$f_c = \frac{c_o^2 (1-\sigma)}{1 \cdot 8 c_c h \sqrt{(1-2\sigma)}}$$

where

f_c = critical frequency,

c_o = sound velocity in air,

$$c_c = \sqrt{\left\{\frac{E(1-\sigma)}{\rho(1+\sigma)(1-2\sigma)}\right\}}$$

= compressional wave velocity in wall material

E = Young's modulus of wall material,

σ = Poisson's ratio of wall material,

ρ = density of wall material,

h = thickness of wall.

Of the parameters involved, c_o, the sound velocity in air, cannot be altered. With a sufficiently wide choice of materials the wave velocity, c_c, and Poisson's ratio, σ can vary widely, but between most practicable materials the variation is limited. The most important parameter is usually the thickness, which can vary over a range of 50:1 between conceivable walls.

Practical implications for single-leaf walls

Some typical values of critical frequency are shown in Table 1. The critical frequency of a brick wall is not easily determined and the properties of brick walls are variable, but it appears that the critical frequency of a one-brick (215 mm) wall is usually in the neighbourhood of 100 Hz. A one-brick wall thus satisfies the condition of having a critical frequency near or below the lower limit of the practical range. In practice the insulation given by a brick wall near 100 Hz is better than might be expected from the idealised curve of Fig 4 because of effects related to

Table 1 Approximate critical frequencies

Material	Thickness *mm*	Surface mass *kg/m²*	Critical frequency *Hz*
Brick	215	400	100
Brick	102·5	200	200
Lead	18	200	15,000
Plasterboard	10	9	2,900–4,500
Plasterboard	20	18	1,400–2,300
Glassfibre reinforced gypsum	10	18	2,000
Plywood	10	7·5	1,300
Steel	3	25	4,000

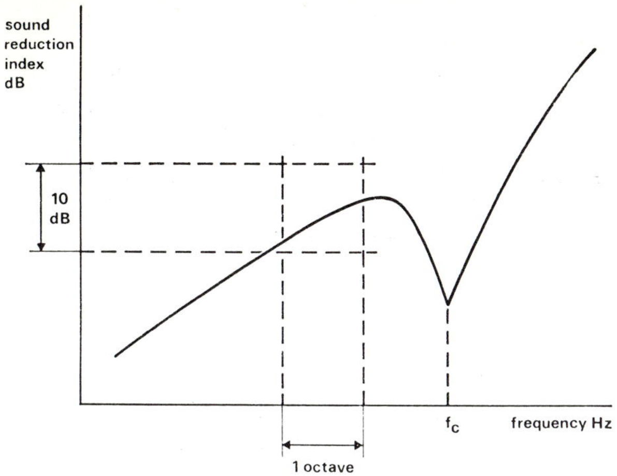

Fig 4 Typical form of insulation curve for a single-leaf wall

the limited area of practical walls. These arise because near the critical frequency, 100 Hz, the sound wavelength in air and the bending wavelength in the wall are both equal to about 3·4 m, and are therefore comparable with the linear dimensions of a normal room. Another departure from simple theory is that the insulation given by a one-brick wall levels off near 1,000 Hz instead of continuing to increase, because new modes of transmission arise in a thick wall. In practice this is not serious because the insulation is adequate.

Concrete walls—including those made of lightweight concrete—fall into the same category as a one-brick wall, having a critical frequency near 100 Hz if they are thick enough to provide the mass required.

With thinner brick or concrete walls insulation is poorer not only because the mass is reduced but because the critical frequency goes up to a frequency at which its effects are more serious.

Since brick and concrete walls operate mainly above the critical frequency, the insulation they provide is affected by the amount of damping present. It is therefore theoretically possible to improve the insulation such walls provide by modifying the materials used to increase their internal damping. There has in fact been research in University and in Industrial Laboratories on the use of polymers to increase the internal damping of concrete.

It is logical to consider the possibility of going to the other extreme, ie using walls with a critical frequency near or above the upper end of the frequency range at 3,150 Hz. Table 1 contains examples of panels with a critical frequency above 1,000 Hz. However, the mass required for a single-leaf wall with a high critical frequency to meet the party-wall grade can be calculated, and is just over 200 kg/m². This is about half the mass of a one-brick wall. It does not appear feasible to combine the required mass with a high

critical frequency in a practicable panel. Materials which would provide the required combination in a homogeneous panel exist, but they are expensive; the most suitable is lead, which (Table 1) would have a critical frequency of 15,000 Hz with a surface mass of 200 kg/m². With stiffer materials, a sophisticated sandwich construction would be necessary. For practical purposes, single-leaf walls with a high critical frequency can be ruled out; to obtain practical walls with a high critical frequency it is necessary to use two or more leaves separated by a wide gap.

Flexible double-leaf walls

The basic idea behind a double-leaf wall is shown in Fig 5. If three rooms were arranged in a row with successive pairs separated by walls giving 25 dB insulation, then the insulation between the end rooms would be 50 dB. It is reasonable to hope that insulation of the same order can still be obtained if the middle room is reduced to the sort of air gap that is practicable within a party wall. In contrast, if the two original walls could be joined together into a single wall without lowering the critical frequency, then, according to the mass law, the total insulation would be only

$$25 + 6 = 31 \text{ dB}$$

Practical experience has confirmed that a double-leaf wall with flexible (high critical frequency) leaves can provide satisfactory insulation with much less mass than is needed with a single-leaf wall. Walls already in use provide satisfactory insulation with a total mass of 60 kg/m²—only 15% of the mass of a one-brick wall. However, a double-leaf wall performs better than a single-leaf wall of the same total mass only above a certain frequency which depends on the cavity width. This frequency is the 'mass-spring-mass' vibration frequency f_0 of two rigid bodies of

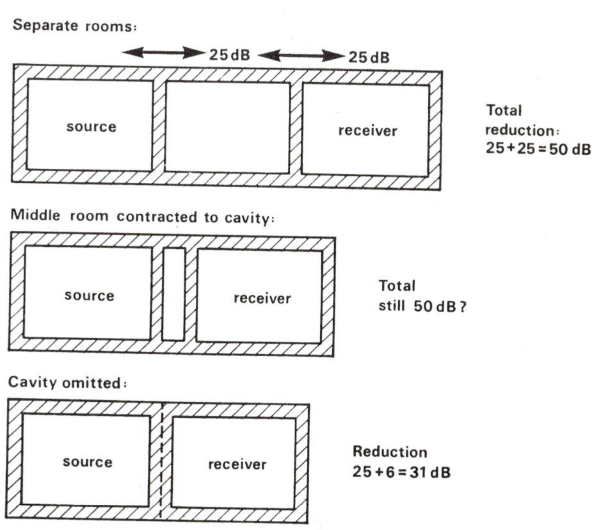

Fig 5 Basic idea behind a double-leaf wall

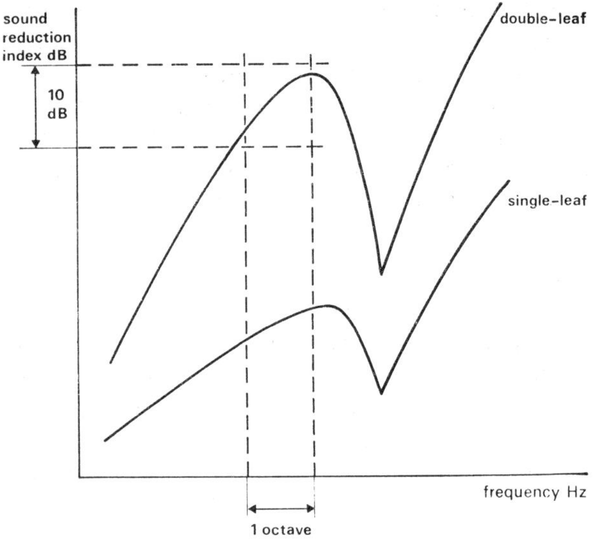

Fig 6 Comparison of shapes of typical insulation curves for single and double-leaf walls

the same mass as the leaves connected by a spring with the stiffness of the air in the cavity:

$$f_0 = \frac{1}{2\pi} \sqrt{\frac{\gamma P}{d}\left(\frac{1}{m_1} + \frac{1}{m_2}\right)}$$

where P = air pressure (approx 10^5 N/m^2)

γ = ratio of specific heats (approx 1·4)

m_1, m_2 = masses of leaves per unit area

d = cavity width

To gain advantage from double-leaf construction, it is essential that the cavity is wide enough for the 'mass-spring-mass' frequency to be below 100 Hz, and desirable that it is as wide as possible. Thermal double glazing with a narrow cavity does not provide good sound insulation because the 'mass-spring-mass' frequency is well above 100 Hz. A sharp dip in the insulation curve usually occurs near the 'mass-spring-mass' frequency. A similar effect can occur in more complicated arrangements such as dry-linings on masonry walls.

It is sometimes said that a double-leaf wall 'beats the mass law' Whether this is so is essentially a matter of definition and it is more convenient to regard single and double-leaf walls as obeying different kinds of mass law. A double-leaf wall provides nearly twice the insulation which each leaf would provide on its own and the effect of changing the mass of both leaves is twice the effect of a similar change in the mass of a single-leaf wall; thus if the total mass is doubled the insulation is increased by approximately

$$2 \times 6 = 12 \text{ dB.}$$

It will be recalled that for single-leaf walls insulation increases by approximately 6 dB per octave except where it dips near the critical frequency. For double-leaf walls the corresponding slope is 12 dB per octave. Consequently the curve of insulation against frequency for a double-leaf wall has the typical shape shown in Fig 6.

Because of the steep slope, it is usually found that while it is difficult to obtain sufficient insulation with lightweight double-leaf walls at the lowest frequencies, there is then a wide band of frequencies in which the insulation is well above what is required. Above this region, there are again problems because the dip near the critical frequency is even more pronounced than with a single-leaf wall. At low frequencies, the only solution, after choosing the largest practicable value for the cavity width, is to have sufficient mass, but having dissimilar leaves and using an absorbent curtain or filling of a material like glass fibre helps to overcome problems arising from the critical frequency dip. If the leaves have different critical frequencies, the dip, although broader, is not so deep. Absorbent material in the cavity is useful because the thickness needed to make an effective contribution decreases with frequency and reaches a practicable value at the higher frequencies. However, even in the critical frequency region absorption serves only as a supplement to reflection, which still makes the major contribution.

Although double-leaf lightweight walls can provide satisfactory insulation, it is not yet possible to predict with much confidence whether a given design will prove satisfactory. In particular, more needs to be known about the factors which can spoil insulation. There is a large difference between the vibration levels in the two leaves when a wall is behaving satisfactorily. As a result, insulation can easily be spoilt by solid bridges. With lightweight panels, studding is usually needed to increase static stiffness. It is almost certain that insulation will be inadequate unless the two leaves are attached to different studs. With low-rise housing, it is possible to build structurally detached houses with a gap of 300 mm or less between them, but in high-rise buildings some solid connection at the edges of a party wall is unavoidable. Study of a similar problem for windows has shown that mounting the panes in a neoprene gasket goes a long way towards providing the desired isolation between the leaves. There is good reason to expect that similar mountings can be useful in double-leaf walls.

Stiff double-leaf walls

The use of cavity constructions has been discussed at some length for flexible lightweight walls, because with such walls a cavity is essential. With stiff, heavy walls the situation is different. In practice, the choice is between a single-leaf wall and a double-leaf wall of similar total mass, eg a one-brick wall or two half-brick leaves. As was seen earlier, the loss in insulation which results from using a thinner wall is more than would be expected from considering the mass alone, because the critical frequency goes up into a region where its effects are more serious. This shift in critical frequency tends to offset the advantages of having a cavity. For practical arrangements, the two effects roughly cancel out and there is little to choose between single and double-leaf constructions. As with lightweight walls, ties between the leaves are liable to spoil the insulation, and it is advisable to keep to designs which experience has shown to be acceptable.

Floors

A heavy concrete party floor insulates against airborne sound on the same principle as a single brick or concrete wall: its effectiveness depends on the critical frequency being near the lower end of the practical frequency range, and the mass required for a given level of insulation is the same as with a wall. To obtain satisfactory insulation with less weight, it is again necessary to use a double-leaf construction. The obvious practical difference as compared with walls is that the floor must be stiff enough not to deform too much under loading. This requirement rules out having a critical frequency near 3,150 Hz and consequently the aim must be to make the floor as stiff as possible in order to have a critical frequency near or below 100 Hz. For ceiling panels which have

only their own weight to support, a critical frequency near 3,150 Hz should be acceptable. The best known lightweight construction is a timber-joist floor combined with a pugged ceiling. Although the details are complicated, this conforms roughly to the basic principle of combining a low critical frequency floor and high critical frequency ceiling.

Flanking transmission

The subject of flanking transmission is a complex one, because this type of transmission does not depend solely on the properties of the flanking wall, but also on the properties of the party wall, on whether the flanking wall is continuous, and on how firmly it is bonded to the party wall. The principles are most fully understood for heavy walls which operate almost entirely above the critical frequency.

The simplest case is that of a single-leaf external brick wall. It is convenient to consider first a hypothetical arrangement in which the external wall is vibrationally isolated from the party wall (Fig 7a). In considering direct transmission through a wall operating above the critical frequency, it was seen that most of the vibrational energy of the wall is associated with the 'coincidence' effect and is contained in free bending waves. In the arrangement of Fig 7a these waves can travel up and down the external wall without hindrance. Consequently the vibration level is the same in the parts of the wall adjacent to the receiving room as in the parts adjacent to the source room. As compared with the same construction used as a party wall, there is an inflow of energy over only half the wall instead of the whole, but energy is dissipated over the whole area. Thus the vibration level over the whole wall is half what it would be in a similar party wall. The radiation into the receiving room for a given vibration level is the same as if the wall were acting as a party wall. Consequently, for approximately square rooms, transmission via the external wall is about half (in decibel measure 3 dB lower) that which the same wall would transmit if acting as a party wall. With similar party and external walls, total transmission is then nearly 2 dB more than direct transmission.

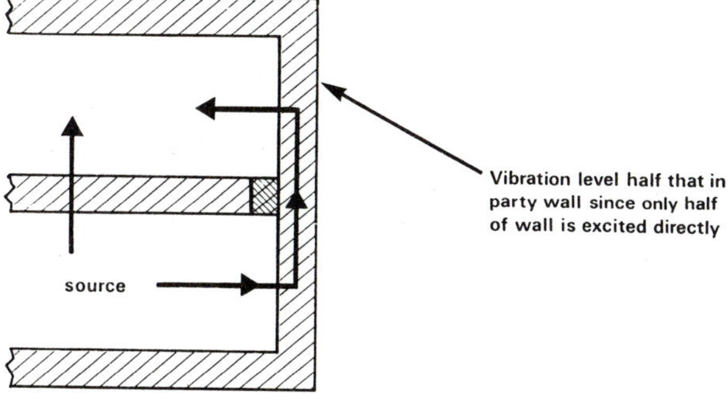

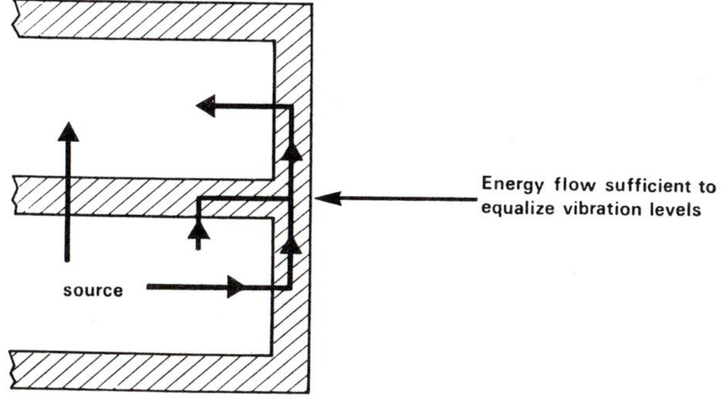

Fig 7 Flanking transmission with heavy single-leaf party and external walls

If, as is more usual, the party wall is bonded to the external wall (Fig 7b) there is extra restraint at the junction and energy is less freely transmitted between the two parts of the external wall. However, the energy levels in the two parts of the external wall will still be virtually the same if a packet of energy traverses a section of the wall a fairly large number of times before being dissipated: this situation is approached for practical values of damping. A complicating feature is that energy flow between the party wall and the external wall is of the same order of magnitude as that between the two parts of the external wall. If the walls are of the same thickness,

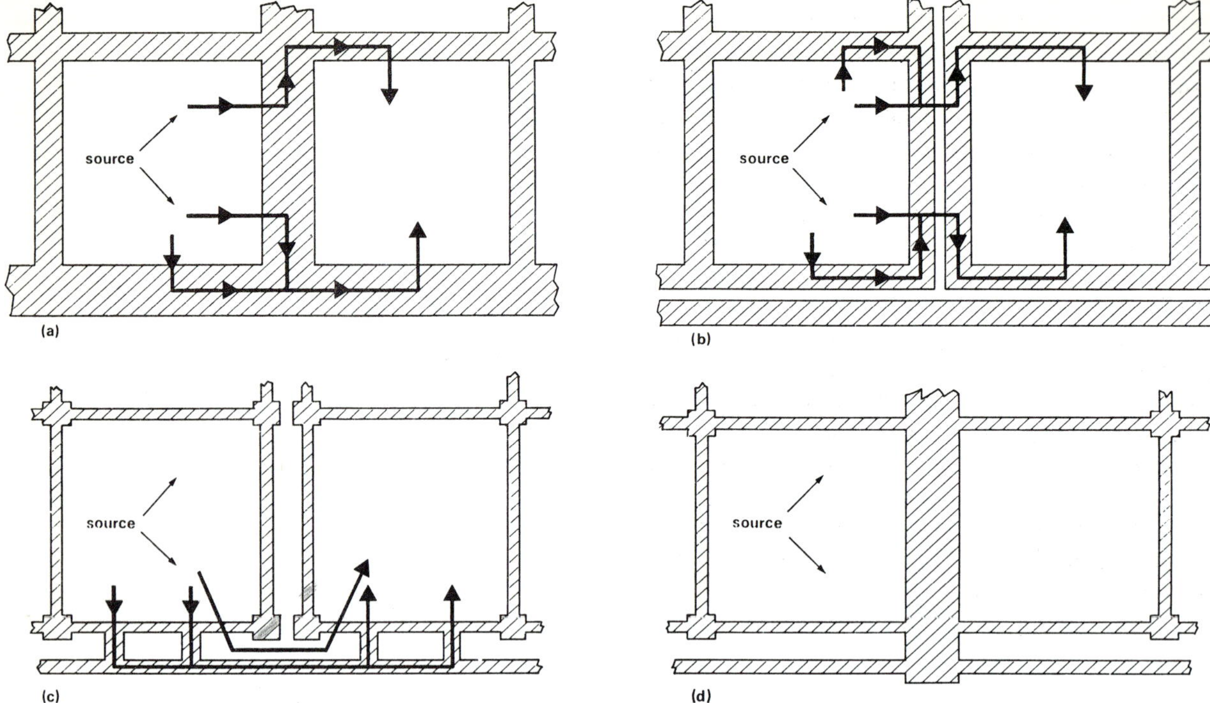

Fig 8 Principal flanking paths: (a) heavy single-leaf party and external walls; (b) heavy double-leaf party and external walls; (c) lightweight double-leaf party and external walls; (d) heavy single-leaf party wall, lightweight double-leaf external wall (flanking negligible)

the result is that the energy levels in the party wall and the two parts of the external wall become nearly equal. To produce this result, there must be a net flow of energy from the party wall to the external wall. As far as the party wall is concerned, this loss of energy has the same effect as additional damping in the wall, so that direct transmission is less than if the two walls were vibrationally isolated. But because of the additional energy received by the external wall, flanking transmission becomes nearly the same as direct transmission instead of being 3 dB less.

This result holds if the external wall has the same thickness as the party wall. Clearly flanking transmission depends on the properties of the wall in a very similar fashion to direct transmission. Thus if the external wall is much thinner than the party wall, flanking transmission is likely to exceed direct transmission.

Flanking transmission also occurs via internal partition walls; indeed, the fact that these are usually thin suggests that they are likely to contribute more than external walls. However, the position is more complicated because resistance to transmission across the junction with the party wall is usually high enough to produce a significant lowering in the vibration level of the partition wall. At present, only actual measurements can determine the precise importance of internal partition walls.

Possible flanking transmission paths for a number of types of construction are sketched in Fig. 8. Two paths are indicated for a wholly lightweight construction (Fig 8c): one via the air in the cavity, and one via ties across the external wall. The danger of a serious amount of transmission by the latter path is reduced by the fact that, except over short distances (as across the ties), structure-borne transmission depends on resonant vibrations. In discussing direct transmission, the only resonant vibrations mentioned were free bending waves above the critical frequency. Actually natural vibrations of panels below the critical frequency can increase direct transmission and contribute flanking transmission, but these vibrations radiate poorly, and it should be possible to produce designs with which their effects are unimportant.

Impact insulation

In several respects, the requirements for impact insulation are the same as for insulation against airborne sound. The importance of mass may be deduced in the same way as for airborne sound (Fig 2): the energy radiated by a floor depends on the vibration amplitude, and Newton's law of motion indicates that the vibration amplitude is inversely proportional to the mass of the floor whether the exciting force is air pressure or the impact of a solid object. It is also clear that, as for airborne sound, a useful part can be played by a cavity construction which ensures that the vibration level in the radiating leaf is much lower than in the leaf directly excited. For best results, transmission through solid connections must be less than the unavoidable transmission through the air in the cavity. This is clearly not practicable when the upper leaf is supported on the lower. Instead, to raise impact insulation to a satisfactory level, use is commonly made of the lesser degree of isolation between the leaves provided by a floating floor. In this, the leaves are separated by a resilient layer, usually glass wool. A complicated theory is needed to give a satisfactory explanation of the behaviour of floating floors, but there is a fairly simple criterion for the minimum thickness of a resilient layer of given stiffness: it needs to be thick enough for the 'mass-spring-mass' frequency of the system floating layer/resilient layer/structural flow to be below 100 Hz.

Further reading
BS 2750:1956 Recommendations for field and laboratory measurements of airborne and impact sound insulation in buildings (with 1963 amendment) British Standards Institution.
National Building Studies Research Paper No. 33 Field measurements of sound insulation between dwellings: P H Parkin, H J Purkis, W E Scholes HMSO 1960

Digests:
102 Sound insulation of traditional dwellings—1
103 Sound insulation of traditional dwellings—2
128 Insulation against external noise—1.
129 Insulation against external noise—2
187 Sound insulation of lightweight dwellings

Sound insulation of traditional dwellings—1

This Digest and the next are mainly concerned with the design of new dwellings—houses and flats—of traditional construction but the principles discussed are equally applicable to the conversion of existing buildings.

Part 1 deals with: requirements of building regulations; explanation of the grading system; sound insulation of party walls in traditional construction; sound transmission between rooms in the same dwelling; improvement of existing dwellings.

Part 2 deals with: floor constructions, concrete and wood joists.

Requirements of building regulations

The building regulations for Scotland (1963) and for England and Wales (1972) incorporate mandatory requirements controlling sound transmission between dwellings. There are some differences in the standards of performance required in Scotland and in England and Wales, but in no case is the required performance below Grade I. Lower grades are, however, still of interest, for example, in the improvement of existing dwellings or in defining performance in other types of building for which there are no mandatory requirements for sound insulation.

The grading system

Party-wall grade. This grade is based on the performance of the one-brick party wall. It reduces the noise from neighbours to a level that is acceptable to the majority; a lower standard certainly could not be justified on present evidence. A higher standard is not yet practicable, mainly because at this level of insulation flanking transmission is usually about equal to direct transmission and there is little to be gained from improving only direct transmission.

Grade I. This is the highest insulation that is practicable at the present time *vertically* between flats. It is based on the performance of a concrete floor construction with a floating floor, which gives the best floor insulation obtainable by normal structural methods. Noise from the neighbours causes only minor disturbance; it is no more of a nuisance than other disadvantages which tenants may associate with living in flats.

Grade II. With this degree of insulation the neighbours' noise is considered by many of the tenants to be the worst thing about living in flats, but even so at least half the tenants are not seriously disturbed.

Worse than Grade II. If the insulation between flats is as low as 8 dB worse than Grade II, then noise from the neighbours is often found to be intolerable and is very likely to lead to serious complaints. With better insulation than '8 dB worse than Grade II' the likelihood of complaint decreases gradually, but when there are also other reasons for dissatisfaction serious complaints about noise may occur if the insulation is worse than Grade II.

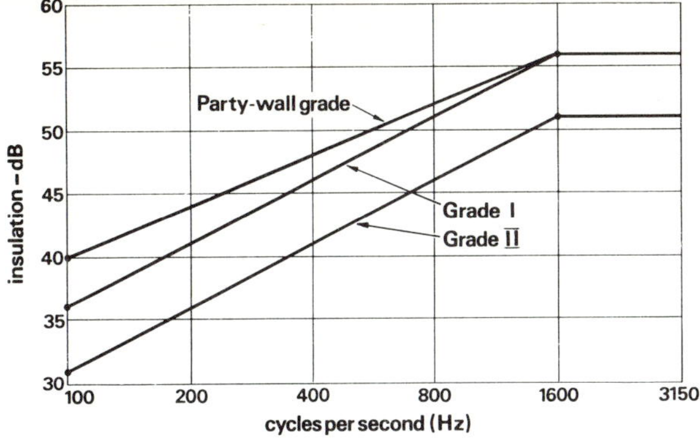

Fig 1   Grade curves for airborne sound insulation

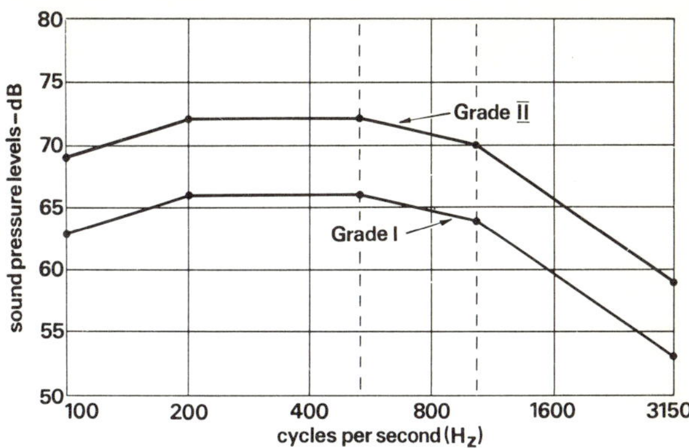

Fig 2   Grade curves for impact sound insulation

The grade curves

The levels of airborne and impact sound insulation that satisfy the party-wall grade and Grades I and II are given in Figs 1 and 2. To qualify for a particular airborne sound grading, the insulation should be *not less than* the value shown at each frequency in Fig 1. To meet an impact grading, the measured noise levels produced underneath a floor by a standard impact machine operated on the floor should not *exceed* the value shown at each frequency in Fig 2.

Wall constructions are required to meet the grade for airborne sound only; floor constructions, in order to be classified under a particular grade, must satisfy that grade for both airborne and impact sound insulation.

Measured insulation curves seldom follow the grade curves exactly and measurements satisfying grade requirements over most of the frequency range may very likely fall short at one or two frequencies; it is not intended that constructions should be condemned or graded down on account of minor faults that may have little significance, and in practical grading assessments a suitable tolerance is allowed. For strict grading purposes, such as conformity to Building Regulations, tolerances must be accurately defined, and it is also necessary to correct, or normalise, measurements to standardised conditions; without this correction a construction would, for example, vary in its insulation value depending on whether it was measured in an occupied or an empty dwelling, because of the differing amounts of absorption present.

All the gradings in this Digest and in Digest 103 are based on measurements made in accordance with BS 2750:1956, normalised to 0·5 sec reverberation time in the receiving room. The permitted tolerance for compliance with a particular grade is a total adverse deviation from the grade curve at all $\frac{1}{3}$-octave frequency bands between 100 and 3150 Hz of not more than 23 dB. The performance of party wall and party floor constructions is given in terms of their sound insulation grading relative to the grade curves.

Wall constructions

The sound insulation gradings of various forms of construction for party walls are shown in Table 1. The values given are based on the assumption that the remainder of the construction is traditional, with all structural elements firmly bonded together.

Another assumption is that the planning is traditional; that is, the size and layout of the rooms is normal or average for dwellings. Large rooms separated by small areas of party wall create favourable conditions for sound insulation; conversely, small rooms, with large areas of separating wall, are unfavourable. At present, no practical guidance on the effects of planning on sound insulation is considered feasible because of the complications of flanking transmission.

Party walls in houses
Solid walls

The basis of the party-wall grade is the one-brick wall, plastered on both sides, weighing not less than 415 kg/m² (including plaster*), but any solid walling material of this weight, plastered on both sides, is likely to meet the grade.

On concrete of open texture, eg no-fines concrete and many lightweight concretes, the plastering must not be omitted from any part of the surface; flues from fireplaces, etc, should be lined or rendered so as to seal off all air-paths through the porous material.

Attempts to seal porous concrete walls with linings of wallboard on battens, plaster strips or dabs, instead of plastering, have usually proved to be unsatisfactory for sound insulation. Non-porous walls lined with plasterboard on battens should be satisfactory if the overall weight is not less than 415 kg/m², but there is some doubt about the performance of plasterboard attached by plaster dabs; this requires further investigation. Dense concrete cast in permanent shuttering of wood-wool slabs, with the surface of the wood-wool plastered, has been found to provide only poor insulation at an important part of the frequency range.

* Throughout this Digest, 'plaster' implies dense, two-coat work at least 12 mm thick weighing not less than 24 kg/m²

Table 1: Gradings of party walls in traditional dwellings

SOLID WALLS		Weight incl any plaster	Grade	CAVITY WALLS, with wire ties of butterfly pattern and plastered on both sides	Cavity width	Weight incl any plaster	Grade
		kg/m²			mm	kg/m²	
One-brick wall	pbs	415	P–w	Two leaves each consisting of: 100 mm brick, block or dense concrete	50	415	P–w
In situ concrete or concrete panels with joints solidly grouted	plaster optional	415	P–w	Lightweight aggregate concrete—sound absorbent surfaces to cavity (see text)	50	300	P–w
175 mm concrete at 2320 kg/m³	without plaster	—	P–w	Ditto	75	250	P–w
175 mm concrete at 2080 kg/m³	pbs	—	P–w	Ditto	50	250	I
Lightweight concrete or other material	pbs	415	P–w	50 mm lightweight concrete at 1280 kg/m³	25	—	II
300 mm lightweight concrete 1200 kg/m³	pbs	—	P–w	100 mm hollow concrete blocks	50	—	II
225 mm no-fines concrete at 1600 kg/m³	pbs	—	P–w				
200 mm ditto	pbs	365	I				
Lightweight concrete, or other material	pbs	220	II				
Half-brick wall	pbs	220	II				

Notes: pbs = plastered on both sides

P–w = Party-wall grade

All weights, thicknesses and widths are minima.

Cavity walls

Cavity walling of two half-brick leaves, separated by a 50 mm cavity, was formerly recommended for party walls on the basis of the higher single-figure insulation. All available evidence now shows that this has no sensible advantage over the one-brick solid wall; indeed, unless the cavity width is maintained as a minimum and wire ties of butterfly pattern are used, the party-wall grade will not be attained.

Materials other than brick (eg concrete or stone) used for the leaves of a cavity wall will attain party-wall grade if the wall is plastered and its overall weight is not less than 415 kg/m².

If light-weight aggregate concrete is used, presenting sound-absorbent surfaces to the cavity, the weight of the wall can be reduced to 300 kg/m² with a 50 mm (minimum) cavity. If the width of the cavity is increased to 75 mm or more, the weight of the wall can be further reduced to 250 kg/m².

In all cases, the requirements already stated regarding plastering and wall ties must be met.

Wall linings

Wall linings on battens or studs can improve sound insulation but their performance is variable because of many factors. Although firm directions cannot be given, some guidance based on present knowledge might be helpful.

No improvement of the wall is of any value if flanking transmission already equals or exceeds direct transmission.

On a solid wall weighing at least 250 kg/m², wall linings can sometimes bring the insulation up to party-wall grade. The weight of the lining may need to be as much as 25 kg/m² but this could depend on the width of the air-space between the lining and the wall. The air-space should always be as wide as possible because of the benefit at low frequencies, where the needs of sound insulation are usually most difficult to meet; an air-space of 150 mm may well be necessary. Because isolation of the lining from the wall is desirable, a lining on independent studding is more effective than one on battens, but lining on battens may occasionally give the small improvement necessary to up-grade a wall that otherwise fails by only a small margin. A moderate amount of sound absorption in the cavity behind the lining (eg a suspended quilt of glass wool or mineral wool) is nearly always beneficial.

Linings over shallow air-spaces (eg on battens) are of little value for insulation of the low frequencies.

Party walls in flats

Although flanking conditions vary between two-storey houses and multi-storey flats, the effect of this on sound transmission in traditional construction is not usually significant. Any of the constructions mentioned in Table I should therefore attain the grading attributed to them if they are built as panels within the members of normal heavy steel or reinforced concrete framed multi-storey flats, provided that the connection between the frame member and the wall panel is made air-tight and rigid by sealing with mortar. The practice of inserting strips of non-rigid material, such as cork or felt, round the edges of the panel to separate it from the frame is not recommended; in certain special constructions such strips can be designed to improve the insulation but in general they are likely to reduce it. In particular, horizontal resilient membranes alone inserted in the walls at floor levels—usually with the object of reducing the flanking transmission up and down the walls—are of no value for the purpose and are best omitted.

A solid wall, weighing at least 365 kg/m², is likely to meet Grade I requirements.

Attention to detail

To achieve the grade that a construction is capable of giving, care must be taken to ensure that the wall construction does not fall below the standard in any area, such as in under-building or roof space, within the thickness of floors and ceilings or behind heating appliances. Eaves and dormer windows may also need special care.

Internal partitions

Although there are no mandatory requirements for sound insulation of partitions, it is desirable to provide a reasonable degree of privacy between rooms. Many dwellings have partitions either of plasterboard-faced panels with a honeycomb core or of plasterboard on each side of 50 mm or 75 mm stud framing; the sound reduction* is about 28–30 dB; this is a low standard of privacy between habitable rooms. Although no systematic investigation has been made of the amount of complaint it provokes, a reduction of about 35–40 dB between most rooms seems acceptable. This can be achieved by a solid partition of brick- or block-work, plastered both sides, weighing 75–150 kg/m², or by a sealed double-leaf dry partition of half the weight with a cavity not less than 50 mm wide. However, this cannot be a universal recommendation because the pattern of living within the home varies widely and it is difficult to determine whether the demand justifies imposing a standard between internal rooms.

Existing houses

It is rarely economical or even practicable to improve the insulation of a completed dwelling. This is because sound insulation is a function of the whole construction, not of the party wall alone and still less of any surface treatment of the party wall. Between well-built houses with one-brick solid party walls

---

* The performance of internal partitions is expressed according to common practice as a single-figure average over the usual frequency range.

and external (flanking) walls, with all brick joints and frogs filled with mortar, the sound insulation should be better than the party-wall grade. This is close to the maximum normally obtainable with ordinary methods of construction. Reducing the direct transmission alone (by insulating the party wall) is pointless because by-passing by flanking transmission leaves the net insulation almost unchanged. It is not often possible to reduce the flanking transmission.

Where the direct transmission is appreciably greater than the flanking transmission, and treatment of the party wall could be beneficial, it will be worth while to engage an experienced consultant. If there is leakage via air-paths, perhaps through underfloor spaces that are linked by gaps around the ends of joists built-in to the party wall, these paths should be sealed before any further treatment is given. Suitable treatment may sometimes take the form of wall linings, though a wide cavity between the lining and the wall behind may be needed to give adequate improvement at low frequencies, as already discussed.

Sometimes the insulation is already up to the party-wall grade, or not far below it, but the householder wants still better insulation. The solution with most chance of success is to increase substantially the weight of the party wall. A further skin of brickwork, half- or one-brick thick, tight up against the existing wall, and mortared to it, may help to reduce both flanking and direct transmission. But this solution calls for new foundations, and it is not possible to say how extensive the heavier wall needs to be to give worth-while results; it might be necessary to thicken up the whole party wall to obtain only limited benefits.

Further reading

National Building Studies, Research Paper No 33, *Field measurements of sound insulation between dwellings*, P H Parkin, H J Purkis and W E Scholes; HMSO: 1960.

BRE Digests

143 *Sound insulation: basic principles*

Sound insulation of traditional dwellings—2

Concrete floors

Various types of floor construction are listed in Table 1 with the insulation grading that each is likely to give in practice for airborne and impact sound insulation. In cases where the gradings under these two headings differ from each other, the lower gradir.g is to be taken as the overall grade; thus, only floors that give Grade I both for airborne and impact sound insulation will give Grade I overall.

Floor finishes

The effect of the floor finish on insulation against impact noise is shown in Table 1; for example, in items (4)–(6) a 'hard' floor finish gives worse than Grade II but a 'soft' finish upgrades the same construction to Grade I. Ordinarily, no floor finish adds significantly to the airborne sound insulation, but the impact insulation of most concrete floors can be raised to Grade I simply by adding a finish or covering that is soft enough. A fitted carpet on an underlay of hair felt or sponge rubber will nearly always give Grade I impact insulation. It is obviously undesirable, however, to rely on a floor covering that is under the control of the occupant of the flat above the floor to provide the insulation required by the occupant of the flat below.

There is no standard grading of floor finishes for impact noise but resilience and thickness both have some effect. The terms hard, medium and soft are used in Table 1 and elsewhere only to indicate the properties of the floorings in relation to impact noise; the following examples should help to interpret this classification:

Hard: *concrete, terrazzo, clay and stone; pitch-mastic and mastic asphalt; magnesium oxychloride; thermoplastic tiles; 2·5 mm linoleum; wood block, parquet and mosaic.*

Medium: *thin carpet without underlay; cork flooring not less than 6 mm thick; wood board and strip; chipboard.*

Soft: *thin carpet with underlay, thick carpet with or without underlay; rubber or plastics on sponge-rubber or felt backing of combined thickness not less than 4·5 mm; cork tile not less than 8 mm thick.*

This list is not comprehensive and in particular many new floorings have not been included, either because no examples have been measured by the Station in field tests or because their long-term behaviour is not known. On constructions which are shown in Table 1 as achieving Grade I only marginally, it would be inadvisable to use floorings of which the performance is in doubt.

Floating floors

Among the constructions shown in Table 1, there are two forms of floating floor, items (2) and (3), for use on any concrete structural base. Each is capable of giving Grade I overall sound insulation, with any floor finish.

The principle underlying the design of a floating floor is its isolation from any other part of the structure. To ensure this, it is recommended that the resilient layer on which it rests should be turned up at all edges which abut walls, partitions or other parts of the structure. Partitions should be built off the structural floor so that the floating screed or raft is self-contained within each room. There must be no continuity between the floating element and the structural base either by fixings or by bridging of the resilient layer.

Concrete screed In this form of construction the screed rests on a resilient layer which is laid over the structural floor. Details of the design of floating screeds—thickness, mix proportions, bay sizes, etc.—are given in Digest 104, *Floor screeds.* A thickness of at least 65 mm is required and if the area of screed is greater than about 15 m² there will be a considerable risk of curling or cracking unless special precautions are taken. In the past, it has been suggested that the screed should be divided into bays, but because of the difficulty of dealing with joints where adjacent bays have moved out of alignment it is now suggested that a floating screed should be lightly reinforced rather than divided into bays. Reinforcement will not, however, eliminate curling.

The concrete screed in its wet state must not be allowed to penetrate the resilient layer, of which the joints are the most vulnerable in this respect. A layer of waterproof paper or plastics sheeting should therefore be used to prevent this. Failure to do so will allow solid bridges to form which would reduce very considerably the effectiveness of the construction for sound insulation.

It is usual to lay wire netting (eg 20–50 mm mesh chicken wire) over the quilt and sheeting to protect them from mechanical damage during the operation of placing the concrete.

Synthetic anhydrite screed A satisfactory material for floating screeds is synthetic anhydrite; shrinkage and curling are practically eliminated even with very large areas. The material is more expensive than an equal volume of concrete but the permissible reduction in thickness (to about 30 mm) helps to offset the higher cost. It is normally placed on a layer of waxed paper which, if adequately lapped, avoids the risk of solid bridging mentioned above. The material has the further advantage of faster drying than concrete screeds.

Wood raft floating floors on concrete (Table 1 item 3) A wood raft floating floor consists simply of wood flooring, fixed to battens to form a raft which rests on a resilient quilt laid over the structural floor slab. The battens must not be fixed to or in direct contact with the slab. For structural reasons, floorboards should preferably be tongued and grooved and not less than 20 mm thick; 18 mm plywood or chipboard of flooring grade is equally suitable. The battens should be at least 40 mm deep (preferably 50 mm or more) and not less than 50 mm wide; they are usually spaced at about 400 mm centres.

The use of chipboard sheets tongued and grooved together and placed on the resilient layer without battens is sometimes suggested; little is known about the insulation properties of this method and there may be difficulty in joining the sheets together to form an isolated raft that is structurally stable. Chipboard, factory bonded to an expanded polystyrene layer and with provision for forming tongued and grooved joints, is also available.

Softwood is generally used for floating floors in dwellings; special precautions are necessary with hardwood because this is usually supplied kiln-dried to a low moisture content. If the moisture content at the time of laying is lower than it will be when the building is in use, the wood will swell as it takes up moisture and the floating raft, being unrestrained, may buckle. Therefore, when using hardwood for this purpose it should be at a higher moisture content than that at which it is usually supplied and species of low movement value should be specified.

Resilient layers The resilient layer is a very important part of floating floor construction for which only reliable material should be used.

Glass wool and mineral wool are in common use; quilts of the long-fibre type have been found the most satisfactory. As a basic guide to a suitable quilt, the long-fibre glass-wool type PF 225 with a nominal thickness of 13 mm has proved satisfactory. The prefix means that it is paper-faced (one side); it has an uncompressed density of 36 kg/m³.

The thickness and density quoted are minimum figures for this type of quilt; increased thickness or density tend to improve the sound insulation performance because the resilience increases. The type 600 resin-bonded glass-wool quilt or slab, 25 mm thick, of density about 100 kg/m³, which is sometimes used under heated floor screeds for thermal insulation, is eminently satisfactory from the sound insulation viewpoint. Long-fibre mineral-wool quilts are likely to prove just as satisfactory for sound insulation in floating floors as their glass-wool counterparts. For design purposes, it may be assumed that the 13 mm '225' quilt compresses in service to 3 mm under wood battens or to 6 mm under a concrete screed and that the 25 mm '600' quilt, normally used under concrete screeds, does not compress to less than 22 mm.

The above quilts are equally suitable for use with concrete-screed or wood-raft floating floors, on

Sound insulation grading of concrete floors between flats

Construction		FLOOR FINISH [1]	GRADE	
			AIRBORNE	IMPACT
(1) Concrete floor 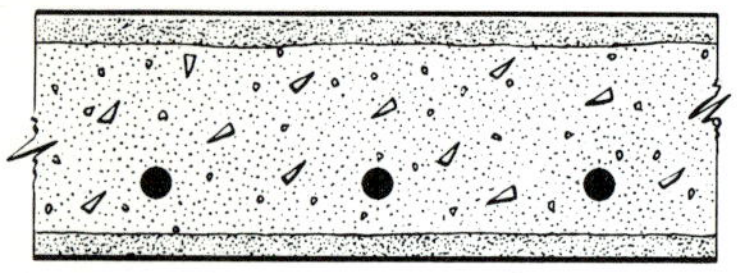	The basic floor construction of items (1) to (5) is assumed to weigh not less than 220kg/m^2 (including any integral screed and plaster finish) and may be dense or lightweight reinforced concrete, hollow concrete beams or concrete beams with hollow clay block infilling	Hard	II	4dB worse than Grade II
		Medium	II	II
		Soft	II	I [2]
(2) Concrete floor with floating screed 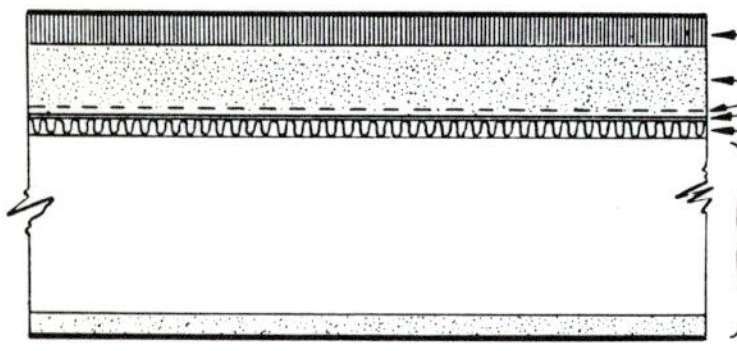	Any floor finish / Screed / Wire mesh / Paper / Resilient layer / Not less than 220kg/m^2	Any	I	I
(3) Concrete floor with floating wood raft 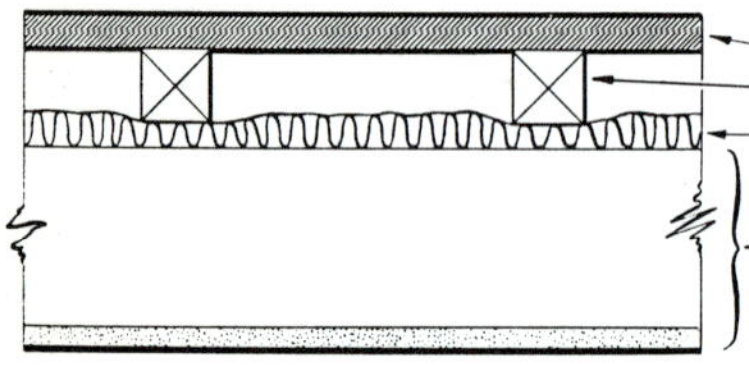	Wood flooring / Battens / Resilient layer / Not less than 220kg/m^2	Any	I	I
(4) Concrete floor with suspended ceiling 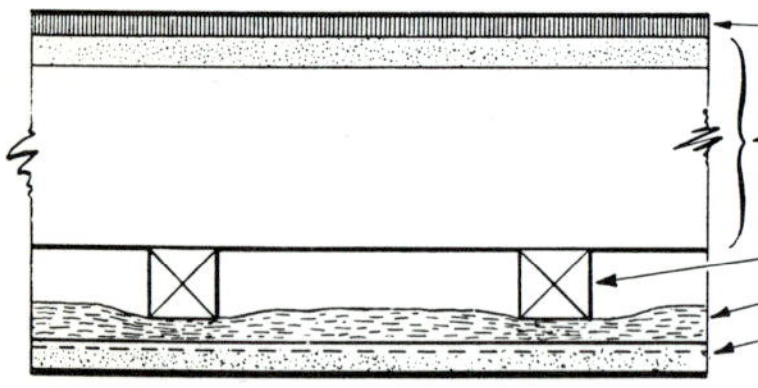	Floor finish / Not less than 220kg/m^2 / Battens wired to slab / Absorbent quilt / Heavy ceiling e.g. plaster on expanded metal lathing	Hard	I [2]	2dB worse than Grade II
		Medium	I [2]	II
		Soft	I [2]	I
(5) Concrete floor with lightweight screed 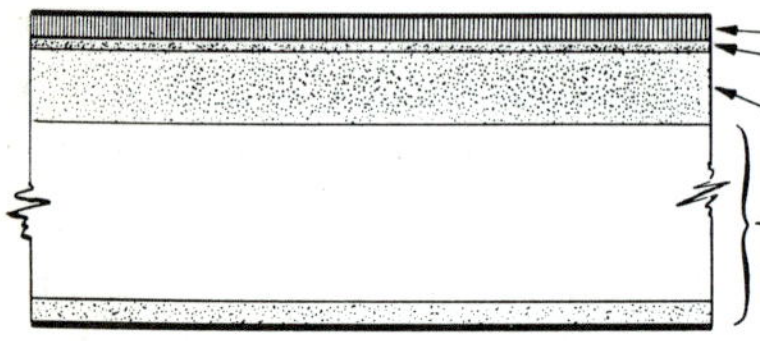	Floor finish / Dense topping (see text) / 50mm lightweight screed / Not less than 220kg/m^2	Hard	I [2]	4dB worse than Grade II
		Soft	I [2]	I [2]
(6) Heavy concrete floor 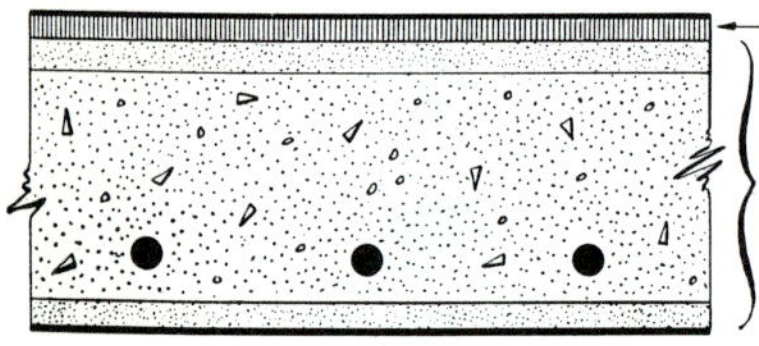	Floor finish / Not less than 365kg/m^2 including screed and plaster	Hard	I	4dB worse than Grade II
		Soft	I	I

Notes: (1) refer to paragraph ' Floor finishes on page 1   (2) the grades shown may be obtained only marginally.

concrete slabs or on wood joists. When laying paper-faced quilts on concrete slab floors, the paper face should be upwards (essential under floating screeds unless additional waterproof paper is used), whilst on wood-joist floors the paper face should be downwards. Although glass wools and mineral wools are the most tried materials for resilient layers under floating floors, other materials can be used if they have adequate resilience in a stable or permanent form. Resilience is best checked by measurement in the field against the Grade I performance standard. Permanence is more difficult to check and is likely to be verified only by experience.

Expanded polystyrene board has been used on a fairly wide scale as an alternative to glass-wool and mineral-wool quilts. Complete reassurance that its creep properties are insignificant is still lacking but after a number of years experience there is no evidence of any serious deterioration in performance. Therefore, although the Station has not itself investigated those properties of the material that affect its resilience, it is suggested as a resilient layer for floating floors provided the following specification is adhered to:

(a) density to be in the range 15–25 kg/m³

(b) nominal thickness to be not less than 13 mm

(c) the board should be pre-compressed to half its initial thickness with a rapid recovery to at least 90 per cent of the initial thickness.

Expanded polyurethane shows some promise as a resilient layer but has not been much used as yet and not fully investigated.

Some other materials, eg cane fibreboards and hair felts, have been found unsatisfactory because of compaction under continuous loading.

Pipes and conduits in floating floors   It is often necessary for services such as electric conduits, gas and water-pipes, etc, to traverse a concrete floor. Whenever possible these pipes should be accommodated within the thickness of the floor slab and integral screed, if any, but sometimes they have to be laid on top of the slab and contained within the depth of a floating floor. This need not cause trouble with a floating screed provided that the pipes do not extend more than about 25 mm above the base, that they are securely fixed so as not to move whilst the floating floor is being laid and are haunched up with mortar on each side to give continuous support to the resilient quilt. When two pipes cross, one of them should be sunk into the base slab. The resilient quilt should be carried right over the pipes. If a wood-raft floating floor is being used and the pipes have to be laid above the slab, the pipes can of course be readily accommodated parallel to and between the raft battens, but in the other direction the battens will have to be notched over them; the battens must be thick enough to allow for this.

Lightweight concrete screeds
Lightweight concrete is sometimes used for floor screeds without a resilient quilt (see Table 1 (5)). In many instances this construction has given Grade I airborne sound insulation, but with no improvement in impact insulation. A soft floor finish (see p. 190) is therefore necessary in order to bring the overall rating up to Grade I.

Although not all the details of the construction have yet been investigated, the essential requirements appear to be as follows:

The density of the concrete screed should be not more than 1100 kg/m³.

The thickness of the screed should be at least 50 mm, exclusive of any dense topping (see next paragraph).

An impervious or airtight layer should be provided above the lightweight screed. The dense concrete topping often required on a lightweight screed to ensure a satisfactory base for the floor finish will serve this purpose.

It is not yet known whether some types of lightweight concrete screed are better than others for sound insulation purposes, assuming that the density requirements are met.

Recommendations for the design and laying of lightweight aggregate concrete screeds are given in Digest 104.

Suspended ceilings
Suspended ceilings are chiefly of benefit against airborne sound and are comparable with lightweight concrete screeds in that they can be used to raise the sound insulation of a normal concrete floor to Grade I, provided that a soft floor finish is also used to give the necessary improvement in impact insulation. One form of construction is shown in Table 1, item 4. Not all the suspended ceilings that have been measured have given a satisfactory improvement of sound insulation and the requirements for a successful system of construction are not all known precisely; the following features appear to be significant:

The ceiling should be not less than about 24 kg/m²; plaster on expanded metal lathing or on plasterboard can provide this weight.

The ceiling should be essentially airtight so as to eliminate direct sound penetration via air-paths, such as would occur with open-textured materials or with open joints.

The points of suspension from the floor structure should be as few and as flexible as possible.

The air space above the ceiling may range in depth from 25 to 300 mm or more—the deeper the better —and should preferably contain sound-absorbent material.

Table 2   Sound insulation grading of wood-joist floors between flats

Construction				Grade	
Floor	Ceiling	Pugging	Walls	Airborne	Impact
Plain joist	Plasterboard and single-coat plaster	None	Thin	8 dB worse than Grade II	8 dB worse than Grade II
			Thick	4 dB worse than Grade II	5 dB worse than Grade II
		15 kg/m²	Thin	4 dB worse than Grade II	6 dB worse than Grade II
			Thick	Possibly Grade II*	Possibly Grade II*
	Heavy lath and plaster	None	Thin	Probably 4 dB worse than Grade II*	Probably 6 dB worse than Grade II*
			Thick	Grade II	Grade II
		80 kg/m²	Thin	Grade II	Grade II
			Thick	Grade II or possibly Grade I*	Grade II
Floating	Plasterboard and single-coat plaster	None	Thin	4 dB worse than Grade II	3 dB worse than Grade II
			Thick	Possibly Grade II*	Possibly Grade II*
		15 kg/m²	Thin	2 dB worse than Grade II	2 dB worse than Grade II
			Thick	Grade II or possibly Grade I*	Grade II or possibly Grade I*
	Heavy lath and plaster	None	Thin	2 dB worse than Grade II	Grade II
			Thick	Grade II or I†	Grade I
		15 kg/m² (as Fig 2)	Thin	Possibly Grade II*	Grade II*
			Thick	Grade II or I†	Grade I
		80 kg/m² (as Fig 1)	Thin	Probably Grade I	Probably Grade I
			Thick	Grade I	Grade I

*Assumed from other measurements   † May give Grade I with very thick walls

Although useful for sound absorption in the acoustic treatment of rooms, ceilings of soft insulating fibreboard are not recommended for sound insulation between rooms because of their light weight and porous nature.

Wood joist floors

The influence of wall thickness   Indirect or flanking transmission always has some effect on overall sound transmission and the performance of wood-joist floors is influenced by the amount of flanking sound transmitted via the walls. If the sound energy passing up or down the walls is greater than that passing through the floor, then the walls and not the floor will control the sound insulation between the rooms. Further treatment of the floor will be of little value unless the sound transmitted via the walls can be reduced correspondingly. This means reducing the vibration of the walls by one of the following means:

(a)   making the walls thicker

(b)   making the floors heavy enough and stiff enough laterally to restrain vibration of the walls.

Concrete floors are heavy and stiff enough to restrain vibration of the walls but most wood-joist floors are not and the maximum net sound insulation is controlled by the thickness of the walls, even though the floor may have potentially higher sound insulation. Therefore insulation values for wood-joist floors can only be given in conjunction with the wall system, as in Table 2.

To be classed as a thick wall system, three or more of the walls below the floor must be at least one brick thick or of similar weight; the walls above the floor need not be so thick. The overall thickness of a cavity wall does not count in this respect, but only the thickness of that leaf which constitutes the vertical flanking path and radiates sound into the room below. Metal anchorages connecting floor joists to external walls are sometimes employed in order to give lateral support to the walls, but the additional stiffness imparted to the walls by this means is insufficient to give any improvement of sound insulation.

Wood-joist floors with thin walls   Unless in conjunction with a thick wall system, most wood-joist floors, even if designed for sound insulation, will fall short of Grade II by at least 2 dB, because of transmission via the walls. To reach a higher standard than this, the floor must be heavy enough and stiff

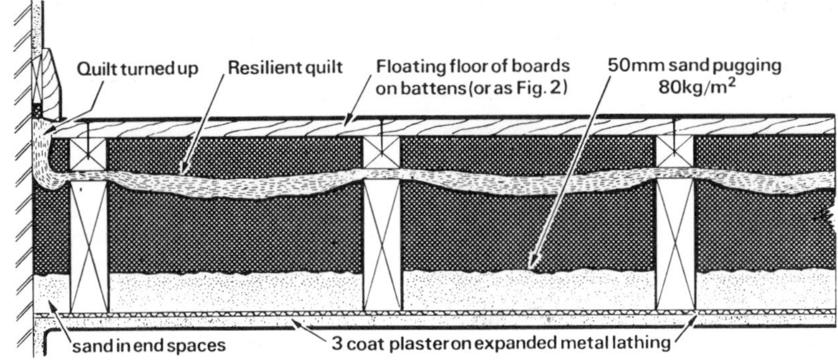

Fig 1 Insulated wood-joist floor with heavy pugging

enough to restrain the walls laterally. The only satisfactory method known at present is to provide a ceiling of expanded metal lath and three-coat plaster, loaded directly with a pugging* of 50 mm of dry sand or other loose material weighing not less than 80 kg/m² together with a properly constructed floating floor. The construction is shown in Fig 1.

The pugging will be most effective if it is supported by the ceiling and not on separate supports (pugging boards). The metal lathing must be securely fixed so as to support the combined weight of the pugging and plaster. The sand must not be omitted from the narrow spaces between the end joists and the walls; it should be as dry as possible when it is placed in the floor and should not contain deliquescent salts.

There is not much evidence at present about the use of alternatives to metal lath and plaster ceilings for supporting the sand pugging. However, it seems likely that strong board ceilings (eg thick plasterboard or asbestos board) will give similar results provided that they are firmly bonded to the walls, the joints between boards are solid and airtight and the combined weight of ceiling and pugging is the same, ie not less than 120 kg/m².

Wood-joist floors with thick walls   When the walls are thick, it is possible to use a lighter form of floor construction than the one just described without falling below Grade II insulation—though the heavier construction is still to be preferred because with thick walls it can be relied on to give Grade I insulation. In the lighter form of construction (see Fig 2) the floating floor remains an essential feature but the pugging is reduced in weight to 15 kg/m²; the ceiling may be of plasterboard with a single-coat plaster finish, but a heavier ceiling is preferred. The pugging should be supported direct on the ceiling

and not independently. Wire netting separately stapled to the joists is sometimes inserted above the ceiling to retain the pugging in position in order to ensure that the floor attains a full half-hour fire resistance; the netting must not be allowed to prevent the pugging from bearing fully on the ceiling. Alternatively, the same fire resistance can be achieved by increasing the plaster finish on the plasterboard ceiling to 13 mm thickness. The pugging material normally recommended is high-density slag wool (about 200 kg/m³), laid to a thickness of about 75 mm. Loose or pelleted mineral-wool puggings of 110–150 kg/m³ density need to be 100–130 mm thick. Other pugging materials can be employed provided they are of loose wool or granular type and the thickness used is sufficient to give the stipulated weight of 15 kg/m²; very light materials such as glass wool or exfoliated vermiculite are not suitable. Old wood-lath and plaster ceilings, which are usually quite thick, can provide very good sound insulation; they are often associated with thick walls, in which case the floors are likely to be Grade II without further treatment. The addition of a floating floor may sometimes result in Grade I insulation; therefore in conversion work, wood-lath and plaster ceilings should be retained whenever possible.

Floating floors for wood-joist floors   The floating floor consists of wood-board or strip, plywood, or chipboard flooring, nailed to battens to form a raft, which rests on resilient quilt draped over the joists. The raft must not be nailed down to the joists at any point and it must be isolated from the surrounding walls, either by turning up the quilt at the edges (which is the better practice) or by leaving a gap round the edges—this can be covered by a skirting.

* 'Deafening' Scotland

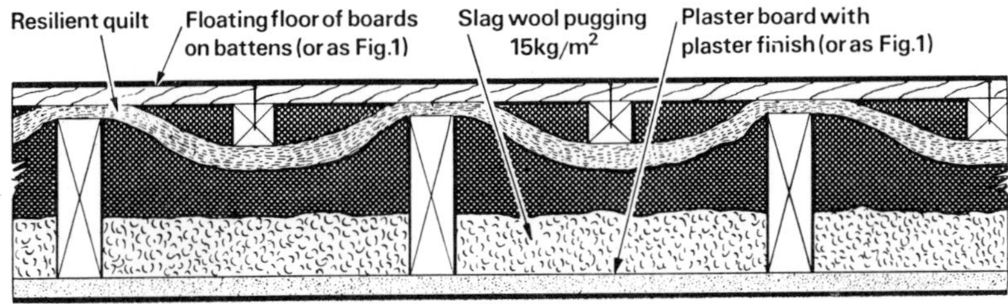

Fig 2 Insulated wood-joist floor with light pugging

Skirtings should be fixed only to the walls and not to the floors. The flooring should be not less than 20 mm thick, preferably tongued and grooved. The battens should be 50 mm wide and preferably 40 or 50 mm deep to form a stable raft. They should be parallel with the joists because battens that cross the joists provide too small a bearing area and overload the resilient quilt; the whole area of the top edges of the joists should share the loads transmitted from the floating raft. If it is essential for the battens to lie across the joists, then specially designed resilient pads must be used instead of the normal quilt or in addition to it.

There are two common methods of constructing the raft; one is to place the battens on the quilt along the top of each joist and to nail the boards to the battens in the normal way, as shown in Fig 1. The other method is to prefabricate the raft in separate panels the length of the room and up to 1 m wide, with the battens across the panels positioned so that they will lie between the joists when the panels are placed in position; the battens of adjoining panels should be staggered on alternate sides of the centre line between the joists and should project beyond the flooring sufficiently as to enable the panels to be screwed together to form a complete raft of the type shown in Fig 2.

Because the flooring is not nailed down to the joists, particular care must be taken to level up joists that are to carry a floating floor; in particular, end joists must not be lower than the others as this produces a tendency for the floating raft to tip and for furniture next to the walls to rock. This could encourage householders to spoil the insulation by nailing down the floating floor.

Partitions on floating floors It has become a practice in some methods of construction, particularly for two-storey dwellings, to build the internal partition walls up to first-floor level and then to construct the wood-joist floor continuously over the whole area of the dwelling, building the upper partitions on top of the flooring. This practice has sometimes been adopted even on a floating floor but is not recommended. Partitions should be supported either on the partitions below or on the floor joists, the floating floor being constructed as a separate independent raft within the confines of each room.

Suspended ceilings An independent ceiling is not very effective as a means of improving the sound insulation of a wood-joist floor. When used in addition to a floating floor and pugging, a suspended ceiling gives little further improvement; to be of much value when used alone, say for improving the insulation of an existing floor, it needs to be so heavy that it might not be practicable to construct it.

Although the airborne insulation at high frequencies could be improved by a comparatively lightweight suspended ceiling, it is at the lower frequencies that wood-joist floors are mainly deficient and more weight is generally required to remedy this deficiency. Suspended ceilings are not of much benefit for impact insulation. If a floating floor is being built it is usually a simple matter to pug the ceiling, and nothing worth while is then gained by adding a suspended ceiling.

Sound insulation of lightweight dwellings

This digest replaces Digest 96 which is now withdrawn.

This digest sets out the Station's present knowledge of sound insulation in dwellings gained from field measurements of industrialised systems of lightweight construction. The trend towards lightweight construction has made it necessary to develop ways of achieving adequate sound insulation other than simple recourse to weight, as in traditional houses. The alternative principles employed, notably structural isolation, are outlined. Practical forms of construction embodying these principles are discussed, against the background of established grades or standards of performance and the requirements of the building regulations. It is suggested that Digest 143 'Sound insulation: basic principles' should be read in conjunction with this digest.

Digests 102 and 103 are concerned with sound insulation between dwellings of basically traditional construction, in which the weight of the structural elements (separating and flanking elements) is by far the most important factor; performance standards or grades relating to walls and floors are suggested, and various traditional constructions are reviewed and graded.

Strictly speaking, no firm recommendation is made in Digests 102 and 103 respecting the choice of a standard, although for walls between houses the fact that only one criterion (the house party-wall grade) is nominated seems to imply that a lower standard would not be considered satisfactory. The British Standard Code of Practice* recommends the house party-wall grade in houses, but selects Grade 1 for walls and floors in flats.

Building regulations

The Building Regulations 1972 require only that a party wall or floor shall, in conjunction with its associated structure, provide 'adequate' resistance to the transmission of airborne sound and that a party floor between dwellings shall also (in conjunction with its associated structure) provide adequate resistance to the transmission of impact sound.

* CP 3: Chaper III: 1972, 'Sound insulation and noise reduction'.

For walls, the requirement is deemed-to-satisfy if measurements made and normalised in accordance with Sections 2A and 3A of BS 2750 : 1956* show the transmission of airborne sound to be reduced by the amounts shown in Table 1. A tolerance not exceeding 23 dB total adverse deviation from the values shown, measured at all sixteen frequencies, is permitted.

Floors must meet the standards of Table 2, again with a permissible total adverse deviation of 23 dB each for airborne and impact sound.

The Building Standards (Scotland) (Consolidation) Regulations 1971, operative only in Scotland, set out in Part VIII precise minimum sound insulation performance standards for new dwellings, equivalent to the BRS house party-wall grade between houses and Grade I between flats.

The two sets of regulations differ in that, for England and Wales all party walls are expected to satisfy the house party-wall grade, while for Scotland this standard is only required between houses, the slightly lower standard of Grade I being acceptable in flats for party walls as well as for party floors.

* BS 2750 : 1956, 'Recommendations for field and laboratory measurement of airborne and impact sound transmission in buildings'.

Prepared at Building Research Station, Garston, Watford WD2 7JR
Technical enquiries arising from this Digest should be directed to Building Research Advisory Service at the above address.

Sound insulation grades

The three grading curves, party-wall grade, Grade I (Airborne sound) and Grade I (Impact sound), are shown in Tables 1 and 2. Grade II curves have no relevance to the present regulations, but they can be derived from the Grade I curves, being 5 dB worse at all frequencies for airborne sound and 6 dB worse at all frequencies for impact sound. In measured field results some spread is inevitable for any given construction, and a tolerance not exceeding 23 dB total adverse deviation from the grade curve measured at all sixteen $\frac{1}{3}$-octave frequency bands between 100 and 3150 Hz is, therefore, permitted in determining compliance with a particular grade.

The term 'party-wall grade' is used in this digest to denote the highest of the BRS grades, which was based on the average performance of normal one-brick solid party walls in traditional houses.

The BRS grades are related to field measurements, not to laboratory test results, which may be very different from field results of the same nominal constructions. The allocation of a grading to a type of construction should, whenever possible, be made on the average results of at least four examples, gradings of single examples being regarded as tentative.

Measurements of sound insulation for grading purposes can only be made between two rooms or enclosed spaces of a reverberant nature. Normally, the rooms will be on opposite sides of the party wall or floor, and it is convenient to quote the results as being the sound insulation of the wall or floor, although what is measured is not the insulation of the separating element alone, but the net reduction between the pair of rooms or dwellings via whatever transmission paths may happen to be present. Sometimes flanking transmission exceeds direct transmission and controls the measured reduction. This is the principal reason for large discrepancies between laboratory and field results for particular separating wall constructions. Measurements are affected by local conditions, such as furnishings, and for grading purposes all results are 'corrected' to conform to standardised furnished conditions, i.e. 0·5 sec. reverberation time in the receiving room.

Lightweight house construction

The most influential factor for sound insulation in conventional residential buildings is the weight of the structure, not only of the party wall or floor but of the flanking walls or floors as well, but it is well known that double-leaf walls can give insulation in excess of their weight contribution. Cavity party walls of plastered lightweight concrete blocks can show a saving of about 40 per cent of the weight of brick walling but this is not enough to satisfy modern demands. Moreover, the current move towards maximum factory prefabrication with rapid

Table 1 Party-wall grade

(Hz)	Minimum airborne sound reduction through wall (dB)
100	40
125	41
160	43
200	44
250	45
315	47
400	48
500	49
630	51
800	52
1000	53
1250	55
1600	56
2000	56
2500	56
3150	56

(Maximum total adverse deviation 23 dB)

Table 2 Grade 1

(Hz)	AIRBORNE SOUND minimum reduction through floor (dB)	IMPACT SOUND maximum octave-band sound pressure level under floor (dB)
100	36	63
125	38	64
160	39	65
200	41	66
250	43	66
315	44	66
400	46	66
500	48	66
630	49	65
800	51	64
1000	53	63
1250	54	61
1600	56	59
2000	56	57
2500	56	55
3150	56	53

(Maximum total adverse deviation 23 dB)

site erection calls for dry-construction techniques as well as for the lowest possible weight. Dry construction adds to the acoustic problem, because it is vital for sound insulation that all air-paths should be eliminated. Wet plastering, which could seal all joints and cracks, is currently out of favour and new developments in industrialised house construction have had to bear in mind the special need for efficient sealing. Sound insulation factors other than weight have had to be explored to the full.

The essential requirements for good acoustic performance of dry lightweight construction are:

(a) an independent structure for each house;

(b) a double-leaf party wall construction;

(c) a certain minimum weight in each leaf;

(d) wide cavity separation between the two leaves.

The houses should be regarded basically as detached houses alongside each other. There is no objection to disguising them as a terraced row by

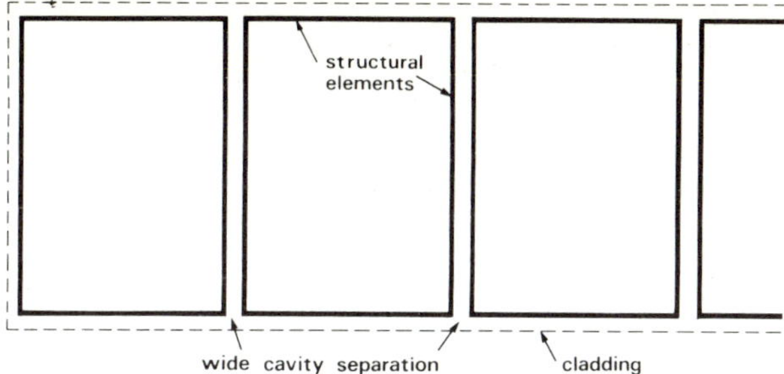

Fig. 1 Schematic plan of 'detached' terraced housing.

continuous claddings of suitable type with minimum acoustic continuity, but the houses should be self-contained structurally, as shown schematically in Fig. 1. They may be skeletal framed structures lined with prefabricated panels inside and out, or they may be constructed of loadbearing panels. The commonest system at present is the timber-framed structure, factory-made in storey-height sections and site assembled, linings being added either in the factory or on site. Typical details of a party wall of this construction are given in Fig. 2; the graph in Fig. 3 shows average performance and variability based on recent measurements of 28 party walls of this type.

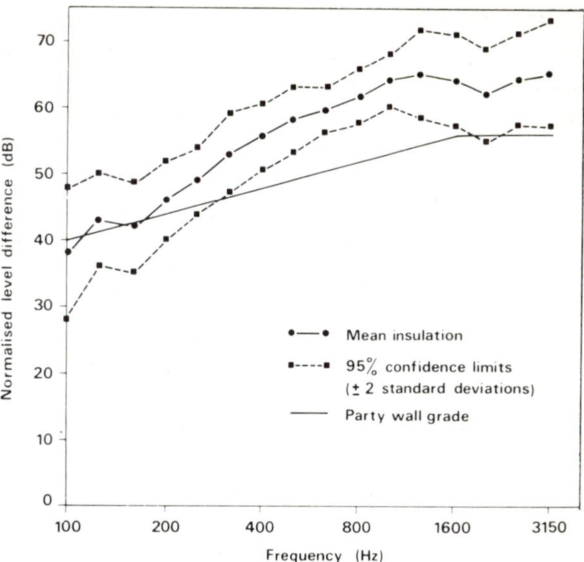

Fig. 3 Typical acoustic performance of isolated house systems.

A precise minimum weight for the leaves of the party wall cannot be given, but about 25 kg/m² is suggested as the minimum for each leaf, i.e. about 50 kg/m² for the double wall. This weight is exclusive of any framework, whether in timber, steel or concrete; it is the continuous weight of the linings or panels. The most common lining the Station has encountered, especially with timber-framed systems, is 40 mm thick laminated plasterboard. This may consist of two layers of 20 mm plasterboard, three layers of 13 mm, or four layers of 10 mm. The most common form is three layers of 13 mm plasterboard, two layers being glued together in the factory and pre-fixed to the timber-framed sections, the third layer being fixed *in situ* with the sheets laid in the opposite direction to the pre-fixed sheets. Recent tests on walls using one layer of 13 mm and one layer of 19 mm plasterboard seem to indicate that this also is satisfactory. This method should help to minimise the risk of joints between panels opening up in course of time and forming air-paths through the party wall.

Other panel materials of equal weight are likely to behave acoustically in much the same way, with minor variations in performance at different frequencies depending on the resonant properties; for practical purposes, the differences between the materials in normal use are not thought to be significant. Considerations such as thermal and moisture movements should be kept in mind, however, because of the serious acoustic effects that could arise from the joints opening up; obviously, the material must be robust enough to withstand mechanical damage.

The width of the separating cavity cannot be fixed very precisely, but 225 mm seems to be about right as a minimum dimension. Making the cavity wider is beneficial. Cavity width is measured between the backs of the lining panels; supporting framework up

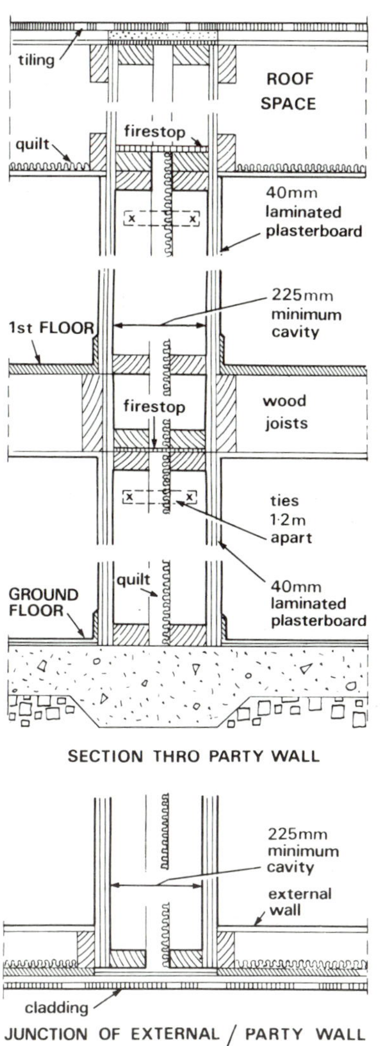

Fig. 2 Typical construction of party wall in isolated house system.

to, say, 15 per cent of the panel area may project into the cavity, providing it does not touch the opposite frame. It is quite possible that the weight of the leaves and the width of the cavity are linked, since both are relevant to the critical problems of sound insulation performance at low frequencies, i.e. the bottom octave (100–200 Hz) of the range we are concerned with. If so, it may be possible to reduce the cavity a little provided the weight of the leaves is increased to compensate; however, it may be necessary to increase the weight by at least 50 per cent each time the cavity is reduced by 25 mm.

Cavities or air-spaces are more effective for sound insulation if they contain a reasonable amount of sound absorbent material, especially if the cavity is a wide one designed to make a significant contribution to the total performance. As most panels will present a fairly hard reflecting surface to the cavity, the insertion of a sound absorbent quilt or blanket is recommended to provide the necessary absorption. Almost any standard glass-wool or mineral-wool quilt 13–25 mm thick will serve, except that any paper covering should be limited to one side of the quilt only. The quilt is usually suspended in the cavity by stapling it to the frame on one side. Its position in the cavity is not important. In platform-type timber-framed construction the provision of two quilts, one on each side of the cavity and firmly attached to the timber frame, contributes to fire-resistance and may dispense with the need for ties bridging the cavity.

Structural isolation

The important matter of the structural independence of adjoining houses needs to be considered in more detail. For effective sound insulation, complete isolation above ground level should be the aim, but this often conflicts with other functional needs, such as stability and fire protection, which may call for bridging of the cavity. Weather protection at the ends of the party wall may also bridge the cavity. Fortunately some forms of bridging, applied with restraint, give the required standard of sound insulation. For example, normal roof tiling can safely be carried across the top of the party wall provided the cavity extends up to the underside of the tiling. Fire protection regulations require the top of the wall to be sealed against the passage of fire from one side to the other; to comply with this, roof tiling needs to be bedded on non-combustible material (e.g. mortar) which is usually supported over the cavity by asbestos-board. This construction is

acceptable acoustically. Similarly, an asbestos-board bridge inserted across the cavity at first-floor level as a fire-stop can be accepted; best results are likely to be obtained if it is fixed to the structural frame on one side only or not fixed at all but simply left resting on the horizontal members of the timber frames.

In two-storey housing, the fire requirement across the party wall is one hour's fire resistance. Within a two-storey house of single occupancy, the required fire-protection standard for the first-floor construction is half an hour. This difference raises a special problem in timber-framed houses of 'platform' construction, because collapse of the floor at the end of half an hour could cause premature collapse of the supporting leaf of the party wall, exposing to the fire the timber frame of the other leaf. Structural stability is thus a contributory factor in fire resistance. A method of safeguarding stability employed in some systems is to tie the timber frames of the party wall together by a row of metal straps 1·2 m. part just below each ceiling level. The straps are usually 40 mm by 3 mm flat steel bar nailed or screwed to the sides of the timber studs. These ties seem to have no serious effect on sound insulation, though this may not continue to be the case if the number of ties is increased beyond the dozen or so used in the houses tested.

Floors in flats

The type of house construction discussed so far can, of course, be used for two- or three-storey flatted dwellings, but in that case the floors also have to reach a certain standard of sound insulation, and requirements for impact sound reduction as well as for airborne sound reduction must be satisfied. If the building is timber-framed or a very light steel-framed structure, concrete floors may not be feasible and a joisted floor may be the only appropriate type.

It is known that a Grade I standard of sound insulation is unlikely to be achieved with wood joist floors in traditional structures (except with very heavy supporting walls) because of flanking transmission. It is not clear whether flanking transmission is also a problem in timber-framed structures. Some tests done by BRE have shown that it is possible to achieve Grade I performance with timber joist floors in timber-framed structures but there is not yet sufficient evidence on which to base detailed guidance.

Wilful damage on housing estates

This digest reports on some of the findings that have emerged from a short exploratory study by the Building Research Station of wilful damage on housing estates. The study was confined to the design and technical aspects, it did not attempt the more difficult sociological and psychological problems. The recommendations made are based on some forty interviews and inspections and are applicable to the design of buildings, equipment and layouts, and the selection of materials, fittings and finishes, both at the inception of a scheme and subsequently when maintenance and repairs are undertaken.

General lessons and recommendations

Among the factors which promote vandalism are boredom and lack of discipline among young people, unsettled conditions of occupancy and absence of pride of possession and community spirit among tenants; these factors seem to be most prevalent in high-rise estates, though the precise effect of density is not clear. These adverse factors can be countered to some extent strategically, for example, by the provision of well designed children's playgrounds and other recreational facilities—though regrettably these in their turn can become targets for vandalism— and by the encouragement of tenants' associations, by good housing management, caretaking and maintenance. Estate layout and building design, particularly the selection of materials and finishes, can reduce the incidence of vandalism by offering greater resistance to damage and less 'satisfying' targets, but no matter to what extent the resistance to actual physical damage is increased, no design can be secure against disfigurement. Technology can, however, usefully reinforce the efforts of the psychologist and the sociologist.

Some general recommendations follow. All are common sense but the survey showed the need to state them because they were not always followed in practice, perhaps because of the need to limit first cost, perhaps because designers did not foresee the conditions that would develop on a particular estate. Some specific recommendations are summarised in the tables on pages 202 and 203.

Good lighting and avoidance of hidden corners

Some of the recommendations of Digest 122 *Security* apply also in the prevention or reduction of wilful damage. Clear visibility is a recognised deterrent; this means good lighting of buildings, particularly public staircases and corridors, garage courts and car parks, and the avoidance of hidden corners particularly in entrance halls and under stairs.

Removal of builders' rubbish and safeguarding of plant

Wilful damage is often at its worst when an estate is first occupied. This may be due in part to the social problems involved in 'settling down' and forming a community. But it may also be associated with the fact that unfinished parts of an estate, contractors' plant and builders' rubbish, are attractive both as adventure playgrounds* and as targets for youthful vandals. Digest 122 gives guidance on security on building sites against theft; adoption of its recommendations is a first step to countering vandalism on estates where building work is still in progress. Ladders should be safeguarded and where practicable unfinished work should be fenced off from the rest of the estate. Broken bricks, timber offcuts, etc, are handy missiles or weapons and should be cleared from a site before families move in.

A good standard of maintenance

Buildings and site works in a dilapidated state breed vandalism. The dilapidation may start with accidental damage, poor initial design or workmanship, failure to complete a maintenance operation or failure to repair promptly a minor act of wilful damage. The responsibility for such failures falls in part on the original designers and in part on those responsible for management and maintenance.

* A BRS study of children's play on housing estates (National Building Studies Research Paper 39:1966) found that 'Building and demolition sites are utilised as adventure playgrounds, for building huts, lighting fires, or for the sort of freedom where "I can fire my catapult at old tin cans", "I like to play on the workmen's shed . . . you have to climb over a big green gate. . . . On the top of the shed is something like green sandpaper."'

Fig 1 Dark finish on light substrate

Fig 2 Lamp cover at kicking height

Avoid:	Alternatives:	To reduce risk of:
Wall finishes		
Soft-textured	Resist damage or marking: cleanable:	Defacement by writing, scratching or carving
Easily scratched, particularly if of colour contrasting with substrate *see* Fig 1	—special paints and glazes, similar in colour to substrate	
Light colours	—ceramic mosaic	*All are vulnerable to aerosol sprays; some authorities therefore prefer cheap finishes, easily renewed*
	—glazed tiles	
Requiring renewal, but too expensive for frequent renewal	Resist damage or marking; difficult to clean:	
	—ribbed metal sheet —roughcast —rough-textured bricks	
Materials		
Glass in large panes and vulnerable positions	Windows, doors, signs, etc: transparent plastics (polycarbonate considered best; acrylic, UPVC and butyrate sheet also available); armour-plate glass	Breakage
		Plastics sheets can be removed; risk of being burnt appears to be less than in lighting fittings
	Opinions vary on the use of glass bricks; though stronger than window glass, they can still be broken and seem to be particularly attractive to vandals	
	Translucent covers to lighting fittings: if inaccessible, use plastics (polycarbonate or butyrate)	*see* Fig 2
		Breakage; *but may become target for catapult, air-gun or stone-throwing*
Plastics covers to lighting fittings near ground	Glass (preferably armour-plate)	Burning
Plastics and asbestos-cement rainwater pipes below 2 m in height	Cast iron	Breakage
External copper and lead piping	Keep pipe runs within the building	Theft
Soft mortar joints, if accessible	Hard mortar	Scraping out
Design features		
Easy access to flat roofs from adjacent walls or lower buildings, projections giving foothold, rainwater pipes, etc. *see* Fig 3	Wide projections at eaves and top edges; parapet walls *see* Fig 4	Damage to felt or asbestos-cement roofs and accidents by falling through the latter
Tile-hanging or glass below ground floor window cill level, particularly if adjacent ground is open for children's play *see* Fig 5	Brickwork, blockwork, etc.	Breakage. accidental or deliberate; removal of tiles
'Up-and-over' garage doors of inadequate strength *see* Fig 3	Good quality doors with lock at floor* Heavy framed, ledged and braced doors with robust hasp, staple and padlock; no glass*	Youths swinging on them; distortion; dislocation of balancing mechanism
	* Alternatively, or in addition, garages to be overlooked from dwellings	

Fig 3 'Up-and-over' converted to 'down-and-under'; accessible roof

Fig 4 Roof access made difficult—though there is still the rainwater pipe.

Fig 5 Damage may be wilful or accidental

Fig 6 Heavy copings

Avoid:	Alternatives:	To reduce risk of:
Design features (Contd)		
Projecting garage door handles about knee-high	Retractable handle for 'up-and-over' doors; or tee handle locked in vertical position	Kicking off, leading to break-in; access to roof
Lever handles to external doors	Spherical knobs	Breakage *but in old people's dwellings lever handles may be needed*
Painted metal posts, rails, etc, in playgounds or public spaces	Non-ferrous metal; galvanised or plastics-coated steel	Chipping and scratching of paint
Common access areas		
Dark corners in entrance halls and open spaces under stairs	Good artificial lighting, day and night if necessary	General nuisance and accumulation of rubbish
	Locked external doors with telephone communication between tenants and visitors	*Limited experience, effectiveness not proved—subject to vandalism*
	Caretaker's office overlooking entrance	
Vulnerable features in lifts *see also* **Wall finishes**	PVC or other sheet flooring	Damage by fouling *though PVC is damaged by hot tobacco ash*
	Metal control buttons	Burning or prising out plastics to dislocate lift service
	Stop buttons	Damage by wedging doors open, eg by milkmen *but not all authorities agree*
	Provision of WC off lift hall	Fouling of lifts *but WC is itself subject to vandalism*
Vulnerable features in visitors' toilets	Specially designed sanitary fittings; pipework and mechanisms concealed; embedded conduit; concealed fixing of fittings and equipment; recessed head or hexagon socket screws	Deliberate or accidental damage; electric shock or blown fuses; theft
Landscaping features		
Excessive amounts of soft landscaping *see Figs 7 and 8*	Roses or other prickly bushes; hard pavings; planted plots raised and with strong retaining walls; paths in the right places	Shortcutting over beds; damage or theft of plants and turf. *Opinions vary on the effectiveness of raised beds*
Cobbles or granite setts	Large slabs or jointless pavings	Lifting and use as missiles
Ranch fencing in white-painted wood	Vertical pales; horizontal boarding suitably strong and unpainted	Damage to boarding or painted finish by climbing
On low walls—thin concrete or stone copings, particularly with overhang; tile copings; brick-on-edge copings without strengthened angles and ends	Large section *in situ* or precast concrete copings, flush on one or both faces and anchored to wall *see* Fig 6; brick-on-edge with *in situ* concrete angles and ends	Removal
Open spaces too large; 'nooks and crannies' in layout	High standard of lighting; all-night floodlighting in extreme cases	Nuisance

Fig 7 Short cuts over flower beds

Fig 8 Raised flower beds

Avoidance of features which attract youthful adventure and vandalism

A number of design features provide opportunities for adventure and for vandalism and should be avoided. Blocks of lock-up garages may suffer badly from attack if they are hidden from general view; abandoned cars are added attractions. So are flat or low-pitched roofs which have easy access; asbestos-cement sheeting is particularly at risk and provides an accident hazard. Monopitch roofs falling to the rear are not easy of access from the front, especially if protected by a wide overhang, but they may need to be protected at the back. Other design features which provide targets are posts, signs, special lighting fittings, etc, which may be unusual in appearance or become objects for climbing and swinging on. Their design should either be kept simple and unattractive for adventure play or they should be strong enough to withstand it.

Caretaking on mixed estates

Housing estates developed with blocks of flats at high density in urban centres usually have caretakers, but caretaking is often less adequate on lower density estates when blocks of flats are mixed with low-rise housing. In these situations the problem of vandalism and wilful damage can be serious, especially where there is a high proportion of landscaped public open space and hidden corners.

Layout

Damage may be inflicted on flower beds and ornamental grass by people taking short-cuts across them instead of following the formal paths provided; fences may also be damaged in an effort to make a route where none was planned. Raised beds may help to prevent this (Fig 8) but the whole problem could be anticipated when the layout is planned so that paths are provided along routes that users are likely to follow.

Maintenance and appearance criteria

The attitudes of housing authorities to the problem of wilful damage can be broadly classified according to whether the criterion of 'maintenance' or 'appearance' takes precedence. Design and maintenance policies vary between the two schools of thought though they share a common interest in cost. The aim of the 'maintenance' group can be expressed simply as to find means of minimising the combined cost of maintenance, repair and supervision. This may lead to a choice of robust finishes, fittings, etc; high initial cost is acceptable if the reduction in maintenance and other subsequent costs is likely to show an overall saving. Unfortunately, when precautions against wilful damage are intensified beyond a certain point, appearance and amenity standards may drop. The aim of the 'appearance' group is, therefore, to steer a middle course between the unsightliness of wilful damage and the possible bleakness of anti-vandal treatment.

Given some initial damage on a new estate, regardless of whether or not it is made good, vandalism is likely to remain a problem for some time. Eventually, though possibly not until the first full redecoration removes all signs of defacement, maturity and a well-maintained appearance may combine to discourage further attack. Persistence in the repair of damage and removal of defacement is essential if headway is to be made against the problem. Deterioration in appearance can arise from causes other than vandalism, such an accidental damage, misuse or ordinary wear and tear, or weathering, decay, corrosion and other forms of failure; if these are allowed to develop, conditions favouring vandalism can become established in an estate previously free from it, or re-established in one from which it has been eliminated. Comparatively minor faults or incipient failure can provide temptation even to well-suppressed vandalistic tendencies.

The probable incidence of wilful damage in a particular situation is not easy to predict. Technical preventive measures might not, therefore, be adopted in new work unless the most adverse conditions are indicated. If, however, the risk is kept in mind during the design stage, it may be possible to build in protection at little or no extra cost or loss of amenity. The design should not preclude the adoption of vandal-resisting measures later, if found necessary.

As with all matters concerning maintenance, good feed-back from the management and maintenance side to the designers is essential if mistakes are not to be repeated.

Local housing programmes

This digest introduces a subject that is new to the scope of the series, drawing for the first time on the work of the Urban Planning Division. It comments on current factors relating generally to the housing situation and highlights demographic and other data that should be taken into account when local authorities, and others concerned with the provision of dwellings, are considering future house-building programmes.

The digest deals with national trends and quotes national statistics from which there will be considerable local variations. Nevertheless, an awareness of these trends is desirable in the planning of local housing programmes.

For a fuller treatment of the subject, see 'Trends in population, housing and occupancy rates 1861–1961' by W V Hole and M T Pountney; HMSO: 1971.

The context

The UK spends some £1500 million a year in building new dwellings and improving existing ones; about half of this is spent by the public sector and half by the private sector. Whilst the numbers in the total housing stock still fall short of estimated need, there is an excess in some areas alongside an endemic shortage in places such as London. But with a slow rate of population growth, even compared with most other highly developed countries, the questions relating to new construction are now *what type* and *where*, rather than *how many* dwellings.

Despite the large volume of housebuilding since the last war, houses built before 1914 still comprise about a third of the present housing stock. Demolition of unfit dwellings increased during the sixties and together with house improvements this has noticeably reduced the amount of unfit housing (estimated by surveys in 1967 and 1971). Slum clearance, as it was known up to the end of the thirties, no longer exists, but with the solution of the problem of dwellings now classified as unfit almost in sight, new problems of caring for and adapting the older parts of the housing stock will arise as changing needs and standards render them obsolete.

Rising standards of living since the war have already changed expectations regarding space, standard of specification in the home—for example, requirements for thermal comfort—housing layout and environmental amenity. There has been a major swing to owner occupation, so that both private and public building are serving rather different portions of the population than they did before 1939. The broad social objectives of the public sector are still to assist those who are underprivileged in the housing market, but the type of assistance that public authorities are now empowered to provide and the composition of the underprivileged group have changed.

There is increasing pressure on local authorities to consider housing in their areas in the round, public and private sectors together meeting the overall needs. But as the useful life of a house in this country is generally assumed to be longer than 60 years—the period over which public housing is financed—an awareness of recent changes in need, and the direction of likely future trends is crucial to this broader approach to planning.

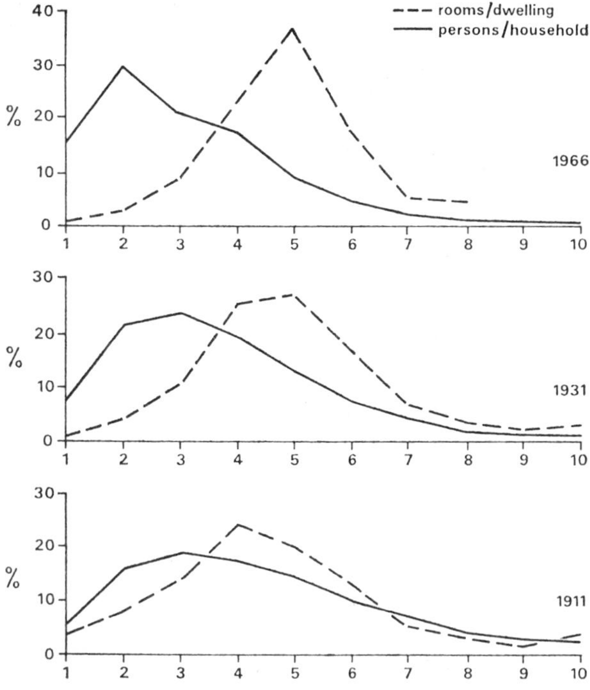

Fig 1   Changes in household size and house size

The present position

The broad aim of housing policies is to produce a situation in which the needs of all households are matched by the available housing stock, with sufficient 'slack' in the system to allow both new households and those wishing to move to choose from a reasonable range of dwellings. Ideally the phrase 'matched by the available housing stock' would cover such matters as number and size of rooms, type of dwelling (house, flat, maisonette, etc), location and type of tenure and cost. Too little is known about the relationship between these factors and the way in which the desires of households vary according to their social and economic characteristics to enable households to be matched closely to housing stock. The three graphs in Fig 1 show the distribution of household sizes and of the housing stock by number of rooms per dwelling for the years

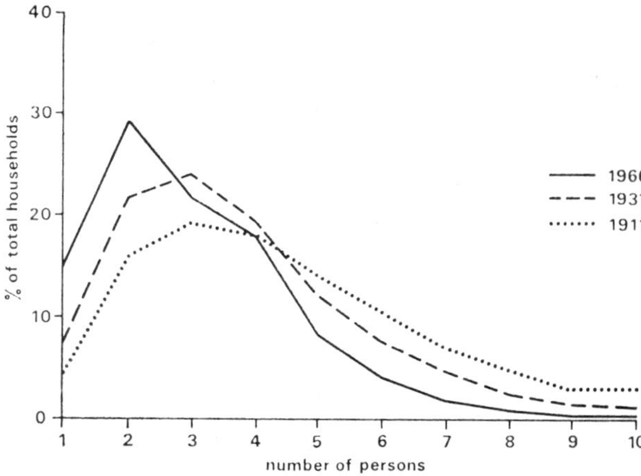

Fig 2   Size of private households; England Wales 1911–66

1911, 1931 and 1966. They show that on the simple criterion of rooms per dwelling there has been a growing mismatch between the range of household sizes and the housing stock. Even allowing for a rise in the demand for space, it can be assumed that there is at least some rough correlation between the size of household and its space needs (see Table 3) but it is evident that an increase in the number of small households has not been accompanied by an increase of small dwellings.

Perhaps more important, however, is the conclusion that housing has, over these 55 years, concentrated more on five-room (three-bedroom) dwellings. Hence there is less flexibility in the housing stock either for a household to move to a dwelling of different size as it needs change, even if its economic circumstances permit this, or for the stock to meet future changes in overall demand for space. This concentration on building three-bedroom houses can be explained partly on a value-for-money basis; the cost per bedspace is less in a five-person than in a one-person dwelling, so there is an economic inducement—to obtain the most space for the financial outlay. Added to this, during much of the past, the public sector has been primarily concerned in housing families with children.

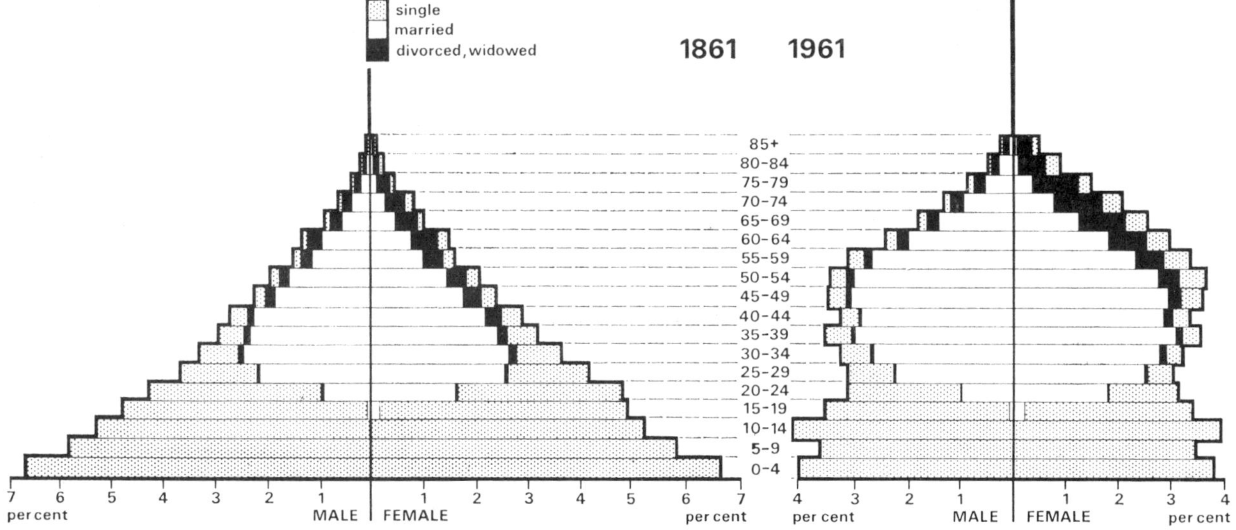

Fig 3   Population of England and Wales, 1861 and 1961 : age, sex and marital status

While the private sector builds in response to its assessment, adequate or not, of demand, the overall planning for need in an area in relation both to existing stock and new building lies with the public sector. Part of this planning must have regard to overall balance of different sizes and types of dwelling within the housing stock and it should take into account demographic and social trends.

Demographic changes

The shift in balance of large and small households over the last 50 years has already been mentioned. In order to make the point more forcibly the household size curves from the three graphs in Fig 1 are brought together in Fig 2. It will be seen that the proportion of one and two person families has increased at each of the three dates illustrated—as it has consistently over the entire period.

This trend towards an increasing proportion of small households reflects a series of demographic changes which have had and are continuing to have important effects on the nature of housing need. The changes can be highlighted by comparing the distribution of the population by age, sex and marital status, in 1861 and 1961, shown in Fig 3. The broad base of persons under 15 years old in the 1861 pyramid, the result of high birth-rates, is one obvious reason why households were larger then. But increasing life expectancy has radically altered the balance in the older age groups from 7 per cent of persons over 60 in 1861 to 17 per cent in 1961. The gap between life expectancy of males and females has also widened, now standing at 69 for males and 75 for females; more and more the housing needs of the very old are those of old women. Since the last war a growing proportion of the population over 15 are married, which has contributed to the increasing rate of household formation. During this period, average age at first marriage has been consistently dropping and is now 24 for men and 22 for women. This is likely to affect both the financial resources available to a couple for renting or purchasing their first home, and the age of the wife when child-bearing is completed.

The effects of declining family size and increasing life expectancy on the relative length of different stages of the family cycle are shown in Fig 4. The picture is essentially a simplified one, based on a series of averages, but it nevertheless brings out two points of significance. The first is that the Victorian wife would have reached 40 before bearing her last child; her post-1930 counterpart is 10 years younger. The second, even more significant, is the marked increase in the length of the final stage, partly due to earlier completion of the family and partly to longer life expectation. Not only has this final stage lengthened on average, but the percentage of couples

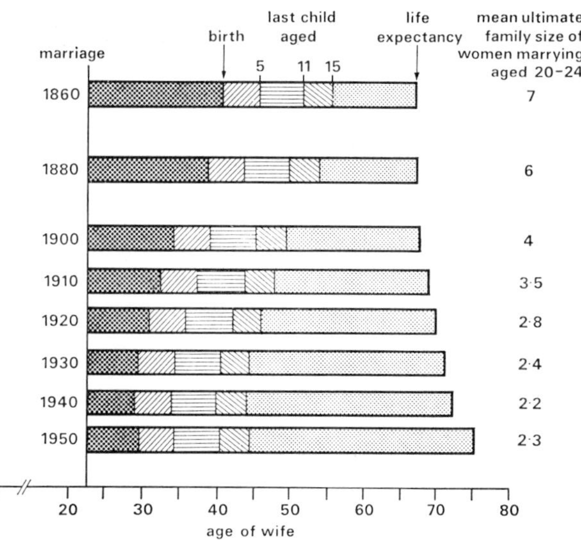

Fig 4 Length of stages of family cycle for successive marriage cohorts, 1860–1950

who might expect to survive to enjoy it has increased. Thus only 52 per cent of women aged between 20 and 24 who married in 1871 might have expected their marriage to continue unbroken by death for 30 years; the corresponding figure for those who married in 1911 was 71 per cent.

More important, with the movement of children away from the parental home now likely to begin when the wife is in her early forties, the couple have an extended period, beginning well before retirement, when they are again alone. This and subsequent widowhood for one partner, usually the wife, have been the major factors in the growth of one and two person households. Added to this, younger single people have tended to form separate households alone or with a friend. Of all households in England and Wales in 1966, 61 per cent contained no dependent children; not surprisingly, dependent children are most likely to occur in households of four or more persons. Much of the housing need in this country is therefore that of small adult households, although a great deal of the housing stock has been designed for a family containing children of pre-school or school age.

For families, the cycle of growth and decline in the size of household is important in terms of changing needs and the extent to which those needs can be met, particularly in so far as the stages are linked with financial prosperity or stringency. In broad outline beginning with marriage—and assuming that the wife is at work (see the next section)—there may well be a period of relative prosperity. With the arrival of children a more stringent period sets in while the demands of a growing family are met from a single wage. The ability to achieve larger accommodation at the time when demands on space are at their maximum may therefore be severely curtailed. Some time later, often when the wife returns to work, the family as it declines in size will enjoy its period of

Age group	millions			per cent		
		projections			projections	
	1961	1981	2001	1961	1981	2001
0–14	12·3	*13·4*	*14·8*	23·3	*23·2*	*23·5*
15–29	10·4	12·8	*13·6*	19·7	22·3	*21·5*
30–44	10·6	10·9	13·0	20·0	18·9	20·7
45–59	10·6	9·5	10·8	20·0	16·4	17·2
60–74	6·8	8·0	7·3	12·8	13·8	11·5
75+	2·2	3·1	3·5	4·2	5·4	5·6
all ages	52·9	57·7	63·0	100·0	100·0	100·0

Table 1 Age distribution of the population of the UK

1971-based mid-year projections
Numbers in italics are projections of those as yet unborn

greatest prosperity, probably continuing until retirement, when financial restriction may again be experienced. For this reason, many old persons are in poor quality housing. Types of small adult households other than married couples vary widely as to economic resources. Some, where all members are employed, are comparatively affluent, others such as students or women living alone have not only restricted finances but limited access to mortgage facilities. Such households may not be eligible for housing by the public sector and the decline of privately rented accommodation makes their position even more difficult. While various types of purpose-built accommodation are provided by the public sector for old people, there is little specific provision by either the public or private sector for other small adult households of working age, suited to their finances or their pattern of living.

The modern trends towards a high proportion of married persons in the population and marriage at an early age show no signs of change. The trend to complete child bearing in the early years of marriage also appears well established (nowadays nearly 70 per cent of all births are to women under 30). However, with over 90 per cent of married women below the age of 35 now practising some form of birth control, both the number and spacing of children is increasingly a matter of personal choice. A change in the pattern of spacing children (interval between marriage and first birth and between first and subsequent births) would lengthen or shorten various stages of the family cycle and this could have a more profound effect on housing need than minor variations in average family size. This must remain a matter for speculation; but as those who will be of an age to marry and build families in the next ten years, and those reaching later stages of the family cycle in subsequent years, are already born, it is possible to form a reasonably reliable picture of their relative distribution in the future population. Owing to the rise in the birth rate between 1956 and 1964, the number of persons aged 15–29 who are most likely to be marrying and producing children ten years hence will be proportionally greater than in 1961 (see Table 1).

Social changes

An increasing proportion of married women are now in paid employment. As recently as in the 1930s the proportion was only 10 per cent; by 1951 it was 25 per cent and it had risen to 38 per cent by 1966. However, their participation in the labour force is markedly affected by the presence of children, as Fig 5 shows; the peak age for working mothers is in their forties, a time when their children are growing up and family dispersal is approaching. The forecast for 1981 by the Department of Employment reflects the view that older married women today belong to a generation that for the most part has not been accustomed traditionally to seek employment outside the home; in future this age band will be drawn from a more emancipated generation with different attitudes to jobs.

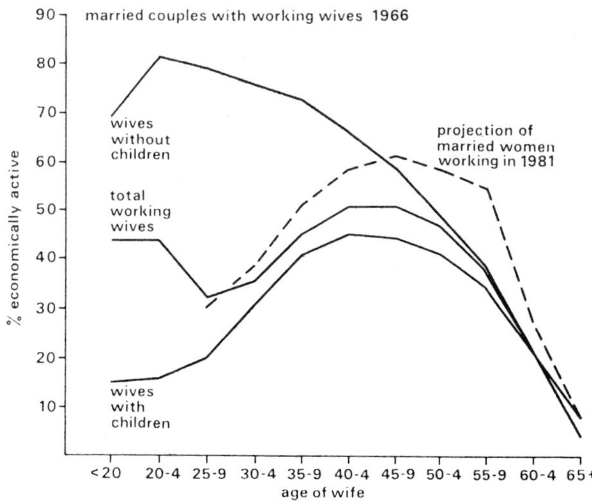

Fig 5 Percentage of wives, with children and without children, who were economically active

Although as yet only some building societies are prepared to take into account a proportion of the wife's income in granting mortgage facilities, the wife's income can contribute in other ways to improving living conditions in the home, through the purchase of furniture and furnishings, labour-saving appliances, convenience heating. At least in the middle income ranges much of the increasing affluence of the fifties and early sixties has been channelled in this direction. In the future there is likely to be a demand from those families with working wives who have reached the reasonably prosperous later stage of the family cycle for comparatively small but high specification dwellings: fully automatic heating, more generous space in service areas for household appliances, easily cleanable surface finishes.

Those who are purchasing houses through the aid of building societies are predominantly youthful: in 1970 27 per cent of borrowers from this source were under 25, 43 per cent between 25 and 34, and 19 per cent between 35 and 44. For some at least, home ownership begins with marriage and this in turn is likely to affect the nature of housing demand when they reach the later stages of the family cycle, by which time they will have a well-established equity in the housing market.

Before 1914 only about 10 per cent of housing was owner-occupied and the status of owner-occupier was associated with a person of some means. By 1947 this proportion had grown to 27 per cent; a further 12 per cent was publicly owned, but privately rented property still dominated the scene at about 60 per cent. Since then, much of this privately rented property has been absorbed into the owner-occupied group and some has been demolished; due to this, and to continuing building in the private sector, 52 per cent of dwellings in England and Wales were owner-occupied in 1970, 28 per cent were publicly owned, and private renting and other sectors had dwindled to 20 per cent. With this major swing to owner occupation it is to be expected that house owners include a much wider range of incomes. Fig 6 shows that, on the basis of

total household income, persons who rent privately are the poorest group, only 50 per cent having incomes of £25 per week or more, compared with 55 per cent of those owning outright, 60 per cent in publicly owned housing and 90 per cent of those in the process of buying. This implies that, as the supply of privately rented property decreases further, many who are now in this group would not be in a position to contemplate purchasing a house, and amongst persons who own their homes outright, many could not afford extensive house improvements or even a reasonable level of maintenance.

Summary

This digest has raised problems and offered only guides to solutions. The main points may be summarised as follows:

1 Those who provide housing—whether local authorities or private developers—should try to form a realistic picture of the households for whom they will be providing in the future. In particular they should be aware of national trends in household formation and size, and consider their local significance. Without necessarily conducting surveys of their own, much information on local characteristics of households and dwellings is available in the national census, and other published sources on housing statistics, and the family expenditure surveys offer a guide at least to regional trends.

2 The housing stock is becoming increasingly inflexible, both as a result of its polarisation between owner occupied and publicly owned housing, and the concentration on building three-bedroomed dwellings. If in the future a family of five or more persons came to expect more generous space standards, there would be little scope to realise this wish, if present trends continue.

3 All the current evidence points to an underprovision of dwellings suited to the needs of one and two person households. Small households of adults are now typical of the British scene; while these include many of the old people in this country, half of the one and two person households in 1966 in England and Wales consisted entirely of adults below pensionable age. Put another way, the small households of younger persons represented over a fifth of all households.

4 From the more affluent of these small adult households of working age, particularly those consisting of an older couple with a working wife, there is likely to be a demand for a high specification dwelling. The less affluent of such small households, faced with a decreasing supply of accommodation for renting privately, are likely to become an increasingly underprivileged group in the housing market, unless local authorities assume greater responsibility for housing them, and assisting them in other ways.

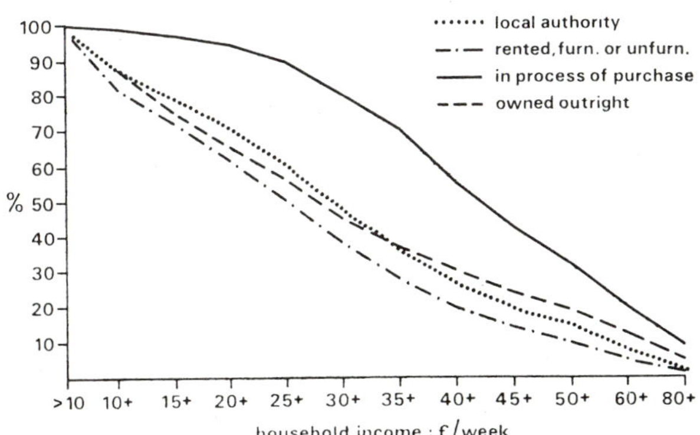

Fig 6 Cumulative distribution of household incomes and tenure: 1970

A note on standards of occupancy

Historically, one of the main concerns in improving housing conditions has been the reduction of overcrowding, which was most likely to be found in slum areas. Definitions of overcrowding have varied in housing literature, but one commonly employed has been two or more persons per room (this definition is more stringent than that of the 1957 Housing Act, which in assessing overcrowding, does not include infants, and counts a child under 10 years as 0·5 persons). Over the last half century, there has been considerable reduction in the proportion of households with the highest rates of room occupancy, as is indicated by the last line of Table 2. Defined as two persons or more per room, overcrowding has ceased to be a national problem, although it still occurs in certain specialised, but comparatively small, areas. The worst housing conditions are now associated with the sharing of facilities in houses that are inadequately converted for multi occupancy.

It will also be seen from Table 2 that small households have progressively bettered their position, but that after a degree of worsening or at best no improvement, only the decade 1951–61 has seen lower occupancy rates for the larger households. The situation in 1961 is in part a reflection of the increase both in small households and in dwellings of five rooms, depicted in Fig 1. Recently, concern has been expressed at 'underoccupancy' in parts of the housing stock. This in turn raises the question of what, in the nineteen seventies, might be considered a desirable standard of occupancy, since something only marginally better than overcrowding is unlikely to represent contemporary needs.

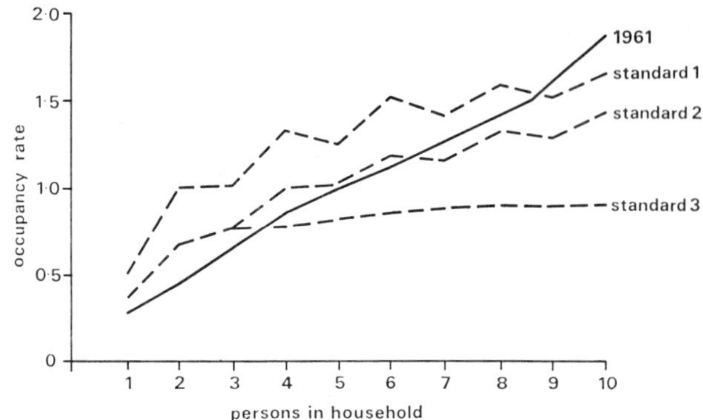

Fig 7 Room occupancy rates for standards of Table 3; rate achieved in 1961

While not selecting a single standard as desirable, a useful though crude means of assessing standards of occupancy can be devised. This takes into account separation of function (eating, sleeping, etc) and degrees of privacy for sleeping. The occupancy rates and numbers of rooms for different household sizes resulting from three such arbitrary standards are set out in Table 3. This indicates that, for a given degree of separation of function, small households will necessarily have lower occupancy rates than larger ones, a point of considerable importance in discussions of underoccupancy.

It is instructive to compare the occupancy rates represented by these three arbitrary standards with the average achieved by each size of household in 1961. It is evident from Fig 7 that households of up to three persons had lower occupancy rates than even the most generous of the three standards (and hence had an excess of space) and those of seven or more persons were deficient even on the basis of

Table 2 Persons per room, 1911–61

Number in household	Persons per room				
	1911	1921	1931	1951	1961
1	0·35	0·34	0·33	0·30	0·27
2	0·48	0·50	0·48	0·49	0·46
3	0·65	0·68	0·67	0·69	0·64
4	0·88	0·85	0·85	0·87	0·82
5	1·00	1·02	1·03	1·03	0·97
6–7	1·30	1·26	1·27	1·25	1·17
8–9	1·55	1·58	1·58	1·57	1·47
10 and over	1·83	1·73	1·67	1·90	1·82
mean for all households	1·10	0·91	0·83	0·74	0·66
Persons per room:	Per cent of all households				
—more than 1·5	16·9	16·3	11·42	5·1	2·8
—more than 2·0	5·4	*	3·9	*	*

*figures not available

Table 3 Room occupancy rates, for different household sizes, for 3 hypothetical standards of space

Persons in household	STANDARD 1 1 living room, no more than 2 persons per bedroom		STANDARD 2 2 living rooms, no more than 2 persons per bedroom		STANDARD 3 2 living rooms, couple share bedroom, all others separate bedroom	
	Rooms required	Occupancy rate	Rooms required	Occupancy rate	Rooms required	Occupancy rate
1	2	0·5	3	0·33	3	0·33
2	2	1·0	3	0·66	3	0·66
3	3	1·0	4	0·75	4	0·75
4	3	1·33	4	1·0	5	0·80
5	4	1·25	5	1·0	6	0·83
6	4	1·50	5	1·2	7	0·86
7	5	1·40	6	1·16	8	0·88
8	5	1·60	6	1·33	9	0·88
9	6	1·50	7	1·29	10	0·90
10	6	1·66	7	1·43	11	0·91
Mean for all households*		1·37		1·10		0·83

*Calculated on the assumption that each household size is equally represented.

In the 1911–61 censuses, the kitchen was counted as a living room only if used for eating; on this basis most modern council housing would be of Standard 2. This should be borne in mind when comparing the standards in Table 3 with those in Table 2.

the second standard. In 1961, 66 per cent of households consisted of up to three persons, and 3 per cent of seven or more persons. Even if the generous space standards now experienced by small households represent in some instances their perceived need for space, there remains the question of whether the type of dwelling in which this space is provided is the most suitable for their way of life.

There are statutory powers for dealing with overcrowding where it still exists. But the increasing proportion of persons in the population living at low occupancy rates implies both a trend to more spacious living conditions, and in some cases underutilisation of the existing housing stock. In allocating tenancies, or considering underoccupation in public housing, local authorities might wish to take note of the generally lower occupancy rates prevailing in the second half of the twentieth century.

Numbers of rooms are only a rough measure of spaciousness, which depends also on room sizes, but the latter are rarely available from large-scale survey data. The number of rooms needed to enable households of various sizes to achieve a given standard of space, such as one of the standards of Table 3, can be used to assess the overall 'fit' between house sizes and household sizes in an area, and the Registrar-General's census of population provides a ready source of data for this purpose. Table 3 shows how separate consideration of each household size helps the interpretation of these data, although average occupancy for all households is no guide to housing conditions in an area.

Further reading

W V Hole and M T Pountney *Trends in population, housing and occupancy rates 1861–1961*; HMSO: 1971

Central Statistical Office *Social trends,* Nos 1–3; HMSO: 1970–1972 (sections on housing and population)

D V Glass *The components of natural increase in England and Wales* pp 186–200 of the first report from the Select Committee on Science and Technology, Population of the United Kingdom; HMSO: 1971

Registrar-General *Sample census of England and Wales 1966* (household composition tables and housing tables) HMSO

Office of Population Censuses and Surveys, Population projections No 2 1971–2011 HMSO: 1972

Registrar-General *Statistical review of England and Wales*; HMSO: annually

Department of the Environment *Housing statistics,* Great Britain; HMSO: quarterly

Department of Employment *Family expenditure survey*; HMSO: annually (see 1970 edition)

W V Hole *The effects of current and future social changes on house design* BRS Current Paper CP 8/71

W V Hole *Housing standards and social trends*; Urban Studies Vol 2 No 2, 1965

*Since 1972 *Housing and construction statistics*

8 Organisation and Site Control

Accuracy in setting-out

The growth of partial or complete prefabrication of building structures places greater emphasis than ever before on the need to attain predictable standards of accuracy in the setting-out and erection processes. The degree of accuracy which can be achieved in construction has a significant bearing on design: in extreme cases, the feasibility of a structural system may depend on the standard of site accuracy that can be safely assumed.

This Digest concentrates on accuracy in the setting-out process, which BRS studies have shown to be critical for the avoidance of subsequent problems during erection. The data included should assist constructors in choosing the techniques necessary to attain the specified accuracy in setting-out, and designers by showing the level of accuracy to be expected in given circumstances.

What determines accuracy in setting-out ?

Ultimately, the standard achieved in setting-out depends on the accuracy of :

Linear measurements

Establishing and transferring horizontal levels

Establishing and transferring verticals

Delineating angles

Delineating straight lines

For each of these operations, a variety of methods may be used, depending on type of site, type of structure, the requisite degree of accuracy, and to a limited extent personal preference.

Accuracy in linear measurement

Given a tape measure, few would doubt their ability to measure distances with precision, but in practice serious errors occur. Studies have shown that, even with experienced site personnel, errors of 25 mm in a taping length of 30 m are possible when using steel tapes. Errors with fabric-based tapes are potentially very much greater and the use of fabric tapes cannot be recommended for any setting-out process.

Taping errors may arise in any of the following ways:

Reading errors are closely allied to the quality and fineness of tape graduation; with most tapes they are unlikely to be significant when compared with other sources of error. The new metric tapes to BS 4484: Part 1: 1969 are graduated in units of 5 mm, except at beginning and end, where 1 mm divisions appear. The 5 mm divisions may be subdivided by eye; the reading error should not exceed 2 mm.

Calibration errors BS 4035 required that tape graduations be correct to within $\pm \frac{1}{16}$ in (± 1.5 mm). Although, as far as is known, no tapes were specifically manufactured to this standard, calibration errors can be ignored if reputable steel tapes, in good condition, are used. Random checks have shown that errors are unlikely to exceed 3 mm at any point marked on the tape. BS 4484: Part 2 will specify an appropriate graduation accuracy for metric tapes and it is hoped that tapes will be available to meet this standard.

Tension errors are the largest source of error in taping. Steel tapes normally available for use on site are graduated fully supported and at a constant tension of 44.5 N (equivalent to a pull of 4.5 kg on a tension balance). Site studies showed that a range of 9 N to 175 N might be applied in practice to a 10 mm wide tape fully supported, resulting in errors of up to 10 mm in a 30 m measurement; in catenary measurements an even greater range of tension was used, from 10 N to 330 N. Some form of tension control is therefore essential if accurate tape measurements are to be made; the optimum form is discussed after the effects of sag have been considered.

Tape sag Wherever a tape is unsupported, it sags into a catenary between the two ends. Because the length of the arc is greater than the length of the chord, which is the distance to be measured, the sag effect will cause an over-estimate of distance. In practice, however, the reverse may be true since some operators faced with a sagging tape try hard to eliminate the sag by increasing the tape tension to the extent that the effect of stretch predominates. This tensile stretch effect can be put to good account if properly controlled: for a given tape cross-section, there is a single compromise

Table 1   Steel tapes: combined effect of tension and catenary errors

Maximum likely errors, in mm, in setting out a distance of 30 m in a single measurement. (Errors for shorter distances always less.)

Applied tension	10 mm wide tape		13 mm wide tape	
	Tape fully supported	Tape in catenary	Tape fully supported	Tape in catenary
Range observed in site experiments	−3 to +10	−13 to +15	−2 to +7	−20 to +12
44.5 N (calibration tension)	0	−12	0	−28
70 N (compromise tension for 10 mm tape)	+2	−3	—	—
105 N (compromise tension for 13 mm tape)	—	—	+3	−3

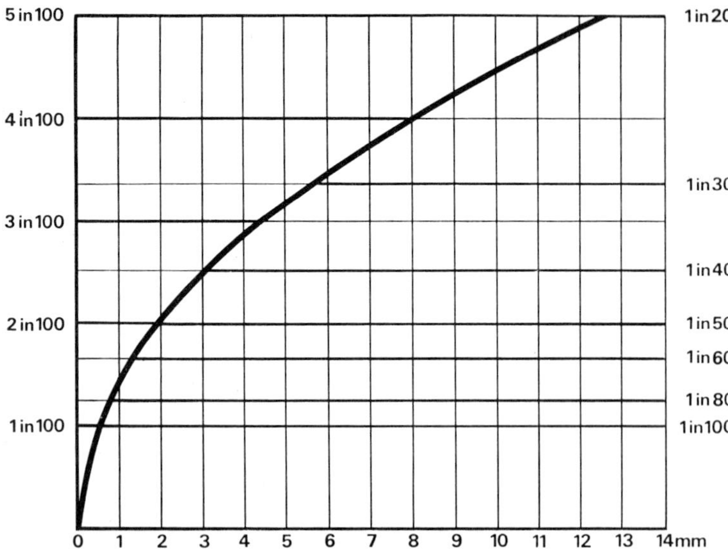

Fig. 1 Correction for slope (per 10 m measured)

tension at which the amount of stretch will nearly compensate for sag when the tape is in a catenary and which will give similarly close accuracy if the tape is fully supported. The appropriate tensions for etched tapes of standard thickness* are 70 N for a 10 mm wide tape and 105 N for a 13 mm wide tape (7·1 kg and 10·7 kg respectively on a tension balance). **Table 1** shows the errors arising when tapes are used in a variety of conditions, i.e. fully supported and in catenary, with and without control of tension; it shows the improved overall accuracy obtained by the use of the compromise tensions stated above.

Differences in level between measured points cause errors which are significant even on quite gentle slopes. Over short distances, these can usually be corrected by holding the tape approximately horizontal but over greater distances more accurate results will be obtained by measuring on the slope and applying the appropriate correction taken from **Fig. 1**.

Thermal movement can cause significant errors in very hot or cold conditions. Tapes are normally calibrated at 20°C and corrections should be made for any considerable departure from this temperature. In dull weather, air temperature is a reliable guide to tape temperature but in bright sunshine tapes—particularly dark coloured ones—quickly reach temperatures above the local shade temperature. When a tape is used in catenary, the effect of this is only marginal: as a rule, the tape will reach only about 5°C above local shade temperature. On a flat surface, however, an excessively high temperature may be reached, dependent on the temperature of the surface. In such a situation, the effect can be largely eliminated by raising the tape on small wooden blocks to allow some air circulation around it. When the tape temperature is higher than the calibration temperature of 20°C, 1 mm in 10 m should be *deducted* from the tape reading for every 10°C difference; when the tape temperature is below calibration temperature, an equivalent *addition* should be made. When measuring existing work, the correction will have the opposite sign.

Accurate taping procedure

If no correction is made to tape measurements other than for slope, the maximum likely error is about 25 mm in a 30 m length, and there is a fair probability of an error of 15 mm. For closer accuracy, further corrections must be applied. For most purposes, use of the appropriate compromise tape tension will bring the accuracy up to the required standard, but for the most accurate work correction for tape

* Heavily chrome-plated tapes and other abnormally thick tapes will require greater tension.

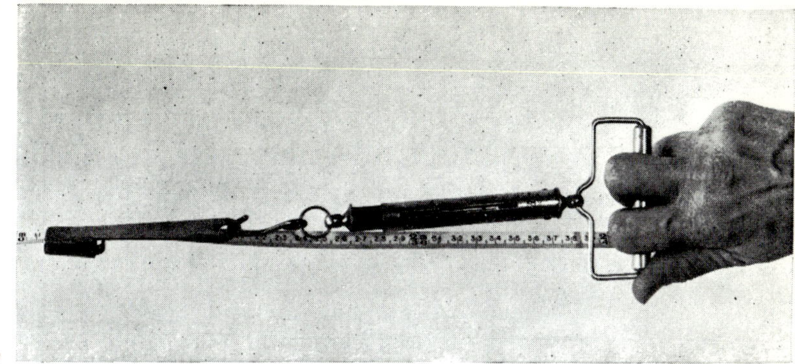

Fig. 2 Tension applied through spring balance

temperature may be necessary if the ambient temperature is markedly different from 20°C. With all corrections applied, and the appropriate compromise tension in use, measurements should be accurate to within ±6 mm in a 30 m length, with smaller errors at shorter distances. To achieve this standard:

1 Use a tape in good condition.

2 Check ring fixing at start of tape to identify zero of graduations (practices differ between makers).

3 Estimate tape temperature, and apply correction if necessary.

4 Estimate or measure slope and apply correction if necessary.
(**Fig. 1**)

5 Apply appropriate compromise tension.

6 Read tape to nearest mm, subdividing the 5 mm graduations by eye and avoiding parallax as far as possible.

Tension control

At the present time, the usual way of applying tension control is through a spring balance, of which some designs (**Fig 2**) have a device for quick attachment to the tape. Skill is needed to apply the correct tension and to read the tape and spring balance simultaneously; to make the operation easier, and therefore more likely to be carried out in site conditions, a constant tension handle has been devised at the Building Research Station (**Fig 3**). This was demonstrated during the 1968 BRS Open Days when visitors were able to measure a distance of 30 m in catenary, first without the tension handle—applying whatever tension they thought fit—then with the tension handle. The comparison of results given in **Fig 4** shows the value of tension control. The BRS tape tension handle is the subject of a patent application; it is hoped that it will be produced commercially.

Accuracy in horizontal level

The accuracy achieved in establishing and transferring levels depends on the quality and condition of the instruments and on the skill of the operatives. **Table 2** lists the more usual instruments used in site work, and gives an estimate—based on BRS and user experience—of

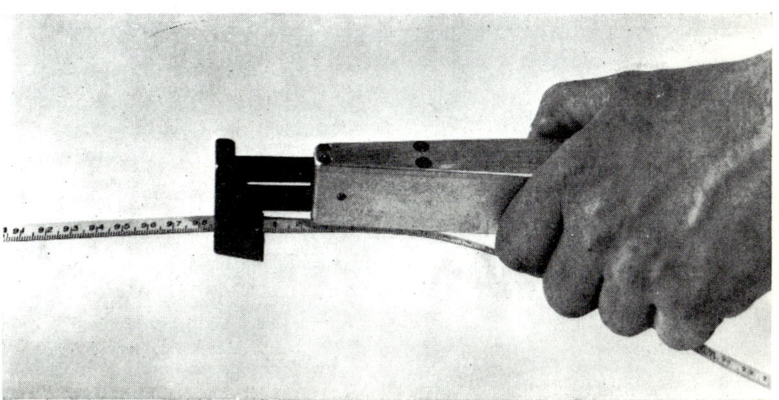

Fig. 3 BRS constant-tension tape handle

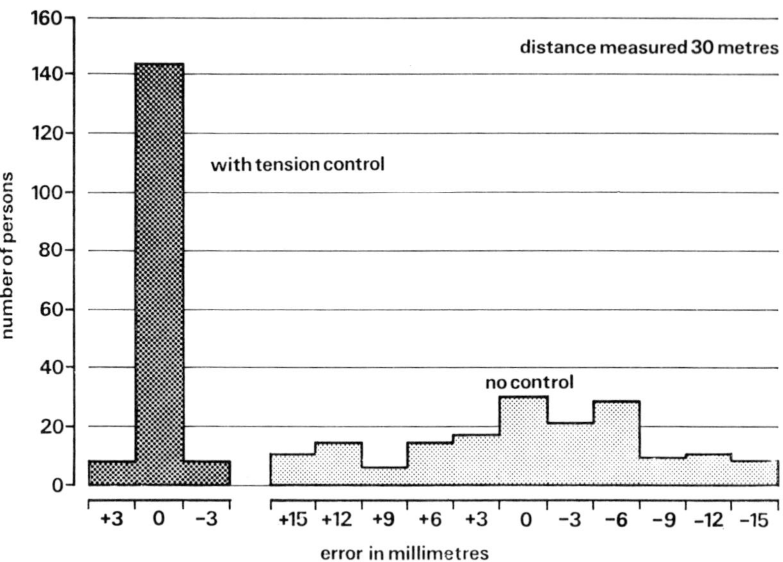

Fig. 4 Effect of tension control

the accuracy which is likely to be achieved in practice with instruments in good condition, and operatives of average competence. The best equipment, used with care and skill, is capable of greater accuracy.

Checking optical levels It is dangerous to assume that because an expensive optical level is being used it will necessarily give the accuracy indicated in **Table 2.** A recent survey of optical levels in use on sites showed 1 in 10 to be seriously out of adjustment, one being no less than 75 mm out in a sight of 30 m. All optical levels should therefore be checked regularly and adjusted when necessary. Trained surveyors know and observe such routines as a matter of course, but much setting-out on sites is necessarily in the hands of people who have had little or no formal training in surveying.

A check which is simple, quick and sensitive is illustrated in **Fig 5.** The level should be set up at point *B*, midway between pegs *A* and *C*. Sighting from *B*, the levels of *A* and *C* are adjusted to give equal intercepts on the staff, i.e. so that the instrument sees them as being at the same level. The instrument is then set up anew at *D*, the distance *CD* being equal to *AB* and *BC*, and sights are again taken on *A* and *C*. Any significant difference in the new levels of *A* and *C* indicates instrument error. For good quality optical levels, *AB*, *BC* and *CD* should each be about 20 m. For simpler self-levelling instruments, having no internal magnification system, a shorter distance is advised, say 7–10 m.

Errors in staff reading With the introduction of the new metric staff, to BS 4484 : Part 1 : 1969, errors in staff reading should be minimised. The staff is graduated in 10 mm intervals and test experience has shown that these can generally be subdivided by eye to the

Table 2   Horizontal Level

Method	Probable accuracy	Suggested useful range	Comments
Water level	±5 mm independent of distance	Limited to length of tube supplied—generally <15 m	Sensitive to temperature. Useful for transferring levels quickly round corners
Light self-levelling level	±8 mm in 30 m	30 m per sight	Particularly easy to use, giving adequate accuracy for much building work : with attachments, can be used to set out standard falls
Optical levels	±5 mm to ±2 mm per single sight, depending on quality of instrument	50 m per sight	Universal in application, for all precise work. Some types can set out falls

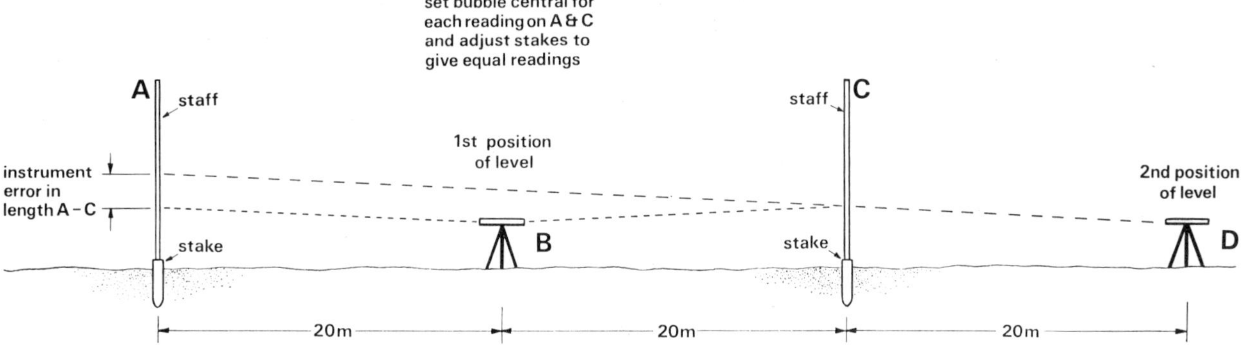

Fig. 5 Checking level for accuracy

nearest 2 mm. For closer accuracy, rarely needed in building work, a parallel plate micrometer may be used for precise subdivision of the scale.

The staff used with the light self-levelling level is intended to be read by the staff man and the figuring should, preferably, be similar to that on an ordinary rule.

Precautions with water levels Theoretically, water levels are very accurate, but in practice errors can arise if the two vertical arms of the water tube are at different temperatures. The difference between sun and shade temperatures can give rise to significant differences in the levels of the liquid in the two arms: the error will be shown up by an inability to obtain the same level when the two arms are compared side by side. A similar effect occurs if any air is present in the water. Before using water levels, therefore, check that the levels in the two arms are indeed the same and that no bubbles are present in the filling water. Keep the instrument out of direct sunlight except when actually being used.

Accuracy in verticals

The range of instruments available for checking and transferring verticals is less comprehensive than for levelling. Those instruments available, and the accuracy it is reasonable to except from them, are given in **Table 3**. Until recently, there was a gap between the capabilities of the plumb-bob and those of the theodolite, or specialised optical plumbing instrument; this has now been filled by a light self-levelling optical plumb, which is based on the light self-levelling level. It is a significant addition to the range, since plumb-bobs have

Table 3 Verticality

Method	Probable accuracy	Suggested maximum range	Comment
Spirit-level	±7 mm in 3 m	3 m	Craftsman's tool, only suitable for rough work up to storey height: unsuitable if component edge is bowed
Plumb-bob	±5 mm in 3 m	3 m	Suitable for general work up to storey height, in still weather
Plumb-bob, heavy, immersed in liquid to damp movement	±5 mm in 10 m	10 m	May be used over greater heights in totally wind-free conditions
Light, self-levelling optical plumb	±6 mm in 15 m	15 m	Used as a conventional optical plumb or in the hand against walls
Theodolite (with diagonal eyepiece)	±5 mm in 50 m	50 m	For multi-storey work. Range limited by visibility of target: tall buildings may be plumbed in two or more stages
Optical plumbing device	±5 mm in 100 m	100 m	Comments as for theodolite, but accuracy is greater

severe limitations in windy conditions, yet on many sites the use of the highest precision instruments is not warranted nor is the experienced staff available to ensure their correct use.

Accuracy in angular measurements

Methods of setting out angles are listed in **Table 4**, with an estimate of probable accuracy.

Using steel tapes Taping methods are most suitable for use on terrain which is relatively flat, and where there is no need to project the angles into trenches, etc. Theoretically, any angle can be set out by using steel tapes to construct triangles in which the three sides bear the appropriate ratios. In practice, only 90° angles are normally set out in this way, using a triangle with sides in the ratio 3:4:5; because the three sides are not very different in length, taping errors tend partially to cancel out and the degree of accuracy obtained is quite high. Corrections must always be made for sloping ground; accuracy will be further improved if all taping corrections are applied and if the tape tension is controlled as described earlier (see **Table 4**).

Using optical instruments For setting out right-angles, an inexpensive tripod-mounted optical square is available. It consists of two small telescopes, the axes of which are at 90° to each other and which can be swivelled in a vertical plane, along with means for centring over a reference point. The instrument can be used equally well on flat or sloping ground, and can project angles into trenches. It gives an accuracy comparable to that obtainable with steel tapes. Optical limitations make sighting over distances in excess of 30 m difficult, and in poor lighting conditions sights of not more than 15 m are preferable. Theodolites provide a higher standard of accuracy—largely determined by their fineness of calibration—and their superior optics enable angles to be projected distances in excess of 50 m; furthermore, any desired angle can be set out. Theodolites calibrated to 20 seconds should be adequate for most site work, but should for preference be fitted with optical plummets; plumb-bobs are a significant source of error in windy conditions. One-second theodolites—which are normally fitted with optical plummets—offer greater accuracy for critical work. The accuracy of an expensive theodolite should not be taken for granted; methods of checking are described in any good text-book on surveying. Routine checks by instrument engineers should also be established practice. The BSI PD on accuracy in building gives guidance on ways of minimising instrument errors in practice.

Table 4 Setting out angles

Method	Angle	Probable accuracy		Suggested useful range	Comments
Steel tape	Generally 90° (using 3:4:5 triangle)	Corrected for slope only: with all corrections and tension control:	±15 mm in 30 m ±6 mm in 30 m	30 m	Unsuitable for uneven terrain, or setting angles in trenches
Tripod-mounted optical square	90° only		±10 mm in 15 m	15 m in all conditions 30 m in good light	Useful on uneven ground, and in trench work. Quick and simple to use
Theodolite graduated to 20 sec, plumb-bob	Any		±10 mm in 50 m	In excess of 50 m	Universal instruments. Precision depends mainly on fineness of calibration. Optical plummet an advantage in windy conditions
Theodolite, graduated to 1 sec, optical plummet	Any		±3 mm in 50 m	In excess of 50 m	

Straightness of line

Where a theodolite is available, there is no difficulty in delineating any required line on site, and in fixing any desired number of intervening points, but on the majority of sites only tapes and levels are available and lines are set out by stretching builders' twine between established points. Tests have shown that even a modest breeze produces sufficient sideways bowing in the twine to introduce serious errors and if enough tension is applied to eliminate most of the catenary the twine may break or a holding peg may be displaced. Better results are obtained with synthetic lines (polypropylene stranded twine has been used) which being stronger can be used much thinner, thus requiring less tension to eliminate the wind error.

Setting-out large or complicated site plans

For large or complex developments, an alternative to the traditional tape and theodolite methods of setting out plans, road alignments, drainage intersections, etc., is the use of angular co-ordinates provided by two theodolites to locate each desired point. Accuracy is improved and cost may be significantly reduced, but knowledge and skill are necessary.

Fig. 6 shows three permanent survey stations, *A, B, C,* which are first established on the site and accurately surveyed. Then, in order to establish the position of any desired point *X*, visible from all three stations, angular co-ordinates *XAB, ABX, XBC, BCX,* are calculated—usually by computer; the point *X* is set out using, for example, co-ordinates *XAB* and *ABX,* and checked by reference to *XBC* and *BCX.* In practice, more than three stations will normally be required in order that all the desired points are visible from at least three stations. If care is taken in siting the stations so as to ensure that they are not disturbed during construction, points can be rapidly checked at any time. The accuracy achieved depends on the quality of theodolites used and on the competence of the operators; in the best conditions a skilled surveyor using a theodolite graduated to one second should obtain an angular accuracy of $\pm$ 5 seconds on each sight, i.e. $\pm$ 2 mm error at a range of 80 m. Thus, even on long single sights up to about 500 m positional accuracy to within $\pm$10 mm should be possible. This is more accurate than suggested in **Table 4**, where only average operative competence was assumed.

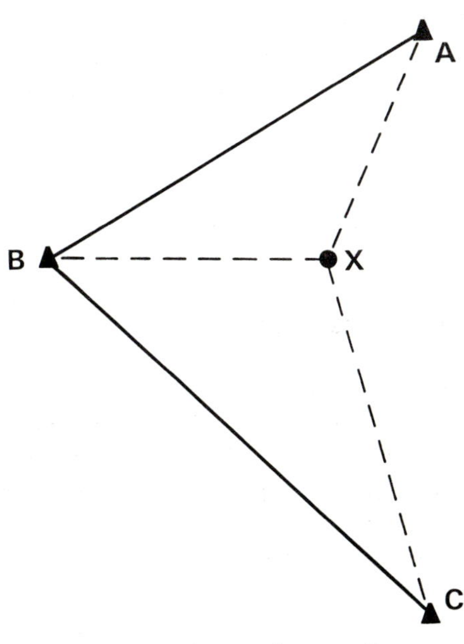

Fig. 6 Setting out by angular co-ordinates

Use of grids in setting out building plans

Where the design of the structure has been based on a grid, use of a matching site grid will minimise errors, but it is essential that the site grid be established with sufficient accuracy. Careful use of steel tapes, following the principles outlined earlier, can establish grids without any use of optical devices, though on sloping ground fewer errors will arise if optical instruments are used. The best instrument for setting out large building plans is the theodolite, since it can project grid lines and angles accurately over considerable distances: furthermore, it can be used to check perimeter grid measurements by projecting measurements from one side of the site over to the other. If the working grid is relatively closely spaced, errors in location of structural elements are further reduced because offsets are smaller.

Demolition and construction noise

Excessive noise can damage people's health or hearing and can cause disturbance to living or working environments.
This digest describes briefly some of the legislation and codes of practice aimed at controlling the levels of noise emitted from sites, and discusses methods of estimating and reducing plant noise.

Legislation

Legislation for the protection of people and the environment against noise has recently been extended by two Acts of Parliament.

The Health and Safety at Work etc Act, 1974,[1] provides for the protection of workers against risks to their safety and health, including the risk of hearing damage. On construction and demolition sites this mostly concerns plant operators or drivers, but men working close to plant must also be considered.

The Control of Pollution Act, 1974,[2] gives local authorities powers to protect the community against noise. Under Part III of this Act, a local authority may specify its own requirements to limit noise on construction and demolition sites by serving a notice that may in particular:

a. specify the plant or machinery which is, or is not, to be used;
b. specify the hours during which the work may be carried out;
c. specify the level of noise which may be emitted from the premises in question or at any specified point on those premises or which may be so emitted during specified hours; and
d. provide for any change of circumstances.

In specifying their requirements the local authority shall have regard:

a. to the relevant provision of any code of practice issued under this Part of this Act;
b. to the need for ensuring that the best practical means are employed to minimise noise;
c. before specifying any particular methods or plant or machinery, to the desirability in the interests of any recipient of the notice in question of specifying other methods or plant or machinery which would
be substantially as effective in minimising noise and more acceptable to them;
d. to the need to protect any persons in the locality in which the premises in question are situated from the effects of noise.

The Act also provides for a developer or contractor to ascertain from the local authority the noise restrictions that will apply to a contract at the same time as, or subsequent to, a request for approval under building regulations under Part II of the Public Health Act 1936, or in Scotland a warrant under Section 6 of the Building (Scotland) Act 1959.

Codes of Practice

Code of practice for reducing the exposure of employed persons to noise[3] gives recommendations for the maximum safe daily dosages of noise for the unprotected human ear. The accepted limit for an eight-hour exposure to a steady sound is 90 dB(A). If the exposure is for a period other than eight hours, or if the sound level is fluctuating (as is usually the case with building works), the equivalent continuous sound level, Leq, should not exceed 90 dB(A). Practical rules for the calculation of Leq are given in the Code. If the daily dosage exceeds the safe limits, the contractor should provide the operatives with ear protectors. Advice on the selection of protectors is also given in the Code.

British Standard 5228, Code of Practice for noise control on construction and demolition sites[4] outlines the application of The Control of Pollution Act, 1974, and considers the role of the local authority as well as that of the developer, the designer and the contractor. In setting a noise limit, a local authority may specify a permissible noise level at a point 1 m from the exterior surface of the nearest noise-sensitive building. The limit may be in terms of a maximum noise level or of an equivalent con-

---

Prepared at Building Research Station, Garston, Watford WD2 7JR
Technical enquiries arising from this Digest should be directed to Building Research Advisory Service at the above address.

tinuous sound level, Leq, for a given period. Although the Code suggests limiting values of Leq for a 12-hour period (eg 0700 to 1900 hours), these values have no legal significance.

Planning

The developer and the designer have a part to play at the early stages of planning a project. Early consultation should be made with the local authority to ascertain the noise restrictions that will apply. The design of the whole project should then be considered so as to avoid, as far as possible, noisy processes and machines. It may also be necessary to make estimates of the total noise emission from the site to check if proposed construction methods will be acceptable.

Details of noise restrictions should be included in the tender documents, so that the tenderer can take this into account in estimating plant requirements and at the same time consider whether any special noise control measures will be needed.

A more detailed check on noise emission can be made when the contract is awarded. For example, a production programme can indicate the locations and periods of operation of plant as well as the numbers and types of machine needed, so that noise levels can be predicted and compared with the limits set. Such a prediction will be possible only if the actual noise output of each machine is known; this information is also needed to determine whether operatives should wear ear protectors. If prediction shows that infringement is likely, means for controlling the spread of noise or the use of quieter machines or methods should be considered.

The Code[4] gives detailed guidance on the prediction and monitoring of noise as well as on methods for controlling the spread of noise.

In general, an extensive knowledge of acoustics is not needed for the prediction of noise levels on construction and demolition sites; a familiarity with some of the general principles and the terminology is, however, necessary.

Noise limits and measurement

Sound is transmitted as pressure disturbances in the air which vibrate the ear drum to cause the sensation of hearing. A sound is quantified by its sound pressure level: a ratio of the pressure of the sound to a reference sound pressure, usually that of the quietest audible sound. This ratio is expressed in decibels (dB) and is measured with a sound level meter. To represent subjective reactions to noise, the dB measurements have to be modified or 'weighted' to take account of the response of the human ear which varies with the frequency of sound. This 'weighting' is achieved by setting the sound level meter to the 'A' scale; the sound level is then indicated in dB(A). A sound level is only meaningful when the location of the point of measurement is specified; for example, the noise level from a building site measured at a point 1 m outside the nearest occupied buildings, or the noise of a machine measured at a point 1·5 m above ground at a distance of 10 m.

The noise from a machine can also be expressed as a sound power level: a measure of the total sound energy emitted per unit time. For practical purposes the sound power level is calculated:

Sound power level, dB(A) =
    sound level, dB(A) + 8 + 20 log R
where R is the distance in metres from the centre of the machine at which the sound level is measured.

Sound power is a measure of the noise output of a machine (or source) and is therefore not dependent on distance from the source, except that the distance should not be too small because of near field effects.

An equivalent continuous sound level (Leq) is a notional steady level which would, over a given period of time, deliver the same sound energy as an actual fluctuating sound. Instruments are available that indicate Leq directly. If Leq is calculated or measured for a period of 12 hours it is referred to as Leq (12 hr) and the value is given in dB(A).

Limitation of site noise by Leq (12 hr) permits a higher Leq for shorter periods provided the background noise during the remainder of the period is at least 10 dB(A) below the 12-hour limit. In that case an increase of 3 dB(A) can be allowed for each halving of the period. Thus, for a limit of Leq (12 hr) 75 dB(A), an Leq of 78 dB(A) would be permissible for a working period of 6 hours, or 81 dB(A) for a period of 3 hours. If it is known that a particular process will produce high noise levels for short periods, it may be more appropriate to restrict the maximum noise level rather than Leq.

Both codes[3, 4] give further information on noise units and recommendations concerning measuring instruments.

Plant noise

The main problem of noise in construction and demolition arises from the operation of plant and powered hand-tools. Some machines are already produced in a sound-reduced form and methods for obtaining greater reductions and for reducing the noise of other plant are under investigation. It remains to be seen how much noise reduction is physically and economically feasible.

In making predictions of site noise it is essential to have reliable information on the sound output of the plant that will be used. The noise produced by a machine depends on many factors such as the speed of operation and the work performed. Some manufacturers publish noise levels for their machines, but the conditions under which the noise levels were obtained are not necessarily the same for all types of plant. Standard procedures for noise tests representative of plant in operation are currently being developed by the International Organization for Standardization and the European Economic Community. The only document so far published is ISO 2151 — *Measurement of airborne noise emitted by compressor/prime-mover — units intended for outdoor use.** Some countries have already set upper limits for the noise outputs of certain types of plant: EEC Directives on this subject are also in preparation. Until standard test procedures have been adopted and the noise outputs of machines are quoted on this basis, use can be made of published figures such as those in Table 5 of BS 5228. The values in this table are maximum sound power levels taken from limited measurements on site and from various published data. In practice, values could be found which fall outside the ranges quoted. Alternatively, plant owners can make their own measurements.

In noise tests that have already taken place on plant such as excavators, loading shovels and dumpers it has often been the practice to measure noise levels at a distance of about 10 m from the centre of the machine (in standard procedures for testing it is likely that this distance will depend on the dimensions of the plant) and at a height of 1·5 m above ground. Measurements are necessary on four sides of a machine and, perhaps, at points above the machine as well, because differences can occur, due for example, to the position of the engine or the exhaust.

There can be considerable variation in the noise output of a machine during working. Measurements 10 m from a mobile crane, for example, gave levels of: 87 dB(A) hoisting; 83 dB(A) slewing; and 73 dB(A) engine idling.

Estimating noise emissions from sites

If the noise outputs of the machine to be used are known and production has been planned so that the locations of the machines and their periods of operation can be forecast, the resultant maximum noise level can be estimated at a point outside the site where a maximum noise limit may be specified.

In theory, and under specific conditions, the level of sound from a point source of noise is reduced by 6 dB(A) when the distance from the source is doubled. Although construction machines cannot be

*Available in UK from British·Standards Institution, 2 Park Street, London W1A 2BS.

regarded as point sources of noise and conditions on sites vary considerably from those on which the theory is based, site studies[5] have shown that this figure can be used to give sufficiently accurate estimates of distance attenuation in practice.

If the noise output of a machine is given as a noise level measured at a radius x from the centre of the machine, the reduction in this noise level at a greater radius y is equal to $20 \log_{10} \frac{y}{x}$.

If, however, the noise output of a machine is given in terms of sound power level, the noise level at any radius up to 700 m from the centre of the machine can be quickly obtained from the graph, Fig 1. For example, if the sound power level of a machine is 120 dB(A), the noise level at a radius of 30 m is $120 - 37 = 83$ dB(A).

Often several machines will be operated simultaneously and the total noise at a point where a maximum noise limit applies has to be estimated. The first step is to determine the noise level of each machine separately at that point. The following table then shows how to predict the total effect of two sound levels:

Difference between the two sound levels *dB(A)*	Amount to be added to the higher sound level to obtain the total sound level *dB(A)*
0 to 1	3
2 to 3	2
4 to 9	1
10 and above	0

If four machines individually produce noise levels of 70, 75, 76 and 77 dB(A) the total noise is obtained by combining the levels in pairs as follows:

70 and 75 dB(A), combined level is $75 + 1 = 76$ dB(A)
76 and 76 dB(A), combined level is $76 + 3 = 79$ dB(A)
79 and 77 dB(A), combined level is $79 + 2 = 81$ dB(A)

Thus the total noise from the four machines is 81 dB(A).

When estimating the noise level at the façade of a building, the Code[4] recommends that an addition of 3 dB(A) should be made to the levels calculated to allow for the effect of reflection. Where multiple reflections occur, due for example to a recess in a building, sound levels may be increased even further.

The estimation of site noise emission in terms of an equivalent continuous sound level (Leq) is more complex because a more precise forecast of the overall working time of each machine is necessary as well as the fluctuations in level that occur during the working cycle of the machine. The Code[4] deals with methods of calculation in estimating Leq and gives examples of their use. As yet, however, there is no published information on practical experience in the use of Leq for the control of noise from construction and demolition sites.

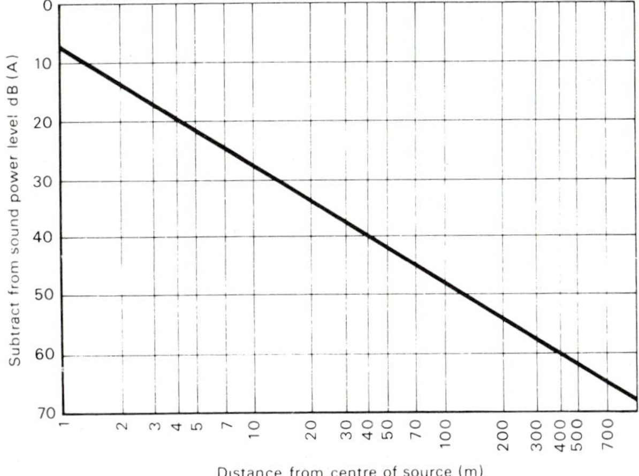

Fig 1 Amount to be subtracted from sound power level to determine the sound level at some distance from the source

Note Sound levels determined by reference to this graph are unreliable at distances less than 3 major dimensions of the source or if the source radiates sound in a marked directional manner

Reducing noise emission

When estimates of site noise emission show that infringement of a noise limit is likely, or when an infringement has occurred, one or several of the following means for reducing noise should be considered:

Siting of plant

By siting plant or operations as far as possible from a noise-sensitive area, or so that mounds of earth, stacks of materials or buildings on site reduce the transmission of sound.

By positioning a machine so that the quietest side faces the noise-sensitive area.

Operation of plant

By operation at a low speed.

By shutting down when not in use.

By keeping engine or machinery covers closed.

By reducing the number of machines in simultaneous operation.

By regular maintenance, notably to avoid rattling noises due to loose fixings or frictional noises due to lack of lubrication and to keep exhaust systems in good order.

Alternative plant or method

By using noise-reduced plant or by modifying the construction method so that noisy plant is unnecessary.

Screening

By erecting a noise barrier between the source of noise and the noise sensitive area (suitable for mobile plant).

By providing an acoustic enclosure for a machine, paying due regard to cooling and ventilation (suitable for a static machine).

By using an acoustic shed in which, for example, a man can work with a concrete-breaker. The operator is then exposed to a greater risk of hearing damage and should wear ear-protectors.

References
1 The Health and Safety at Work etc. Act, chapter 37, HMSO
2 The Control of Pollution Act 1974, chapter 40, HMSO
3 Department of Employment Code of practice for reducing the exposure of employed persons to noise, HMSO
4 BS 5228, Code of Practice for noise control on construction and demolition sites, British Standards Institution
5 Construction site noise, BRE Current Paper CP 57/75

Further reading
Thermic boring, BRE Current Paper CP 58/75
The Nibbler: a new concept in concrete breaking, BRE Current Paper CP 83/74

Structural design
in architecture

*This Digest, arranged to appear at the time of the 1961 Congress of the
International Union of Architects, and stimulated by that event, departs from the
usual pattern of the series. In looking at the processes of structural design, within
the context of architectural design as a whole, the Digest does not attempt to
collate information. Rather it sets out to explore a subject of wide scope on
which different opinions may be held. The discussion, though as factual as
possible, represents therefore a statement of one approach and not a solution
to a problem.*

*At the time of first publishing it was not the practice to attribute Digests to any
single author, but an exception was made in this case since a discussion of this
kind must bear the marks of one man's approach. The Digest was prepared by
an engineer, Mr. R. J. Mainstone.*

Designing any building is essentially a matter of making a long series of
choices—choices about ends and choices about means. The former will
be a dual responsibility of designer and client and will embrace the purpose
the building is to serve, the character it is to have, and any restrictions
within which the designer must work. The latter will be the designer's
responsibility alone and will embrace the detailed planning, the form and
details of the structure and such things as the services to be installed and
the finishes. The completed design will record the designer's choices as a
set of instructions that will enable someone, perhaps as yet unknown, to

The Parthenon. Structure and form indistinguishable and conceived as one in
design. Typical in this of most pre-
nineteenth-century architecture.

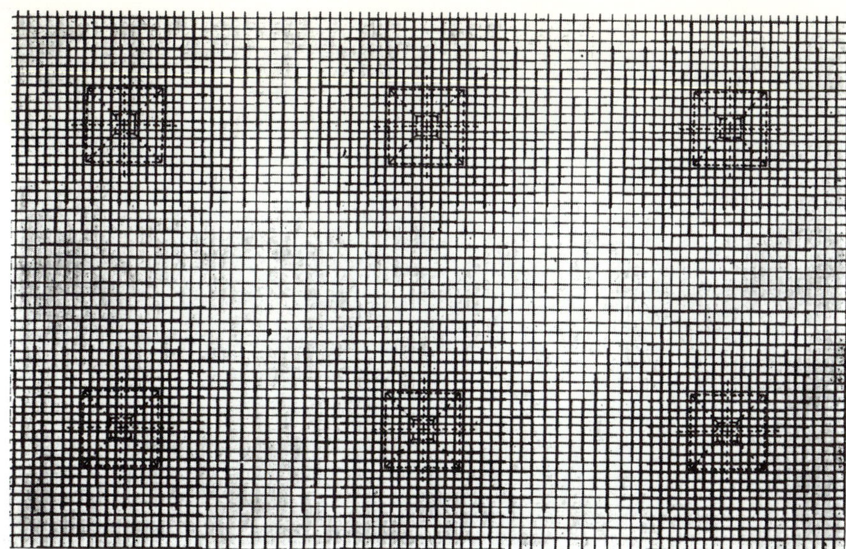

Reinforcement plan of a modern mushroom slab (R. Maillart). Structure now distinguishable from—though not wholly independent of—its formal embodiment and calling, therefore, for separate consideration in design. Typical in this of most recent architecture.

construct the building. It must contain, in embryo, the whole building, though still only as an idea.

It is this process of choosing that it is intended to examine.

It is not peculiar to the architect and the structural engineer. The painter and sculptor, for instance, make and commit themselves to innumerable choices, with each stroke of the brush or blow of the chisel, before a finished work is produced.

What does distinguish the processes of architectural and structural design from those of most other arts is the need for the designer to commit himself to his choices, almost *in toto,* before he can see their final outcome. He does not, as he makes them, see his building grow under his own hand. Only much later can he discover, by simple observation, whether it will serve its intended purpose as well as he expected, whether, visually, it will have the qualities he aimed at, whether it will stand up without moving too much when loads come on to it, and whether it will be practicable to build it with the resources available.

The chief consequence of this distinction—though it is really only one of degree—is to give added emphasis to the process of testing or analysis, which is inseparable from all responsible choice, and to attach a higher value to the sort of knowledge and judgement and imagination that make this analysis possible. Only through adequate knowledge rightly used can the consequences of different choices be foreseen. And only thereby can the choices be guided in the right direction in the initial intuitive phases of design and, when they have been made, subjected to a more formal analysis while it is still possible to modify them if they are found wanting.

Different aspects of architectural design—structural, formal, spatial—differ in that they are directed, primarily, to the pursuit of different ends. In so far as they are pursued separately each must draw on knowledge relevant to its own end. At the same time, each can actually or potentially, draw on different *kinds* of knowledge. And it is these which chiefly affect the ways in which the choices are guided and tested. It is these different kinds of knowledge, therefore, which will now be considered, attention being directed particularly to alternative approaches to the structural side of design.

Basic knowledge

All our knowledge derives ultimately from experience. But that used in architecture can be of two kinds: that derived directly from experience in the field of building and that derived less directly, by scientific enquiry, from experience in fields that are frequently remote from or only marginal to the field of building.

The collapse, in 1940, of the Tacoma Narrows Bridge through aerodynamic instability—a possibility that was not foreseen or considered in the analysis. Showing that the most accurate analysis may be worthless if it ignores a critical type of behaviour.

The reinforced concrete vault of an exhibition hall in Turin (P. L. Nervi). A structure in which the chosen means of construction were stretched near to their limits in vaulting a very large space, so that the form was fairly closely determined by the structural and constructional requirements once the basic choice had been made.

Knowledge of the former kind is typified by the knowledge of the craftsman. It usually embraces both ends and means, and it is the sort of knowledge that is gained through the personal experience of doing something or of being a member of a society that lives and works and worships in certain ways. It can be expressed in the form 'I know what is wanted here' and 'I know how to make it' or, more particularly, 'I know how to make a strong roof for this house with this timber'. Whether it is won laboriously through a long process of trial and error, as it must be in the first place, or gained through following a master's instruction and example, it is essentially a kind of knowledge that is not fully articulate, that stops short of a complete specification of what is known and of an understanding of why it should be so or should lead to the desired result. Not being fully articulate, though, it avoids the difficulties that we are faced with if we try to give precise descriptions of complex skills or of the complex totality of any experience. Knowledge of the latter kind, derived from scientific enquiry, is articulate and is, to that extent, more readily communicable to others. But it is, at the same time, limited to what can be expressed in fairly simple terms. It cannot embrace as a whole the pattern of activities that a building is to house or the qualities that will make it strong and durable and efficient, still less those that will give it the poetry that makes it architecture. It is necessary to isolate these activities and qualities, to break them down into simple constituents and to consider them one by one. And the knowledge so gained is a knowledge of relationships, of patterns of behaviour, of the constancies underlying the ways in which things happen, rather than of the ways in which to achieve particular ends. It is expressible not in the form 'this is the way to do something' but in the form 'if this is done this will be one result'. Its great merit is that, not being tied to the achievement of particular ends, its range of usefulness is infinitely greater than that of the craftsman's knowledge.

The craftsman's approach

The craftsman works by imitation of existing models varied only occasionally by tentative innovation. Such innovations as he does make are directed straight to his ultimate goal—to producing a better hammer or a better jug. Hence the striking 'organic' quality of the traditional forms of such things as a scythe or the hull of a boat.

The architect, or master builder, of the past worked as a craftsman. He served a craftsman's apprenticeship and subsequently he advanced his art by close observation of what he saw being built around him and what he saw of earlier buildings. In these he looked for features that seemed worthy of imitation and for others that seemed to be in need of correction. But there is no evidence that, in so doing, he made any clear distinction

A reinforced concrete coal bunker at the Costillares Institute, Madrid (E. Torroja). A relatively small structure in which purely structural requirements allowed a fairly free choice of the form.

between what we should now call the structure of, for instance, a particular system of vaulting, its form, the visual effects of this form, and its symbolic content or transcendental significance. In fact the evidence nearly all points in the opposite direction. When he approached the limits of what was structurally possible he was inevitably confronted very forcibly with the need to make the building stand up. Yet he seems to have attempted this directly in terms of correct proportioning and disposition of the forms, using his judgement in ways that are not yet fully understood to identify what was significant in them for the stability of the building.

His freedom of choice was always closely circumscribed by the impossibility of stepping safely far outside the bounds of what had been achieved already. His *capriccios* had all to remain on paper. And such freedom as he did have depended a good deal on the relatively simple nature of his aims and on the simple means he adopted to attain them, in particular on his reliance on forms of construction that mirrored his lack of distinction between structure and form by a close identity of the two in the building itself.

The scientific approach

The alternative approach, based on scientific knowledge of behaviour, is of necessity less unified. The designer is obliged to define his ends separately and explicitly in terms of each relevant aspect of behaviour and to pursue them as separate ends while trying to recognise them as merely facets of an indivisible whole. This leads to the splitting up of architectural design into structural design, spatial design and so on, and, incidentally, to this Digest.

In present practice structural design offers, possibly, the best example of this approach. The knowledge of structural behaviour that it uses was not gained from observations that some buildings stood while others fell or threatened to do so. It developed out of scientific enquiry into the principles underlying much simpler though analogous behaviour. The theory of statics considers the nature of the forces and moments acting on any structure or on any part of it and defines exactly the conditions of balance between them, while the theories of elasticity and plasticity define and explore the consequences of different relationships between these forces and the resultant deformations. Only in the large derivative body of theory known as the theory of structures are these basic relationships applied to the particular types of structure used in building. And, whereas the craftsman's knowledge was typified by a caption in Cesarino's edition of Vitruvius that could be freely translated 'in this way you can build a strong tower', that gained from the theory of structures tells the designer no more than 'if you build a tower like this and it is subjected to these loads you can expect that roughly these stresses will be reached at *A, B,* and *C,* and that there will be approximately this deflection at *D'*. It is, for various reasons, always a little uncertain, it takes him a much shorter distance towards his goal, and it has nothing to say about this goal.

More precisely, such knowledge of behaviour is, by itself, inadequate in two ways. It gives the designer no guidance as to what kinds of behaviour may be critical for the structure and it leaves him, when he has somehow identified these, to define in appropriate terms the ends at which he should aim. It is, in other words, up to him to decide whether, taking into account their possible inaccuracies, the predicted stresses in the tower at *A, B* and *C* and the deflection at *D* should be considered as acceptable and whether, if so, this could be regarded as a sufficient guarantee that the tower would be strong enough, or whether the stress at *E* under perhaps a different pattern of loads, or even some deficiency in elastic stiffness, could still lead to failure.

Knowledge of the structural behaviour that is required of a building—of structural ends as distinct from means—is therefore needed as well.

To some extent this also can be scientifically formulated. Patterns of likely loading can, in principle, be specified, at least in terms of their relative probabilities of occurrence. So can the limits of elasticity and strength of the materials. Indeed, at first sight, knowledge of these, coupled with the all-important requirement that the building must stand up, should suffice. The designer soon discovers, however, that there is more to it than this. All buildings left to themselves fall in the end. Much sooner they will distort here, crack there and may even suffer extensive local failures. Some of the movements and failures will be serious, others much less so. A full definition of the end should give all these possibilities their due and guard against each with an appropriate margin of safety.

This is the objective of the current revisions of the principal structural Codes of Practice in terms of the avoidance of 'limit states' such as complete or partial collapse, excessive deflection or local damage. But judgement is still required, both of those responsible for drafting the Codes and of the individual designer. Since, in any one design, it is possible to analyse only a few characteristic behaviours, such judgement is particularly needed in identifying those that may jeopardize safety or in any way interfere with the proper discharge of other functions. But compared with the judgements that earlier designers had to make, those which are now called for are of a more explicit and much more restricted kind, based on a more specialized experience and understanding gained in the light of the new theories.

The reason for adopting this approach—which for want of a better name we must call the scientific approach—is simply that without doing so the designer would be incapable of effectively exploiting more than a small fraction of the opportunities offered to him by the wide range of modern materials and constructional techniques. The geometry of a structure is still a primary determinant of its structural action and therefore of its strengths and stiffnesses, but the wide choice of materials and techniques makes it possible to achieve widely differing performances within a single geometry and thus necessitates an independent consideration of the structure. Provided, though, that the designer gives the structure this independent consideration, his freedom of choice is now potentially so large that it is circumscribed in practice more by his powers of imagination and persuasion than by the physical means at his disposal.

Such, indeed, is the present freedom of structural choice that there is rarely a unique, and therefore determinate, way of meeting a particular requirement. Even when something approaching a unique solution is encountered, as in large bridges and some wide-spanning vaults, there is still more effective freedom of choice than was open to the designers of major structures in the past. The fear that scientific knowledge must destroy the artist's role as a free and responsible creator and turn him into a mere calculator is groundless. Structural 'design' by calculation alone can never be more than the detailing of a design already substantially determined, whether by a specification of its form and type of construction or by a more general specification that it shall be of a particular type and shall, for instance, have the minimum possible weight to carry the loads imposed on it.

In practice, of course, it is neither possible nor necessary to make all the choices that a design entails *ab initio.* To the extent that basic choices of forms and types of construction stem directly from experience that the same forms similarly constructed have previously proved adequate in meeting similar requirements, structural design today still includes elements of what has been termed the craftsman's approach. Even here, though, the correct recognition of similarities and differences can now be ensured only through the insights offered by the scientific approach.

In other aspects of design the balance today between the scientific and the craftsman's approaches is mostly closer to that typical of structural design in the past. The scientific bases for the prediction of relevant performances are less fully developed and there is, in general, much greater difficulty in defining valid criteria of acceptability analogous to the 'limit states' of structural design. The consequent necessity of a greater reliance on the craftsman's direct experience, illuminated only to a restricted extent by scientific insights, inevitably leads to a more restricted freedom of choice or, if this is not accepted, to a greater risk of failure.

Structure in architecture

Structure is an integral part of architecture, and by the master builders of the past (and even the more theoretically minded mechanicoi of the late Hellenistic and early Byzantine worlds) it was never considered otherwise. From this single-minded approach, and from struggles to achieve the utmost possible with the means at their disposal, sprang the remarkable unity of conception of most of their major works—a unity that was reflected also in lesser ones.

Lacking, usually, the stimulus of barely attainable ends, embarrassed, on the contrary, by a bewildering freedom of choice, and forced to think separately of many different aspects of design in meeting the more complex requirements of today, the architect and engineer now find it much more difficult to produce designs with a comparable unity of conception.

Such a unity has been achieved occasionally by engineers working on their own when structural considerations have been paramount and when, because of the magnitude of the project or some restriction of means, there was an adequate stimulus. A structure cannot, in any case, be conceived solely as an abstract system of forces in equilibrium. Its form must be given substance in particular materials put together in a particular way and this substantial form, its constituent materials and the method of construction should always be at the centre of the conception.

More frequently, though, especially when structural considerations have not been paramount, a comparable unity has not been achieved. Either the basic conception of the forms has paid little attention to structural considerations and the disciplines that these may rightly impose or the architect's responsibility for basing his designs also on a careful analysis of other functional requirements has been largely abdicated in unquestioningly appropriating the forms of earlier structural inventions.

A fruitful reintegration seems to call for one or both of two things: (1) a more rational and systematic means for narrowing down the fields for the basic choices of structural form, etc., and (2) better means for bringing together the different aspects of design at the stage when the design is first taking shape.

In truly creative design the bringing together of the different aspects must be at the subconscious intuitive level where ideas are born. And the sort of intuitive mastery needed can be acquired only by much hard work. Perhaps, therefore, the most that can normally be hoped for here is a sympathetic understanding at this seminal level between a number of designers, most of whom remain specialists in their own fields. That should be a prime aim of education, but it is worth consideration also by those researchers who are trying to enlarge the funds of articulate knowledge through which, in part, the intuitions are shaped.

In the design of many buildings, however, the main challenge to creative thinking will lie only in one or two aspects. If these are correctly identified as, for instance, functional planning and environmental control, the primary structural problem will be reduced simply to that of selecting the most appropriate existing structural form and method of construction and proper integration will call simply for planning in detail in such a way as to respect, as far as possible, the limitations of the chosen form and exploit its potentialities. This calls chiefly for a more powerful classification of structural forms and their potentialities and limitations than yet exists.

Such classifications, relating also to plan types, environmental control systems, etc., would also go part of the way towards a rational narrowing of the present freedoms of choice. The complementary need is for a fuller understanding and more adequate specifications of performance requirements, of what it is worth paying to meet particular requirements and of how the price will be determined. The specifications need not be absolute ones. Indications of acceptable limits will usually suffice, together with optima to be aimed at where these also can be stated.

The need for responsible choice by architect and engineer will remain, however, if purposeful structural and architectural innovation are to continue. A principal aim of new developments in design methods should be to make it easier to focus on those choices which, together, are most significant in terms of costs and ultimate user satisfaction.

Further reading

For fuller discussions of some of the topics discussed in this Digest see: R. J. Mainstone, Structural theory and design, *Architecture and Building*, 34, 1959, pp. 106-113, 186-195, 214-221.

R. J. Mainstone, The springs of structural invention, *Journal of the Royal Institute of British Architects*, 70, 1963, pp. 57-71.

R. J. Mainstone, On construction and form, *Program* (Columbia University School of Architecture), Spring, 1964, pp. 51-70.

S. C. C. Bate, Why limit state design? *Concrete,* 2, 1968, pp. 103-108.

R. J. Mainstone, Structural theory and design before 1742, *Architectural Review,* 143, 1968, pp. 303-310.

R. J. Mainstone, L. G. Bianco and H. W. Harrison, Performance parameters and performance specification in architectural design, *Building Science,* 3, 1969, pp. 125-133.

Working drawings

Working drawings should convey the bulk of the technical information about a project. But information is often inadequate, inaccurate or not readily accessible even in drawings which have been carefully prepared. An arrangement of the set of drawings based on a systematic classification of the information to be conveyed makes it easier to locate the relevant drawings and to find, or confirm the absence of, required information.

The recommendations of this digest are based on research fully reported in BRE Current Paper CP 18/73.

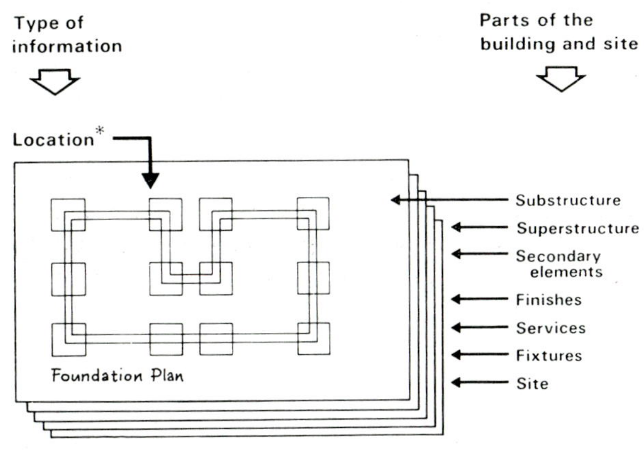

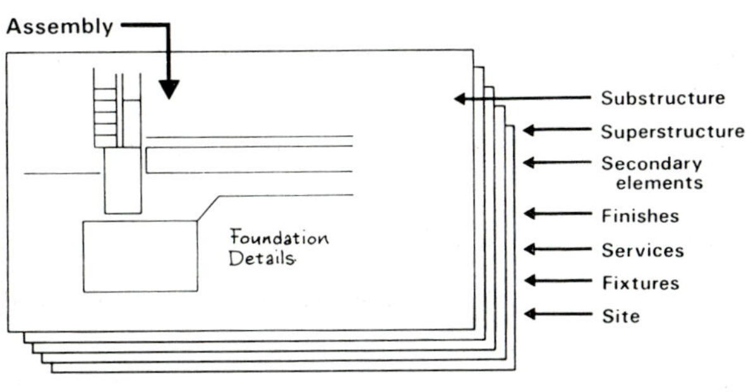

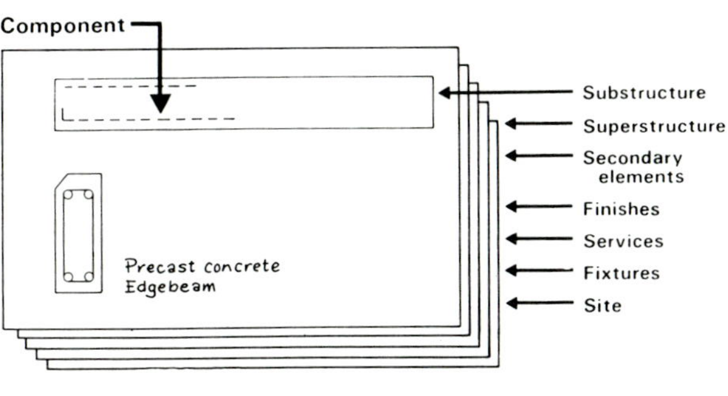

*There will be additional location drawings dealing with the project as a whole

Fig 1   The arrangement of the set

At the drawing board it is all too easy to see a working drawing as an end-product in itself, rather than as a means of communication. Draughtsmen should try to anticipate the use that will be made of a drawing; they should aim to provide 'information for building', not just pictures of a completed building.

The builder needs to know the *shape, size* and *location* of the building as a whole and of all its constituent parts; he must be told the *composition* of the materials to be used and *jointing and fixing* methods.

Information on drawings

Information defining shape, size and location properly belongs on the working drawings. Grids on plans and key reference planes on elevations and sections can improve the quality of dimensional information; when repeated on details, they help to locate them in the building. Composition is best dealt with in detail in bills of quantities or specifications, but drawings and schedules should indicate the materials to be used; graphic symbols may be used for this purpose.

The set of drawings

Two facets should be used to classify information into categories suitable for use in structuring sets of drawings (Fig 1); the first differentiates between location, assembly and component types of information, the second distinguishes parts of the building and site.

Table 1 defines the purposes of location, schedule, assembly and component drawings. Schedules are familiar as collections of repetitive information about parts of a building (eg doors, windows, finishes); they can be included in the 'location' group of drawings but they are more readily retrieved if stored as a separate group.

Table 1 Type of drawing and purpose

LOCATION*	Site and external works:	To identify, locate and dimension the site and external works
*There will be additional location drawings dealing with the project as a whole	Building:	To identify, locate and dimension parts and spaces within the building and to show overall shapes by plan, elevation or section To locate grids, datums and key reference planes To convey dimensions for setting out To give other information of a general nature for which a small scale is appropriate (eg door swings)
	Element:	To give location and setting-out information about one element, or a group of related elements
	Cross-references:	To show cross-references to schedules, assembly and component drawings
SCHEDULE	Element:	To collect repetitive information about elements or products which occur in variety To record cross-references to assembly and component drawings
ASSEMBLY	Element:	To show assembly of parts of one element including the shape and size of those parts To show an element at its junction with another element To show cross-references to other assembly and component drawings
COMPONENT	Element or sub-elements:	To show shape, dimensions and assembly (and possibly composition) of a component to be made away from the building To show component parts of an *in situ* assembly which cannot be defined adequately on the assembly drawing

It should not be assumed that every one of these types of drawing should always be produced. In many cases it will be sensible to combine some types of drawing and even to combine some types of information. For instance, separate component drawings for items produced *in situ* should be avoided if adequate information can conveniently be given on the assembly drawings.

The reason for subdividing location, assembly and component drawings is to achieve smaller, more readily searched groups, each dealing with a clearly recognisable class of information. The groups currently favoured correspond to an elemental breakdown of the building. The elements chosen should represent well-defined parts of the project to which unambiguous common names can be attached. These should not reflect alternative treatments of the same part of the building; it should be possible to retrieve information about, say, an external wall-to-roof junction without knowing whether the designer has opted for eaves or parapet. Similarly, they should not rely on functional distinctions which might not be immediately obvious, such as loadbearing and non-loadbearing walls.

Any of three solutions can meet these requirements:
1 A list of elements specific to a project
2 A list to suit the normal output of the office
3 A universal list, suitable for any project.

The last is preferred because with general usage the elements, and their codes, become familiar to all users, and it will not often be necessary to refer to a key.

As a universal list of elements, Table 2 shows a matrix with code numbers based on CI/SfB Table 1; in the BRE tests, this proved most acceptable. It is adopted in Figs 2 and 4. This list provides alternative levels of detail, so that the drawings for small, less complex, projects can be arranged within the system without excessive fragmentation.

Drawing numbering

Fig 2 shows a simple numbering system which conveys valuable information and is short enough to be easily memorised when following up cross-references. Other information such as the job number, or information relating the drawing to a particular block, zone or room, is usually best placed in a separate but adjacent box.

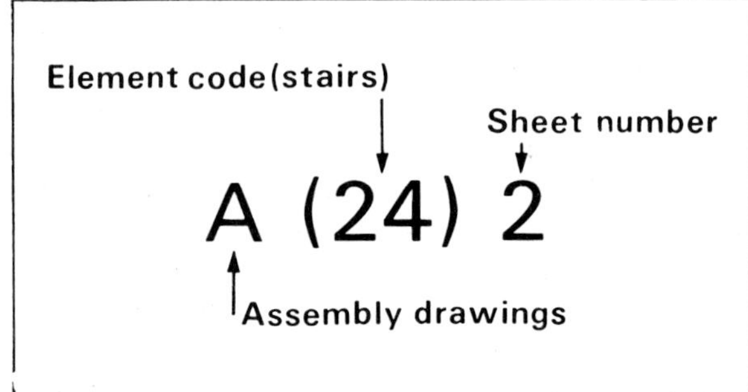

Fig 2

Table 2 Elemental breakdown (from CI/SfB Table 1)

(– –)   Site, project

Substructure	Superstructure			Services		Fittings		Site
(1–) Ground, substructure	(2–) Primary elements	(3–) Secondary elements	(4–) Finishes	(5–) Mainly piped	(6–) Mainly electrical	(7–) Fixed	(8–) Loose	(9–) External elements
(10)	(20)	(30)	(40)	(50)	(60)	(70)	(80)	(90) External works
(11) Ground	(21) External walls	(31) External openings	(41) External	(51)	(61) Electrical supply	(71) Circulation	(81) Circulation	(91)
(12)	(22) Internal walls	(32) Internal openings	(42) Internal	(52) Drainage, waste	(62) Power	(72) Rest, work	(82) Rest, work	(92)
(13) Floorbeds	(23) Floors	(33) Floor openings	(43) Floor	(53) Liquid supply	(63) Lighting	(73) Culinary	(83) Culinary	(93)
(14)	(24) Stairs, ramps	(34) Balustrades	(44) Stair	(54) Gases supply	(64) Communications	(74) Sanitary	(84) Sanitary	(94)
(15)	(25)	(35) Suspended ceilings	(45) Ceiling	(55) Space cooling	(65)	(75) Cleaning	(85) Cleaning	(95)
(16) Foundations	(26)	(36)	(46)	(56) Space heating	(66) Transport	(76) Storage, screening	(86) Storage, screening	(96)
(17) Piles	(27) Roofs	(37) Roof openings	(47) Roof	(57) Ventilation	(67)	(77) Special activity	(87) Special activity	(97)
(18)	(28) Frames	(38)	(48)	(58)	(68) Security, control	(78)	(88)	(98)

Titling and coding

It is first necessary to plan the set of drawings as a whole and to allocate titles which show the purpose and content of each drawing; the elemental break-down will serve as a guide. Element junctions should be considered in turn, to determine which require to be illustrated. Titles should be brief yet should comprehensively identify the sheet content. If the drawing shows a particular feature of an element, this should be stated in the title. If the detail applies at a particular location, this too should be stated (*see* Fig 3).

In wording a title, the coding of the drawing will be a strong consideration. If the draughtsman takes care to exclude from each drawing information that is not relevant to the title, coding should not be difficult. It is not necessary to produce a separate drawing for each element. Most assembly drawings show junctions between elements which have different code numbers. It is usually appropriate to use the higher number: thus, an external wall-to-roof junction should be coded (27), even though it includes external wall information (21). On small projects, and very occasionally on larger ones, it may be appropriate to use main elemental groups (1–), (2–), etc. Different approaches are needed according to circumstances; the most appropriate element code will often result from asking 'Where will the user expect to find this information?' Some of the options are set out in CI/SfB Project Manual, Section 4.2.

Cross-referencing

Although information is easier to find in a structured set, the most efficient way to locate relevant drawings is by cross-reference. Searches for information usually proceed from the general view (location drawing) to the particular (assembly and component drawings). Fig 4 shows how schedules can be used as collecting centres for references. The draughtsman should make a positive effort to provide the references that will aid these searches.

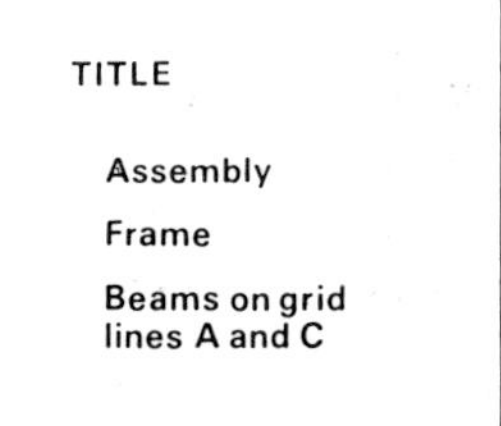

Fig 3

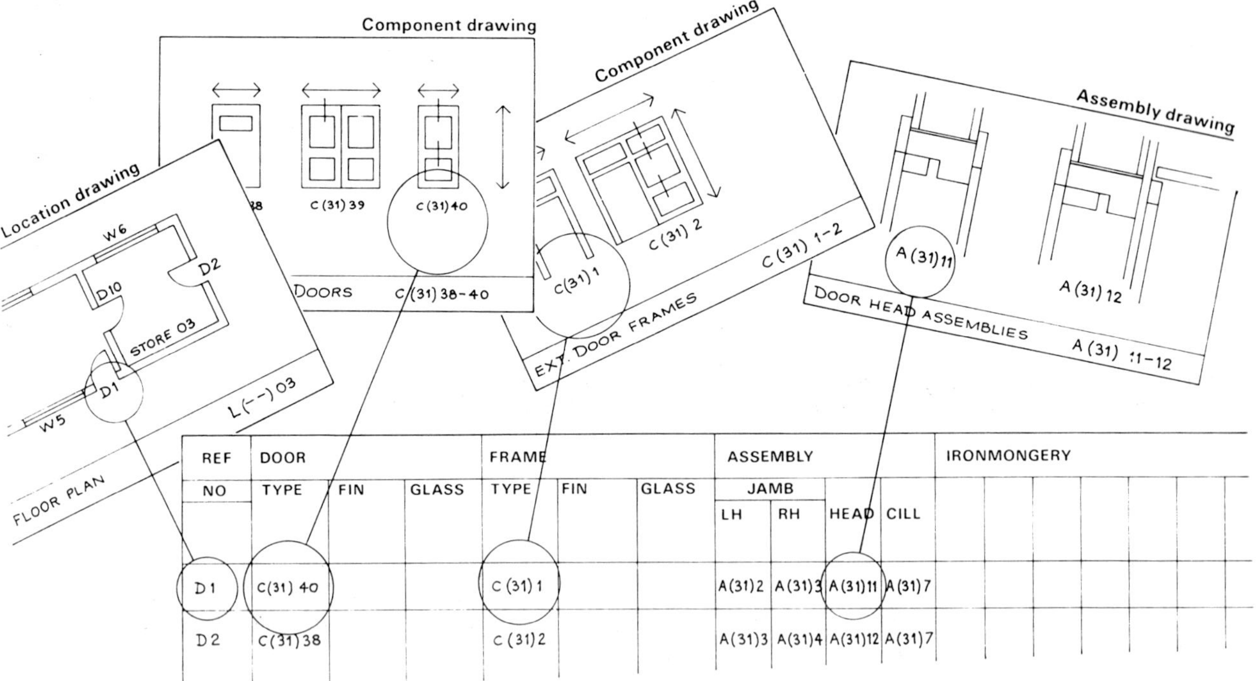

Fig 4 The schedule as a reference centre

The systematic approach

Rules alone will not produce a successfully arranged set of drawings. A positive attitude is required of the draughtsman, but the following points should help:

1 A working drawing is a message from designer to builder, who needs to know not only what the building is to look like but what parts it is made from and how they go together.

2 At the beginning of the working drawing stage, prepare a list of the drawings likely to be required and their titles. Before starting each drawing, consider the title and ask:
What is the purpose of this drawing?
What information will the user expect to find here?
What type of information (location, assembly, component) is it intended to transmit?
What part or parts of the project (elements) is it about?

3 Put the least information on a drawing that will enable it to fulfil its purpose. A cluttered drawing should prompt the questions 'Have I lost sight of the purpose of this drawing?', or 'Should I be producing two drawings instead of one?'

4 On completing a drawing, check that it contains explicit information about the shape, size, location and fixing of all parts. Consider where 'composition' information will be given. Note what references to the drawing will be needed elsewhere in the set.

5 Two or more details may appear on one drawing only if they are of one information type and relate to the same subject. The sheet can then be titled and coded quite specifically.

6 Don't repeat drawings of the same part of the building at several different scales. A small-scale location drawing identifies parts, shows their location in the building and on the site, and gives references to detailed information. A minimum number of assembly and component drawings should be produced at a scale large enough to show the detailed information adequately.

7 Remember the user seeks information from the set as a whole; he starts from the location drawings, so put references on them which will lead to the relevant details. The location, assembly and component groups should be clearly separated. The same sheet size should be used for all the drawings within one group.

8 In addition to a register, a copy of the elements list, eg CI/SfB Table 1 matrix, should be provided. A written explanation of the arrangement of drawings should be issued with the set. All users should be shown how to use the numbering system and how to exploit the structure of the set to locate information.

Project network analysis

The value of project network analysis, also known as PERT, critical path method, etc., for the planning, scheduling and control of building work has in recent years been well illustrated by its application not only to complex large-scale projects but also to medium-sized ones.
This digest sets out the basic principles of project network analysis and shows that its chief merit is a simple logic which imposes a clear and systematic approach to programming, and which is equally well suited to small projects not requiring electronic computers for analysis.

PROJECT NETWORK ANALYSIS is a method of planning, scheduling and controlling projects by recording their analysis in a diagrammatic form which enables each fundamental problem involved to be tackled separately.

It employs a network diagram or flow chart to show the jobs to be done and their inter-relations. From this diagrammatic 'method statement' and estimates of times taken to do individual jobs, it calculates the project duration, the Critical Path (that sequence of jobs which defines project duration) and the float (or permissible delay) for non-critical jobs. A schedule or programme for the whole project is thus devised.

Advantages

1 It separates

(*a*) planning the sequence of jobs from

(*b*) scheduling times for the jobs.

(Compare the simultaneous, and therefore less effective, planning and scheduling of normal bar chart methods.)

2 It shows the inter-relationships between jobs (again, not possible with a bar chart) and enables people to see not merely the overall plan, but the ways in which their own activities depend upon, or influence, those of others.

3 By setting out the complete plan for examination by everyone involved in the project it is easier to assess its soundness and so prevent unrealistic or superficial planning.

4 The effect on the project of alternative methods or individual job times can be examined at the outset.

5 The total requirements of manpower and plant can be readily calculated. By delaying or slowing down non-critical jobs, that is, those not immediately affecting the duration of the project, a schedule may be devised which makes the best allowance for any limitations in available resources.

6 The identification of the Critical Path has two immediate advantages:

(*a*) If the completion date has to be advanced, attention can be concentrated on speeding up the relatively few 'critical' jobs, and

(*b*) money is not wasted on speeding up non-critical jobs.

7 Schedules may be based on considerations of costs so as to complete projects in a given time at minimum expense.

Prepared at Building Research Station, Garston, Watford WD2 7JR
Technical enquiries arising from this Digest should be directed to Building Research Advisory Service at the above address.

Procedure

All projects consist of separate operations or jobs (called here activities). The activities are inter-related—technical or logical considerations will define a certain sequence in which the activities must be performed.

Stage 1—The sequence of activities

The first step is decide which activities are involved, and their sequence. This requires care, thought and a good knowledge of the project in hand. In project network analysis the sequence is presented as a network of which the commonest form is the arrow diagram, each activity being represented by an arrow. Fig 1 shows a typical sequence of activities forming part of a building project.

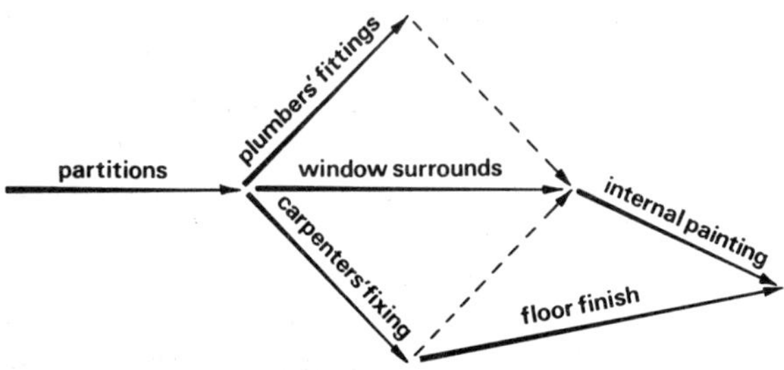

Fig 1

The *duration* of activities is not considered at this stage, so the arrow diagram is not drawn to a time scale. The arrows indicate the direction of time flow only.

The two unnamed broken arrows (called 'dummy' arrows) do not represent real activities; they merely help clarify the sequence by showing, for example, that internal painting follows the completion of carpenter's fixing as well as of window surrounds.

The junction of arrows are called *events*. Sometimes an event corresponds to a well-defined stage, for example the completion of foundations, but usually an event serves no more useful a purpose than the separation of one activity from another.

Only one rule is needed to interpret an arrow diagram: at every event all activities represented by incoming arrows *must* be complete before any activity represented by an outgoing arrow can start.

Fig 1 is thus a more concise representation of the following method statement:

> When the partitions are completed, the plumber's fittings and carpenter's fixing can be started, and also the window surrounds. The floor finish can be started when the carpenter's fixing is complete and the internal painting can commence when carpenter's fixing, plumber's fittings and window surrounds are finished.

The flow chart in Fig 2 is an alternative presentation of the method statement which does not require the use of dummy activities and may be preferred.

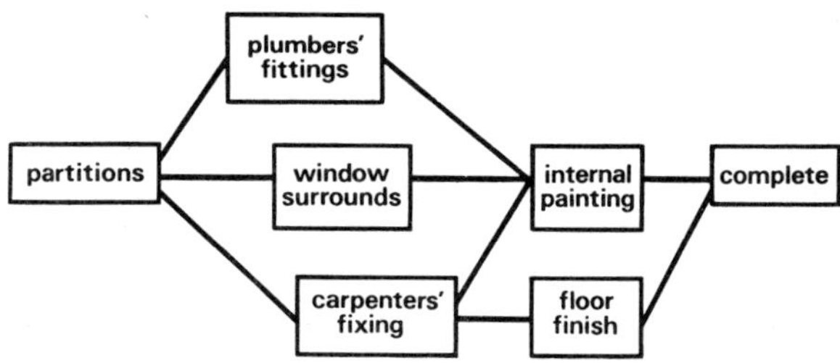

Fig 2

In practice, the necessary operations or activities are listed and the arrow diagram or flow chart drawn by asking of each activity in turn:

> what activities must precede it?
> what activities must follow it?
> what activities can be done at the same time?

Stage 2—Estimated job durations

Once the sequence is known, estimated times (in any convenient units) for each activity are added:

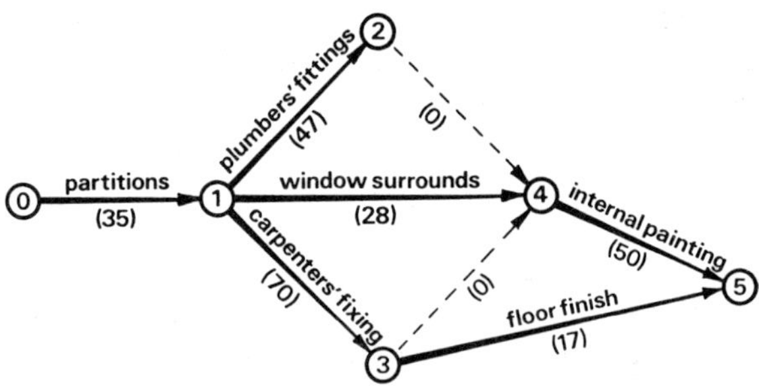

Fig 3   (**N**) = activity duration in hours

This figure repeats Fig 1, but activity durations are added; events are numbered for easy reference and circled to distinguish them from the durations and times written on the diagram during analysis.

Stage 3—Analysis

How soon an event can occur is found by inspecting the alternative paths leading to it. No event can be reached (or new activity started) until *all* preceding activities are completed. Given a start time of 0 the *earliest* an event can occur is given by the *longest* alternative path to it. Table 1 sets out the lengths of the alternative paths to each event; of these the longest is called the earliest time of event.

Event 5 marks the completion of the simple scheme shown in Figs 1 and 3. The longest path to this event (through events 0, 1, 3, 4 and 5) is called the Critical Path since it defines the duration of the project as 155 working hours.

Table 1

Event No.	Alternative paths	Earliest time of event (hours after start)
0	—	0
1	35	35
2	35 + 47	82
3	35 + 70	105
4	82 + 0 or 35 + 28 or 105 + 0	105
5	105 + 17 or 105 + 50	155

In such a simple diagram the Critical Path can be found by inspection but for more complex diagrams it is also necessary to determine the *latest* time for each event to find the Critical Path. The latest time of event is that time which allows all outgoing activity chains to be completed inside the project duration already determined. It is found by working *backwards* through the arrow diagram using similar calculations to those above. The Critical Path then passes through those events whose earliest and latest times are equal. Critical activities associated with these events are indicated by heavy arrows in Fig 4.

The extent to which non-critical activities can be delayed without affecting the project duration is found by comparing earliest and latest times for each event, as in Table 2.

The maximum time available for activity 1,2 is the difference between the earliest time for event 1 and the latest time for event 2, that is 70 hours. But activity 1,2 is estimated to take 47 hours, hence the *total float* for the activity is 23 hours. Total float times for each activity are shown in squares in Fig 4.

A complete project might contain thousands of arrows but the method of analysis is just the same.

However, if the diagram is large or complex, or if it is likely that frequent reanalysis will be needed, it is common to use an electronic computer to carry out the tedious arithmetic. The principles remain exactly as outlined above.

Table 2

Event No.	Earliest time	Latest time
1	35	35
2	82	105
3	105	105
4	105	105
5	155	155

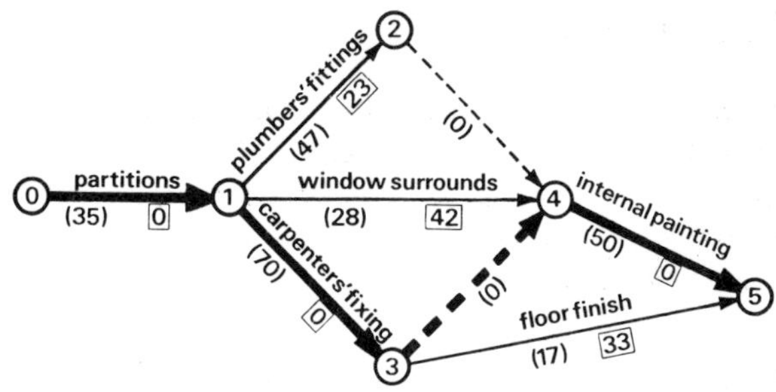

Fig 4

Stage 4—Final scheduling

If the project duration given by the analysis is excessive then either one or more of the critical activities can be shortened or an alternative sequence examined. Re-analysis of the amended diagram will indicate the effects of any changes made. When an acceptable project duration has been obtained the resources required can be examined. The estimated times for activities will be related to their individual requirements for manpower and plant. A total resource requirement (with each activity starting as early as it can) can then be calculated. If jobs in parallel on the diagram require the same resource the *total* resources required may exceed those available. By delaying non-critical jobs up to the limit of their float the need for extra resources can be minimized. Should the schedule obtained in this way still require more resources, at certain times, than are available, then an increase in the project duration is unavoidable.

As a result of this scheduling, a programme is obtained defining the time at which each activity will be performed.

Again, the calculations involved are basically simple but can be time-consuming to carry out; for large networks it is usual to use a computer for such resource allocations.

Stage 5—Project control

PNA is very useful in these planning and programming stages but it should not be considered solely as a planning device.

As work proceeds, PNA can help to control the project by:

1 indicating where management attention should be concentrated—on the critical activities,

2 determining the effects of what has happened to date by re-analysis of the diagram,

3 minimizing the effects of any delays that may have occurred by replanning the remaining activities,

4 carrying out **2** and **3** at regular intervals throughout the project.

Site control and reporting procedures should be devised to provide the necessary information as simply as possible.

The above example, though simple, is sufficient to illustrate the basic method and indicate what it can achieve, without going into the variations and

PNA applied to complete projects

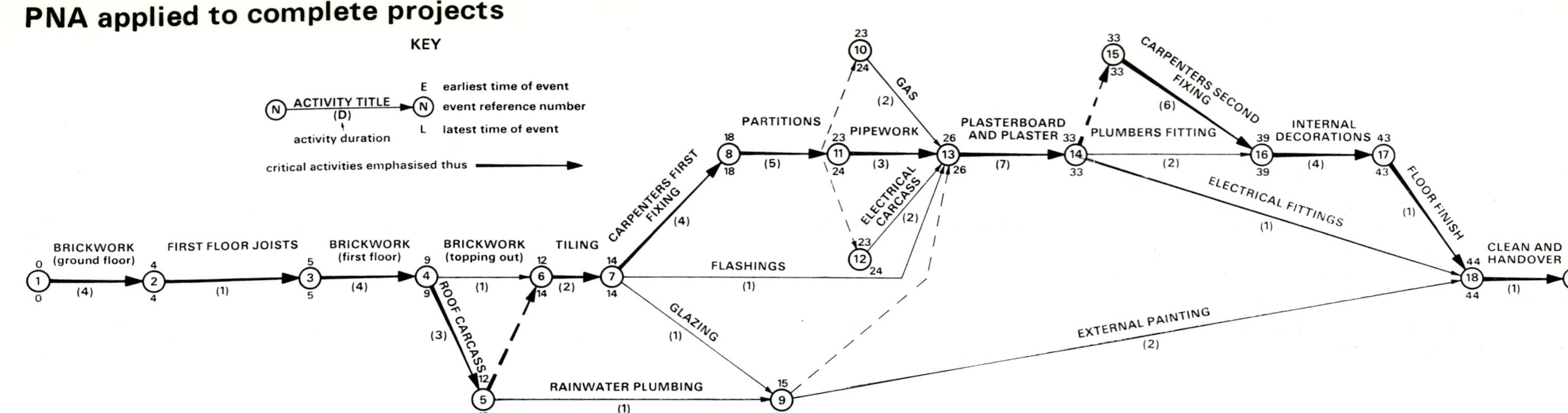

Fig 5 Structure and finishing of a house; network scheduled for event times

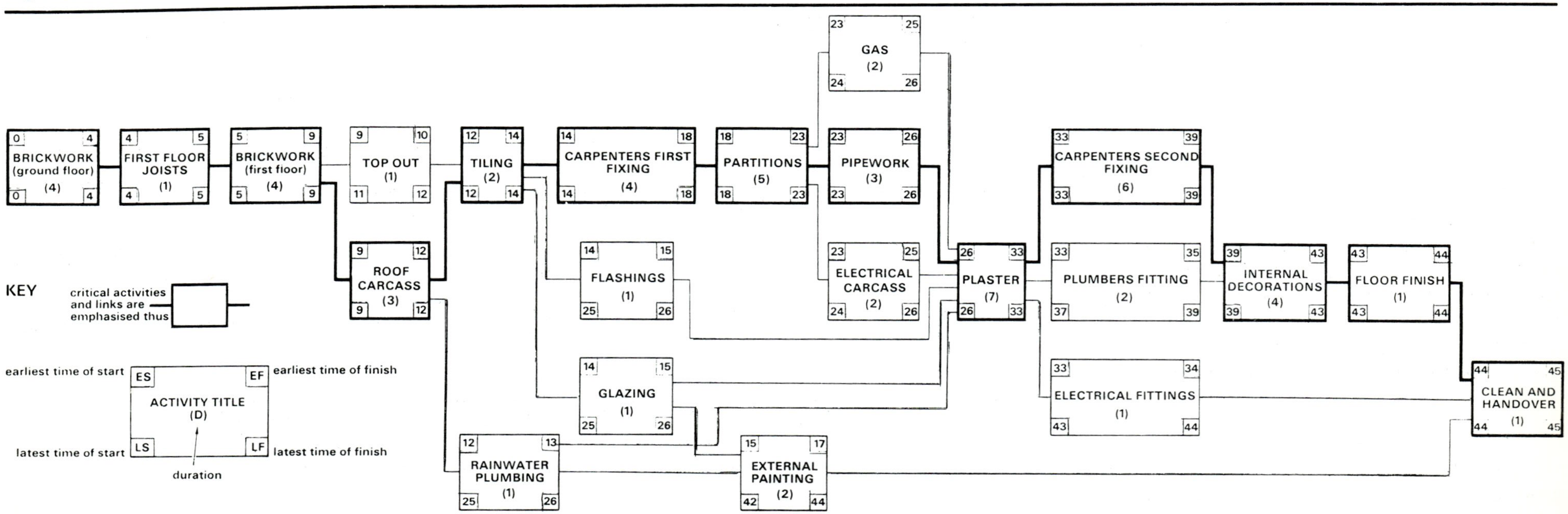

Fig 6 Network for the house, scheduled for activity start times

A House

A modest example, showing the structure and finishing stages of a house, is presented in three alternative forms in Figs 5, 6 and 7.

For the labour force assumed, the earliest completion date is 45 working days from dpc level, this time having been determined by the sequence of critical activities. A different labour force or method of construction may well modify the outlines of these diagrams, without changing the basic PNA approach in any way. Fig 5 shows the arrow presentation; Fig 6 the flow chart form.

Fig 7 presents the schedule in the form of a bar chart. The information presented on this chart could not have been obtained without the use of the earlier network diagrams. This linked bar chart is, in fact, an arrow diagram drawn to a time scale. It may be found more useful to present this information to the executive team in small increments, for example, a fortnightly up-dating of the next four weeks' activity starts.

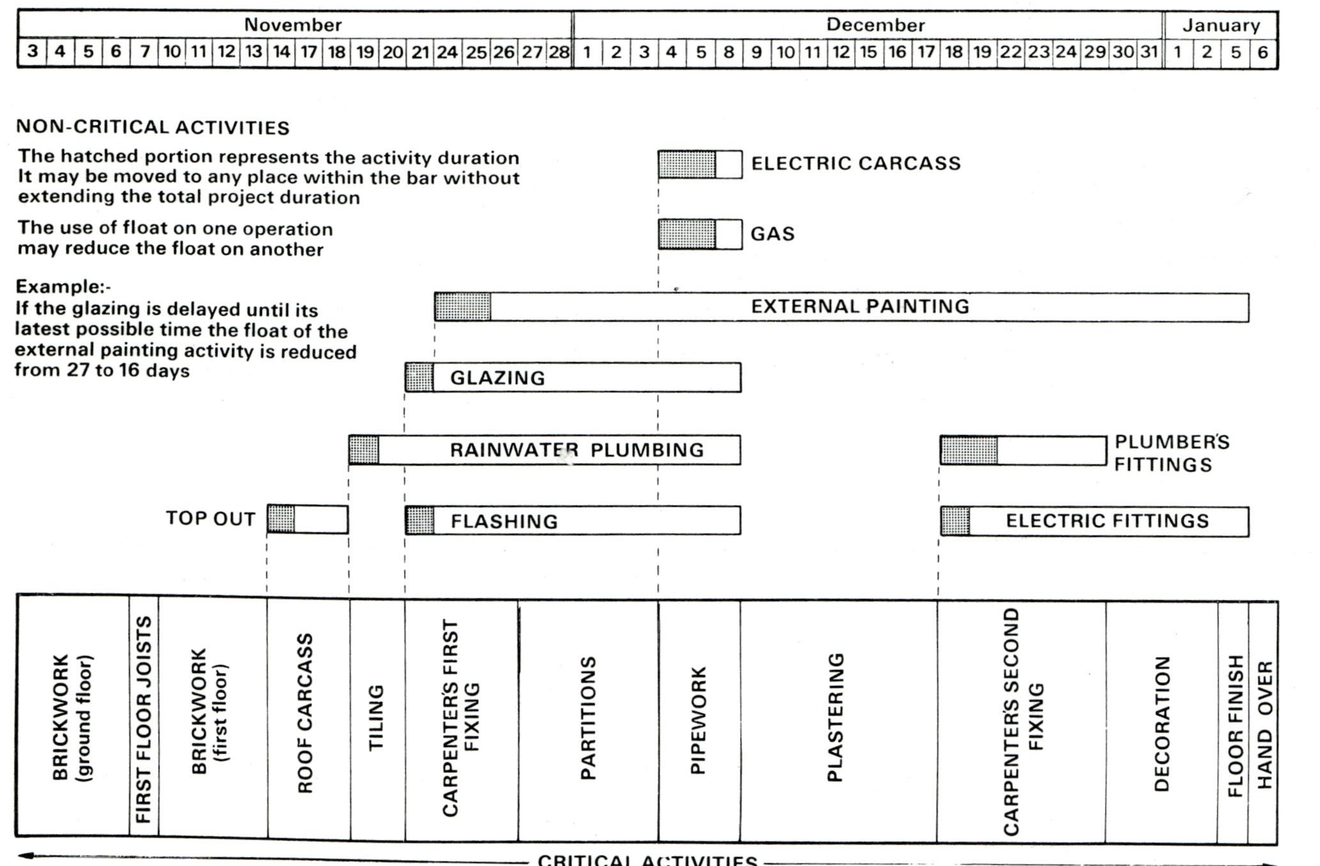

Fig 7

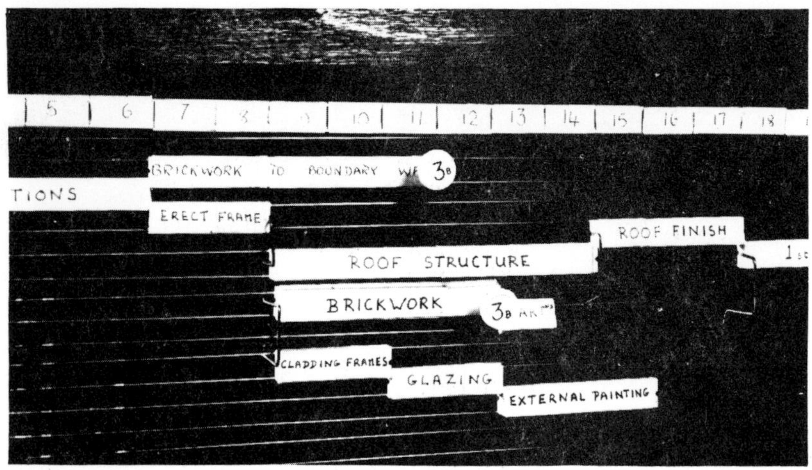

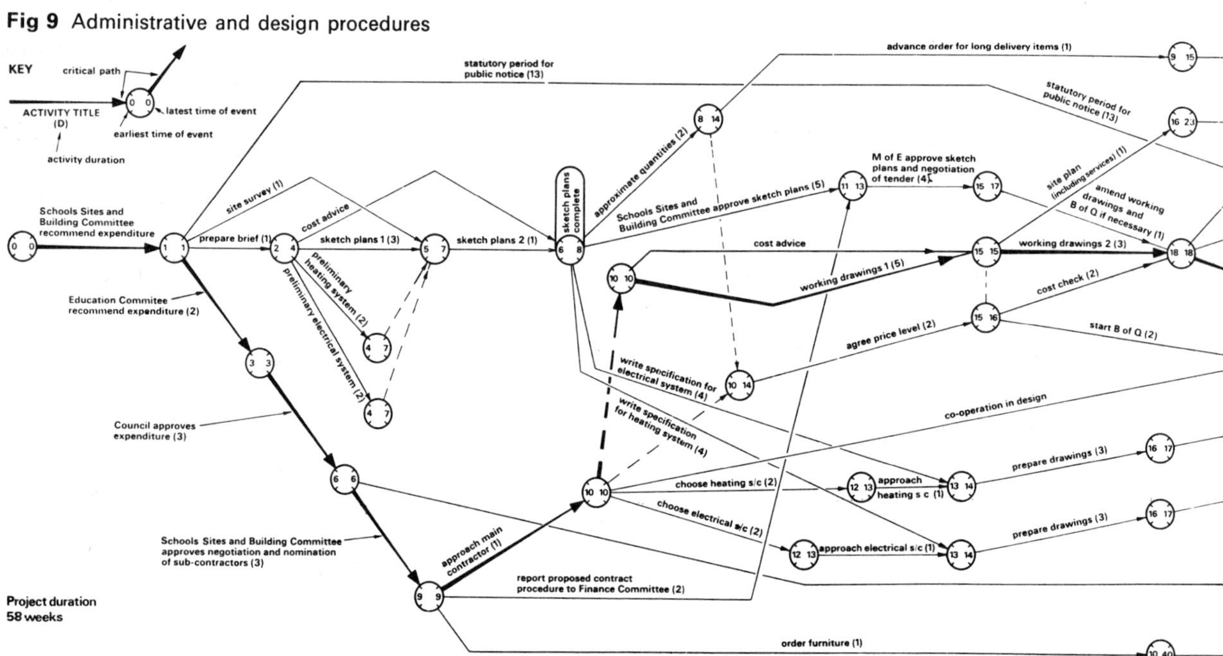

Fig 8 Resources aggregation and activity scheduling. (1) If two brickwork activities each start at the earliest possible time, six bricklayers are needed on days 9, 10 and 11. In stage (2), not shown here, it is demonstrated that by delaying the start of 'brickwork to boundary wall' until day 13, only three men are needed, without delaying completion of the project

refinements appropriate to the nature or size of particular projects and the needs of the planning team.

To recapitulate, PNA enables:

1 All activities to be identified and inter-related in a planned sequence.
2 All activities to be scheduled and their effects upon the duration of the project thereby distinguished as critical or non-critical.

The type of diagram shown in Fig 7 may be set up on a framework of horizontal wires on which slide beads, representing resources, with vertical wires connecting activities which have a precedence relationship. Various solutions of resource allocation to obtain an optimum use of float time may be tried; a simple method of communication to the executive team is a photograph of the most acceptable solution. Fig 8 shows part of such a solution.

Design and planning procedure

PNA need not be confined to site work, as in the previous example. By explicitly presenting the interrelationships of the various activities that make up a project, PNA is especially useful in the design and planning stages.

Fig 9 shows part of the arrow diagram for the administration and design procedures of a four-class county primary school at Barton Seagrave (*see* reference 1).

PNA has also been applied successfully to large-scale planning, such as town planning (*see* reference 5).

Recent research into the feasibility of introducing a production bias to documents used in the preparation of tenders has indicated that a diagram showing a possible sequence of site operations may, with advantage, be produced at the advanced design and quantity measurement stage. It has been found convenient to set up this diagram in flow-chart form, using magnetic tiles, so that changes can be made easily in order to accommodate any differences of approach by the successful tenderer, or to allow for variations in design or materials.

Fig 9 Administrative and design procedures

Conclusions

The basis of PNA is simple to understand and its use for small projects requires no apparatus more complex or costly than pencil and paper.

For the architect, it has an important role to play because the entire sequence of operations has its origin in the design stage. The method has a further attraction to the architect: normal drawingboard practice should enable him to appreciate readily the diagrammatic presentation of the project plan which is central to PNA.

PNA can be used easily for controlling the actual construction process; the level of control can range from that provided by the linked bar chart of a dozen or so activities, to the most detailed time and resource scheduled control system based on a network involving hundreds or even thousands of separately recorded and controllable activities. The success of any control system depends on the accuracy and speed of the information feed-back from the site, the ease with which the control document can be updated and the cost of management relative to the cost of the project. The extra cost of PNA is negligible, but many small building projects are carried out with so little project management that any control tool, even one so simple as a small network, seems out of place. However, larger projects, having full-time supervision will always benefit from the discipline which periodic updating imposes and from the immediately recognisable graphical presentation of up-to-date information of network and bar chart.

Acknowledgement

The project illustrated in part by Fig. 9 was carried out by the County Architect of Northamptonshire County Council and his staff, working in conjunction with staff of the BRS.

References

(a) Available free from the Building Research Station

1 NUTTALL J F, JEANES R E; The Critical Path Method: Current Papers Construction Series No 3

2 JEANES R E; Critical Path Method applied to the overall process of building: Current Papers Design Series No 11

3 NUTTALL J F, AMOS E E; CPM applied to building site control: Current Papers Construction Series No 12

4 BRITTEN J R; The advantage of time-scaled networks and planning frames: Current Papers Construction Series No 41

5 FINE B, BRITTEN J R; Safety nets for urban planners: Current Papers Design Series No 50

(b) Other publications

6 BS 4335:1968 Glossary of terms and symbols in project network analysis: British Standards Institution

7 BATTERSBY A; Network analysis for planning and scheduling: Macmillan.

8 LOCKYER K G; An introduction to critical path analysis: Pitman

9 LOCKYER K G; Critical path analysis: problems and solutions: Pitman

10 Ministry of Public Building and Works; Network analysis in construction design: Research and Development Building Management Handbook No 3: HMSO, 1967

11 SIMMS A G, BRITTEN J R; Project network analysis and critical path: Machinery Publishing Co Ltd, 1969

(c) Publications explaining 'flow chart' approach

12 MULVANEY J E; Analysis bar-charting: Iliffe Books Ltd 1969.

13 CITB; Precedence diagram planning: Pitman 1969 (Learning text).

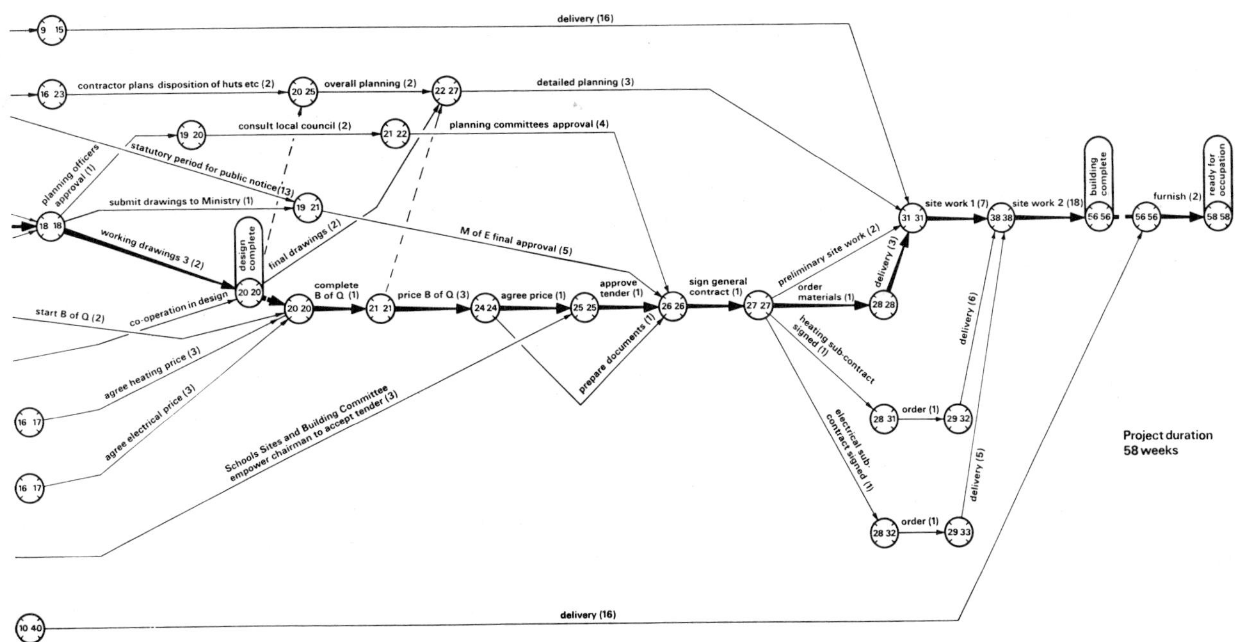

Flow charts to control progress on housing sites

Progress on housing sites can be effectively controlled by the use of a simple display of an outline programme of work-stage completions that is easily kept up to date by recording, week by week, the actual number of completions. Manhour requirements can vary so widely that the preparation of a more detailed programme is rarely justified.

The continually up-dated record of progress makes immediately obvious any need to adjust labour or other resources; it draws early attention to the effects that delays at any stage are likely to have on subsequent stages and gives a continuing and realistic prospect of completion.

An accepted method of programming house-building on any scale is to phase the operations so that the numbers of operatives in the different trades are balanced to achieve continuous production. To meet this need, various devices have been used. A conventional bar chart can be drawn to show a precise sequence of construction and the numbers of operatives required week by week for each stage of work. A more informative picture is produced by 'line of balance'[1] in which the schedules are a series of inclined bar-lines, one for each trade, the slopes of which denote the rate of working; by this technique, the relationship of manning to speed of working and the risks of collision courses and unbalanced manning are vividly displayed. Studies made by the Station have shown that the effort and expense involved in the preparation of a detailed programme may be difficult to justify because some of the implicit assumptions may prove unrealistic; it assumes that the planned sequence will be followed, and that the productivity achieved by the operatives is known beforehand and will remain constant.

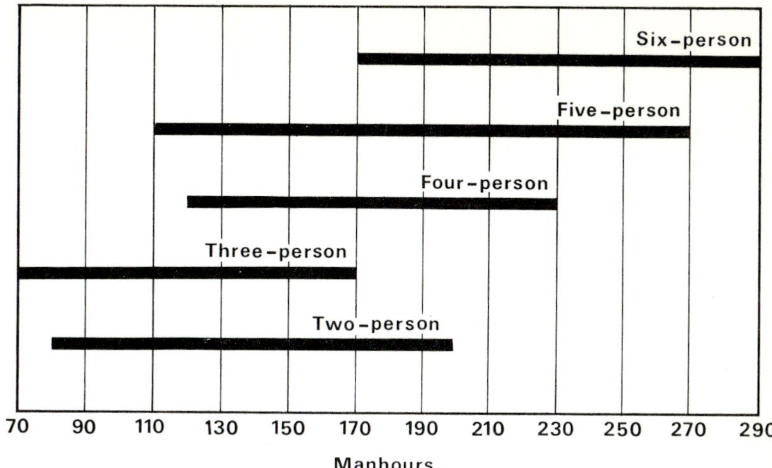

Fig 1 Manhours for brick superstructures of five house types on one site

Fig 1 shows, however, that manhour requirements can vary widely even on the same site and for the same task. Although house size can be seen to affect the *average* manhour requirements for each house type there is a large spread about these averages so that at its worst the two-person type used more manhours for the brick superstructure than many of the other, larger, types. With so little difference in 'apparent' work content between individual dwellings of different size there is little need to consider different house sizes in preparing the programme and with such a wide range in output any attempt at a precise timing of the work must be unrealistic.

Detailed programming presumes closely controlled conditions on site. The pattern observed in practice on the majority of house-building sites is much more a matter of working to an outline plan and adapting to circumstances than any attempt to follow a detailed programme. The most pressing need is for a simple method of recording progress so that major organisational problems are rapidly shown up. Keeping a detailed check on progress is difficult because work tends to be spread over the site but experience on some fifty sites has shown that a sufficient check is obtained by recording how many houses have reached selected stages of completion and displaying this information in the form of a flow chart. A similar basis is adopted for the programme. Only a simple overall plan is required to establish the pace of the work, manning rates and requirements for materials and plant. This is translated into a programme of completions for the selected stages, presented as a flow chart.

Fig 2 illustrates the basic principles of the flow chart; a fuller illustration of the method is given in Fig 4. Fig 2 shows a programme for building ten houses, progress to be checked at five stages of completion. On the horizontal scale, 4 weeks have been allowed for completion

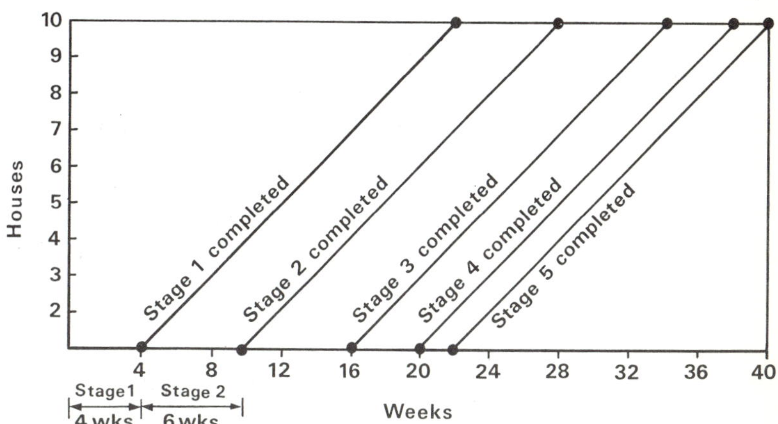

Fig 2 A basic programme

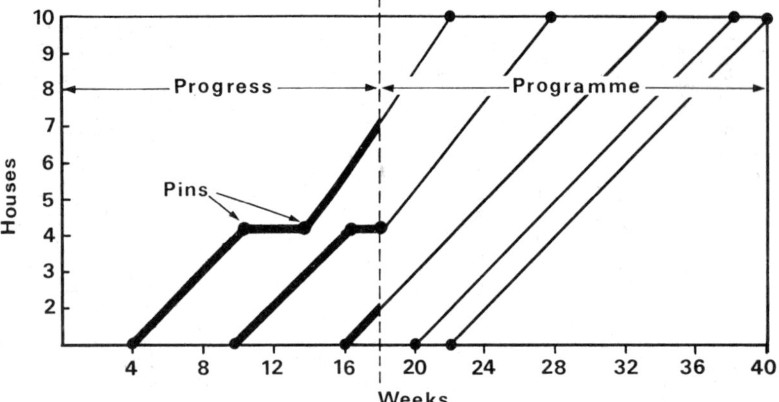

Fig 3   Situation at week 18

of Stage 1, and intervals of 6, 6, 4 and 2 weeks between stage completions thereafter. The slope of the lines represents the required rate of completions, namely one house every two weeks, if the ten houses are to be completed in 40 weeks overall. This initial flow chart is modified in the light of the actual progress. The situation at the end of week 18 is shown in Fig 3 where all before week 18 can now be regarded as 'progress' and all thereafter as 'programme'. Progress on Stage 1 has fallen behind the original programme because of the hold-up from week 10 to week 14, but the rate of completions has been speeded up. If this increased rate is maintained the original target date for overall completion of Stage 1 will be met. The delay on Stage 1 has caused Stage 2 to fall behind schedule. The chart shows that work on Stage 2 must be speeded up if the overall target for this stage is to be met. Although it is not shown explicitly on the chart, the earlier delays may have repercussions on Stage 3; this is a possibility that the site agent will assess in interpreting the situation report provided by the flow chart.

The method of presentation devised at the Station and proved on sites employs a programming sheet fastened to a pin-board. The flow lines are displayed by elastic string and progress is recorded by inserting pins denoting the number of houses completed to the several stages at the end of each week. The elastic flow lines can then be bent around the pins. The use of distinctive colours for the strings and pins of the flow lines is found to make reading very easy.

The actual form and level of detail shown on the flow charts can be selected to suit the requirements of the individual builder but it is suggested that the number of flow lines should be about 20. On one of the sites studied, 36 stages of work were plotted but the lines were found to be too close for easy interpretation. The number was reduced to eight by retaining only the stages that were thought to be proving difficult to control; the new chart was easy to read but the contractor found that too few stages of work were shown for a proper understanding of the production problems as they arose. Other similar experiences have confirmed the recommendation to use about 20 flow lines.

It is convenient to maintain the flow chart on site where charge hands and sub-contractors, for example, can see directly how their work is affecting overall progress. If desired, copies can be kept at head office and in the architect's office so that higher management is kept up to date and so that the architect will be in a position to estimate the likelihood of the contract finishing on time.

Reference

1 Programming House Building by Line of Balance; The National Building Agency, London, 1970

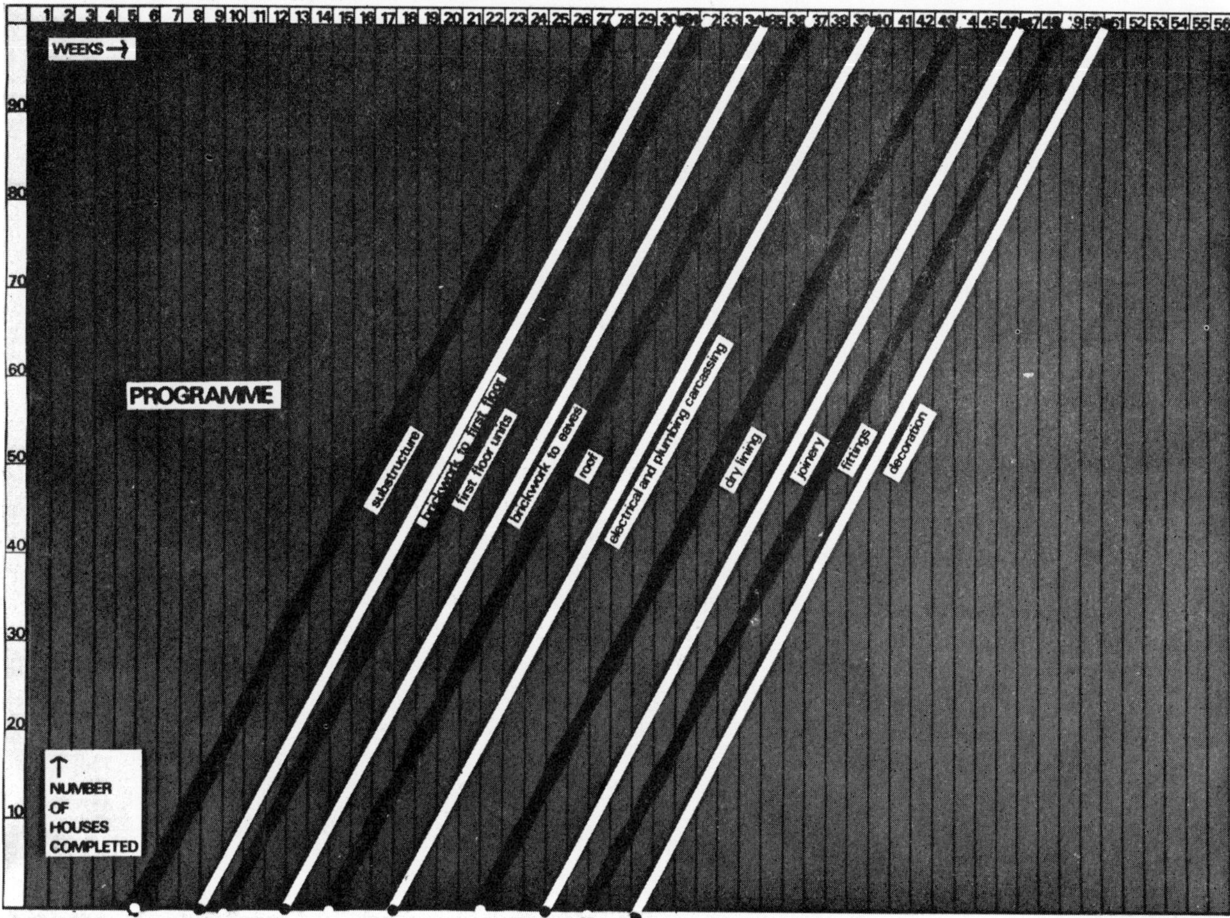

Fig 4a Initial programme

Example

The example shows a programme for the construction of 100 dwellings in 50 weeks. Because of the small scale necessarily used here only ten work stages are recorded and the elastic is wider than would be used in practice.

In general, the tighter the timing between stages, the greater the demand on production control is likely to be. In this case it has been decided to aim at a house-construction time of 28 weeks with intermediate stage completions as shown. This allows 22 weeks between first and last of the 100 dwellings—a rate of about five dwellings per week.

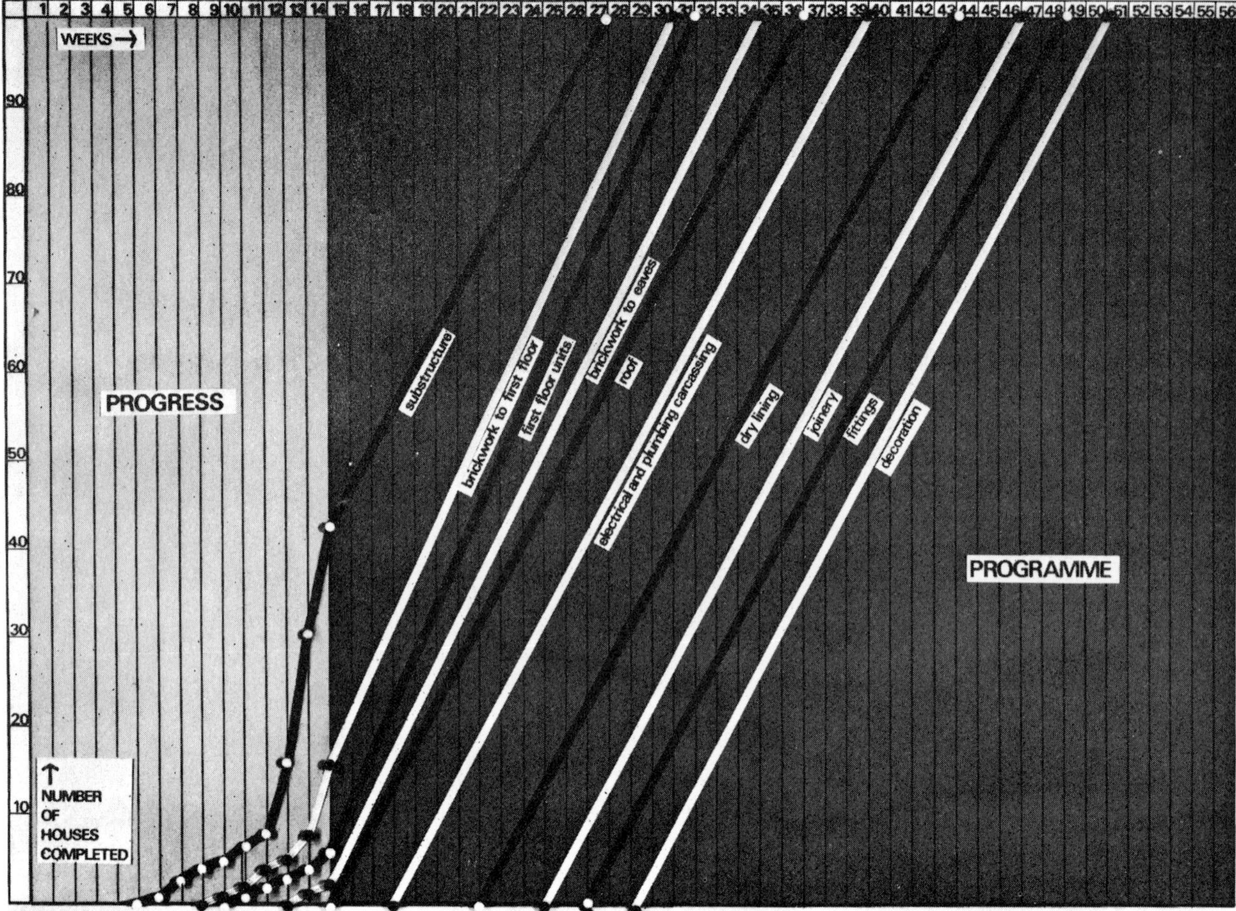

Fig 4b Progress at week 14

Once work begins, pins are inserted on the board each week showing the number of completions of each stage; the use of the elastic string automatically forms the flow lines. Right on programme, the substructure for the first house was completed at the end of week 5 but for the next six weeks delays forced the rate well down behind programme; this in turn prevented the work on the superstructure proceeding at the programmed rate. Additional bricklayers were therefore recruited to speed up the work.

The progress at week 14 shows a satisfactory situation. The rate of progress now being achieved on 'substructure'—the slope of the flow line between weeks 12 and 14—is well above the target rate now required—the slope of the flow line from week 14 to completion —and this means that men can be diverted temporarily to subsidiary work on drainage, not included in the main programme. The current rate of progress on 'brickwork up to 1st floor' seems about right on the evidence so far. It should be possible to adapt to any minor repercussions the original delay may have on subsequent stages and no explicit re-programming is called for.

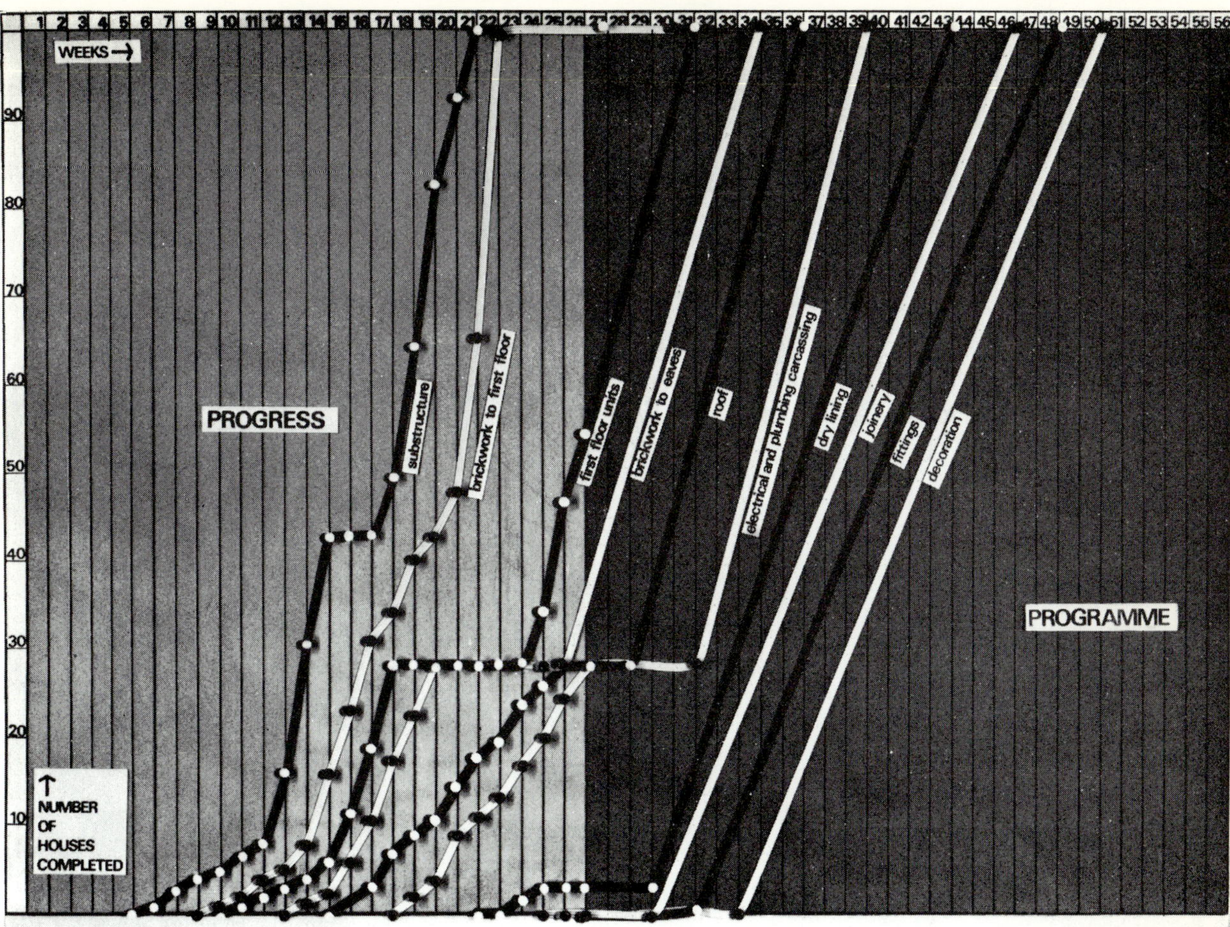

Fig 4c Progress and revised programme to week 26

A failure in the delivery of prefabricated floor units at week 17 has caused a major set-back. When further work on 'brickwork to eaves' became impossible, the efforts of the bricklayers were concentrated on 'substructure' and 'brickwork up to 1st floor' so that these stages were completed several weeks ahead of schedule. Having then run out of work, and the delivery of the floor units still being uncertain, the bricklayers were transferred to another site.

The situation at the end of week 26 is that delivery and installation of floor units has been resumed but the bricklayers have only just been recovered from the other site. Dry-lining has begun but the sub-contractor has withdrawn his men until they can be given a clear run of work. Complete re-programming of the subsequent stages has been necessary to meet the original completion date. Additional resources are required to step up the building rate, since dwellings now need to be completed at a rate of about seven per week.

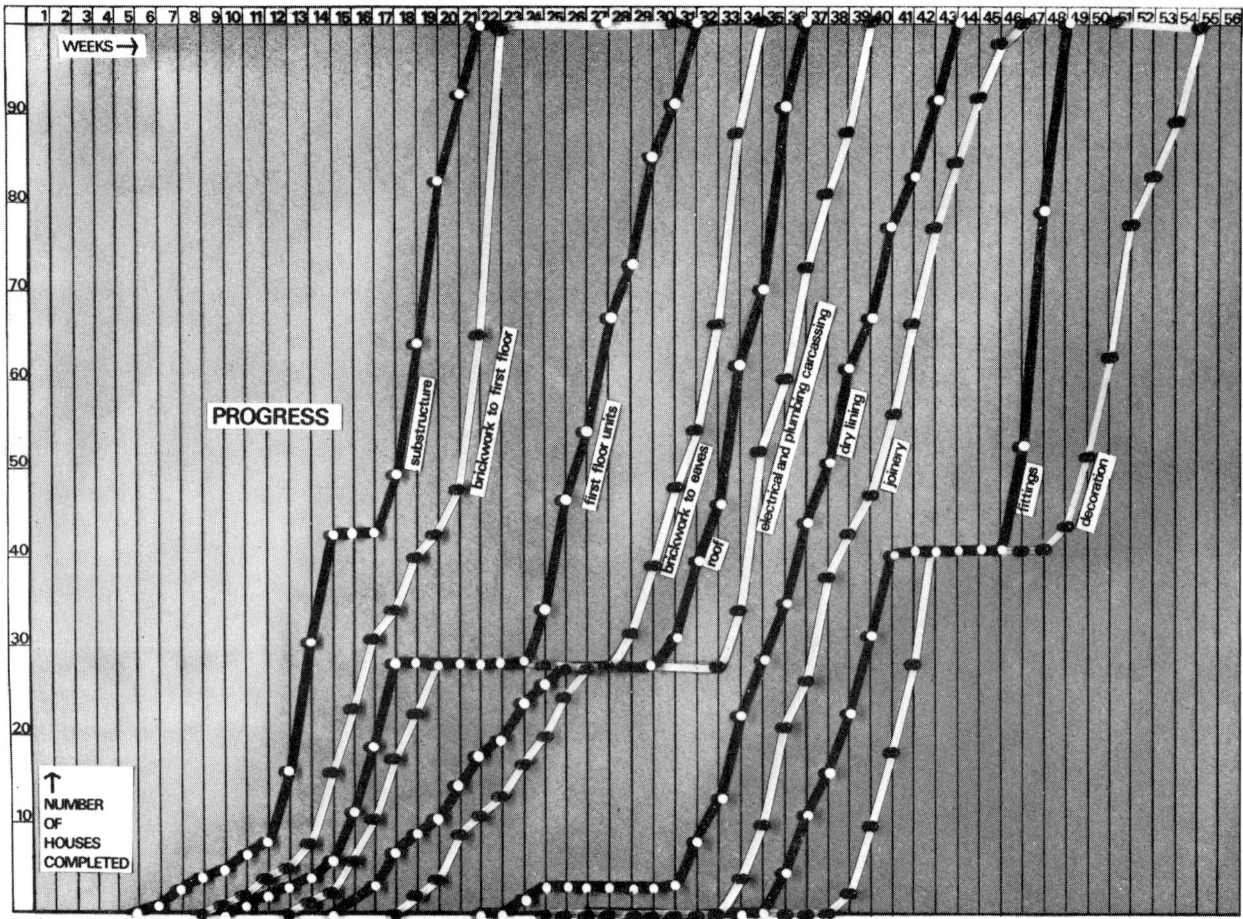

Fig 4d Completion

Additional resources were obtained and progress was well maintained until about week 40, when the sub-contractor installing 'fittings' unexpectedly took his men off the site. Pressure was required to force a restart after five weeks. By intensive working, fittings to the last 60 dwellings were completed in three weeks, putting that stage back on programme. But there was now no hope of the contract finishing on time—60 houses required decoration in three weeks. This would have needed a labour force of about 50 painters, quite beyond the contractor's resources. The work was re-programmed for completion in seven weeks, a total building time of 54 weeks instead of the planned 50 weeks.

European product-approval procedures:1

This digest compares the procedures in various European countries for the approval of new and existing products, ie building units, assemblies and building systems, to satisfy the requirements of building legislation and various non-statutory interests. It outlines the procedures of the Netherlands and France.

For the German Federal Republic, Sweden and Denmark see digest 167.

Procedures for the assessment, certification and approval of building products, to be discussed here, are concerned with: 1—appraisal of suitability for intended use, usually of new products or of established products in a new design situation, and 2—checking the level and consistency of properties relative to standards, normally of established products.

Approval procedures may be applied for purposes of building control or for a variety of commercial purposes. In some European countries, the import of certain items is virtually barred unless there is evidence of authorised independent testing. It is hoped that an outline of the procedures, and an indication of the organisations involved, will not only set the scene for exporters from this country but will also help designers and constructors who need to specify or use products in the countries discussed here, viz the Netherlands and France, and in Part 2, the German Federal Republic, Sweden and Denmark.

Fig 1 shows that whether an established product is to be checked against established requirements (eg Standards) or the performance of a new product is to be assessed, the sequences share the same logical steps: establishing criteria; testing and control; certification; approval or acceptance. The differences are mainly in emphasis.

For an established product, the main concern is likely to be with testing for conformity with requirements (approval testing) or with checking production samples—a programme of quality control. The same establishment may do approval testing and testing for quality control. When conformity with requirements is certified, approval is normally implicit: the emphasis is on testing and control.

For a new product, the emphasis is likely to be on the appraisal stage, which is often complex. An explicit approval for use may be required; this might be from a different body from that producing a statement on suitability for use, that is, stages (iii) and (iv) may or may not be merged.

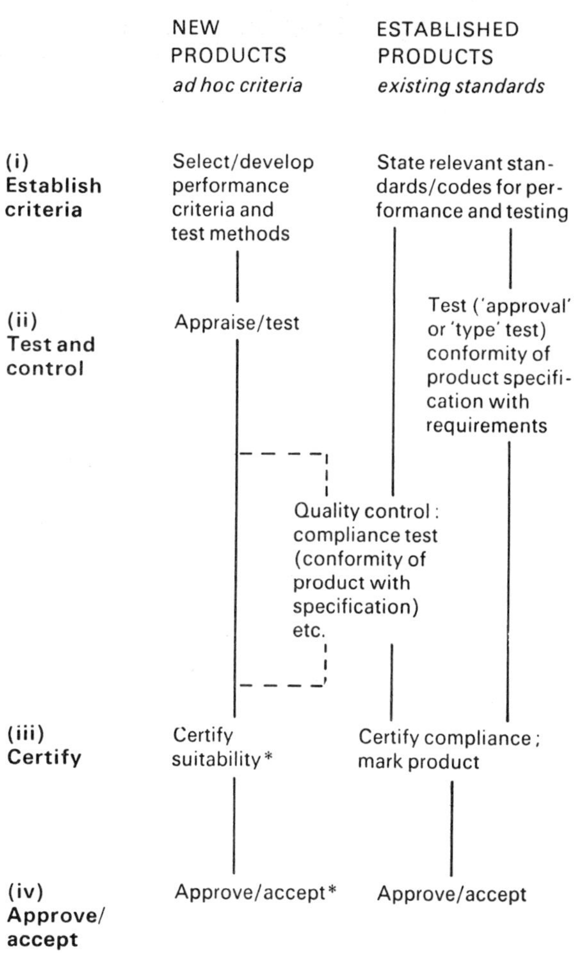

Fig 1   Assessment decision levels

Context	Status	Purpose	Accepting Organisations	Other Organisations Involved
Statutory requirements Building control	Approval	Public health and safety	Building etc authorities	Independent approval organisations
Non-statutory requirements Government building Social housing	Acceptance	Economy, amenity	Public clients	Agrément
Finance, insurance	Acceptance	Financial safeguard	Finance and insurance interests	Standards organisations; professions; technical associations
Industry schemes	Quality assurance	Marketing	Industry	Official and approved laboratories

Fig 2 Status and purpose of procedures and types of organisation involved

Fig 2 classifies procedures according to their status and purpose. Approvals may be required in the context of building control, giving legal sanction to the use of a product as satisfying statutory requirements such as building regulations or measures to ensure the safety of installations. There are other requirements which, although non-statutory, are equally important from the marketing point of view. These are the requirements of 'government building' and 'social housing' which, being major client groups, are important and therefore strong enough to set their own requirements for acceptance; of 'finance' and 'insurance' interests, ie those having a financial interest in the quality of building because they provide either mortgage or insurance cover; and, finally, of schemes set up by industries to ensure consistent quality in their own products, to assist marketing. Although industry schemes are voluntary, they are in some countries integrated into, or adopted as a part of, the building control system.

Figs 3–7, of this and the next digest, outline the various national systems. To the left of each diagram, the relevant groups of accepting and other organisations are selected from the eight shown in Fig 2, with an indication, where appropriate, of only 'minor' involvement. No one country involves all eight groups in their procedures and the changing pattern and number of these reflect the differences between national procedures. The five countries discussed can be divided into two broad groups.

Under the comprehensive systems of building control in the German Federal Republic and Scandinavia, approval procedures are based on the building law, which makes the building authorities responsible for assessing products against legal requirements. Consequently, procedures for product approvals are closely integrated with building control and the approval of new products is part of a wide system of testing and certification. 'General approvals' of new products are issued centrally and valid nationally. A general approval certificate—stating that a product conforms with the applicable regulations—becomes part of the national regulation system, replacing specific approval by the building inspector on site; it relates only to properties that are relevant to the regulations—structural stability, fire safety, hygiene and, recently, pollution control. Approvals are closely linked with product certification and quality control procedures; an important aspect is the initiative taken by industry groups in setting up 'control organisations' and the extent to which their activities have become part of the official system.

In the Netherlands and France, there is no central government involvement in building control and the Agrément System is used to appraise innovations. This is a broad general appraisal of suitability and performance; it is not confined to the requirements of the regulations but extends, for instance, to aspects such as fabrication, installation, ease of maintenance, repair, etc. Agrément is characteristic of countries that do not have centralised direction of building control or national building regulations (UK is exceptional in this). It does not normally do its own testing—in this also UK is exceptional. In these countries where product approvals are not closely integrated with centrally directed procedures, procurement procedures by public clients are prominent.

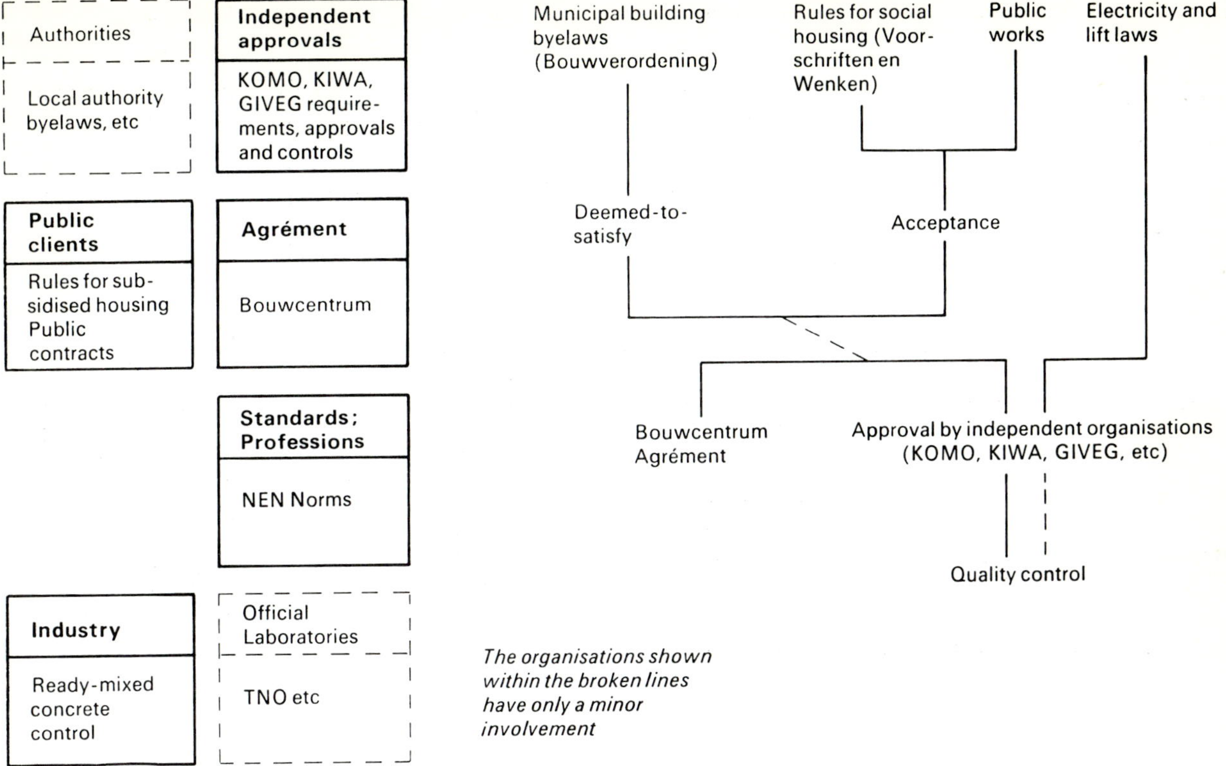

Fig 3 The Netherlands

Netherlands

Building control in the Netherlands is operated independently by the municipalities. Each of these has its own building byelaws, though most of them are based on a national model. About 90 per cent of housing is wholly or partially supported by public funds; both for this and for public works, there is a need for independent quality control and certification, and for an independent appraisal of innovations. Lifts and electrical components are required by law to be approved. For other products, approval is not mandatory but is a condition of acceptance within the different sectors. For example, the requirements for subsidised housing specify some products that must be certified.

The route for approvals and acceptance in these situations, shown dotted in Fig 3 because it does not have the same direct links as other systems to be described, is through specialised organisations that have been set up to provide independent approvals. They are 'approval-and-control' organisations which certify the conformity of products with their own requirements. The best known is KOMO, which deals with general building products, particularly those for which no national standards yet exist but which are purchased regularly by local authorities. It includes representatives of the national authorities with an interest in building, the municipal authorities and various technical and professional associations. It is an entirely administrative organisation, that is, it formulates quality requirements and organises test-ing and quality control but does not itself do any testing.

The other large independent approval establishment is KIWA which was set up by the water supply authorities as the approval body for water supply and drainage components and fittings. It includes representatives of the water supply interests and municipal and regional authorities. It has extensive test facilities and in addition to testing water supply components it does much of the testing for KOMO.

Product requirements prepared by KOMO, KIWA and also GIVEG—the approval body for gas appliances—eventually become NEN Norms.

The Agrément system, operated through the Bouwcentrum in Rotterdam, is concerned with the appraisal of new building products and provides the basis of acceptance of new building systems for subsidised housing. Agréments are now to have a 'deemed-to-satisfy' status in the model byelaws and consequently in most municipal byelaws.

GIVEG *Gasinstituut, Vereniging van Exploitanten van Gasbedrijven.* Gas Institute of the Association of Proprietors of Gas Industries

KIWA *Keuringsinstituut voor Waterleidingsartikelen.* Institution for the Testing of Waterworks Materials

KOMO *Stichting voor Onderzoek Beoordeling en Keuring van Materialen en Constructies.* Foundation for Research Approval and Testing of Materials and Construction

NEN *Nederlandse Norm.* Netherlands Standard

TNO *Toegepast Natuurwetenschappelijk Onderzoek.* Central Organisation for Applied Science

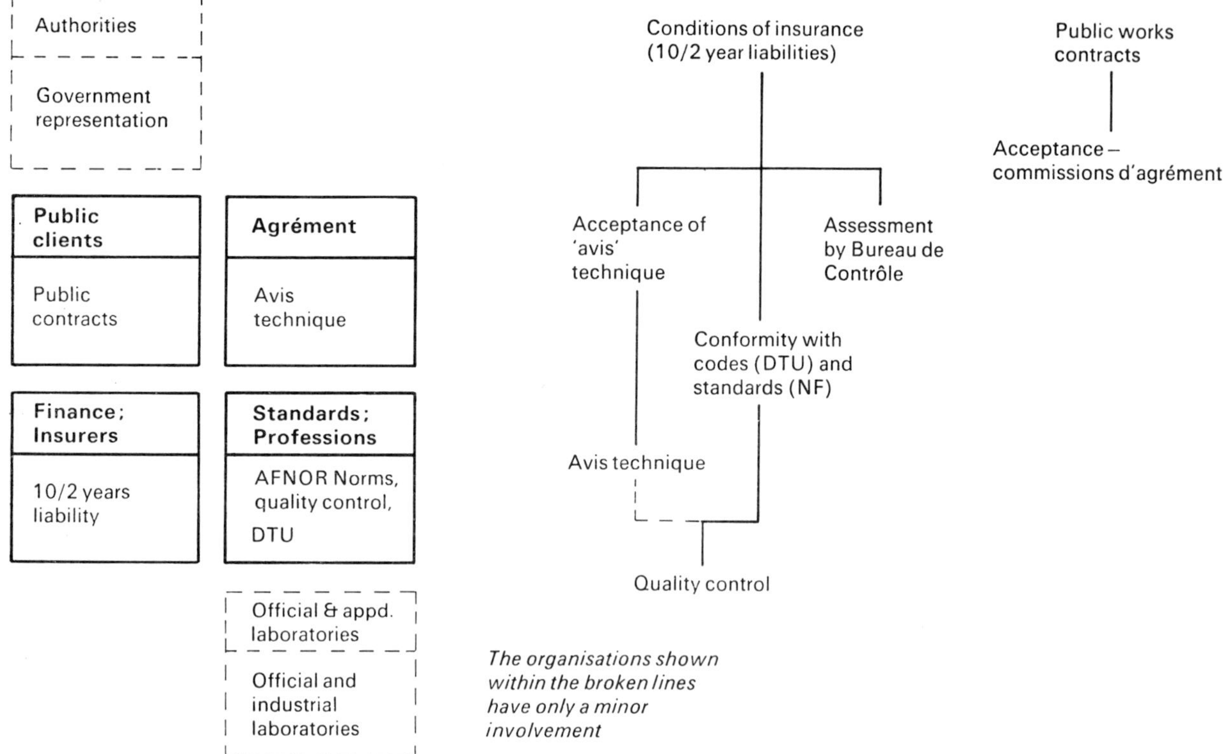

Fig 4 France

France

In France, official building control procedures are not involved in approving design or construction; instead, designers and contractors are liable under the Civil Code for major structural defects and defects in weathertightness (for ten years) and for minor defects (for two years). It is necessary (and for architects compulsory) to insure against these liabilities. For traditional products and constructions, insurance cover is conditional on conformity with established standards and codes but to assess the risk involved in new products and non-traditional constructions the Agrément system was set up. This has been considerably modified in France, where the certificates are now named 'avis techniques' (technical opinions or advice). They are issued by a ministerial commission for which CSTB provides the secretariat and technical backing. The insurers have set up their own procedure for 'accepting' avis techniques; acceptance may be subject to special conditions of construction, or to the exclusion of particular risks, eg weathertightness, under the policy.

Avis techniques are intended for new products or processes for which no standards yet exist but which are in regular production or use. When products or methods of construction for which there is yet no avis technique are to be used, they may be assessed for insurance purposes by one of the three Bureaux de Contrôle—the technical assessment organisations which supervise design and construction. Quality control may be required as a condition of an avis technique.

For established products, covered by a NF, AFNOR co-ordinates quality control procedures, some of which are operated by industry groups. For public contracts, principally for civil engineering works, special commissions (commissions d'agrément) operate acceptance procedures for certain product groups, setting their own requirements for quality and control.

AFNOR *Association Française de Normalisation* French Standards Association

CSTB *Centre Scientifique et Technique du Bâtiment* Scientific and Technical Building Centre

DTU *Document Technique Unifié* Code of Practice

NF *Norme Française* French Standard

European product-approval procedures: 2

This digest deals with the procedures of the German Federal Republic, Sweden and Denmark.

It was explained in the previous digest that in the Netherlands and France there is no direct involvement by central government in building control and that the Agrément system (now modified in France to 'avis techniques') is used to appraise innovations. Also, in these two countries where product approvals are not closely integrated with centrally directed procedures, procurement procedures by public clients are prominent; the relevant 'boxes' in Figs 3 and 4 (in Part 1) show them as having a major involvement.

By contrast, in the German Federal Republic, Sweden and Denmark, now to be considered, product approvals are closely linked with building control. In effect, the building authorities are responsible for checking that products meet the requirements of the building law, and any other legal requirements. It is a coherent system in which the approval of new products is only one aspect; there is also much emphasis on quality control.

The term 'general approval' is applied to the procedure by which the central authority issues a statement that a particular product type (usually, but not necessarily, a new product) complies with the requirements of the regulations, but only of the regulations. In this respect it differs from Agrément, which has a wider scope. General approval is an official licence that a product may be used, under defined conditions, and it is binding on the local building authorities. Quality control is normally a condition of general approval and may also be required for other products as a condition of approval to build.

There are some differences between the three countries in the importance attached to quality control and in the details of procedure, but normally the manufacturer's production methods and his own 'internal' quality control methods must be checked or supervised 'externally'. This involves approving production, especially quality control methods, testing product samples and periodic checking of the factory test equipment; this is undertaken either by an official laboratory (the more usual arrangement for new products) or, for established product groups, by a quality control association set up by an industry group.

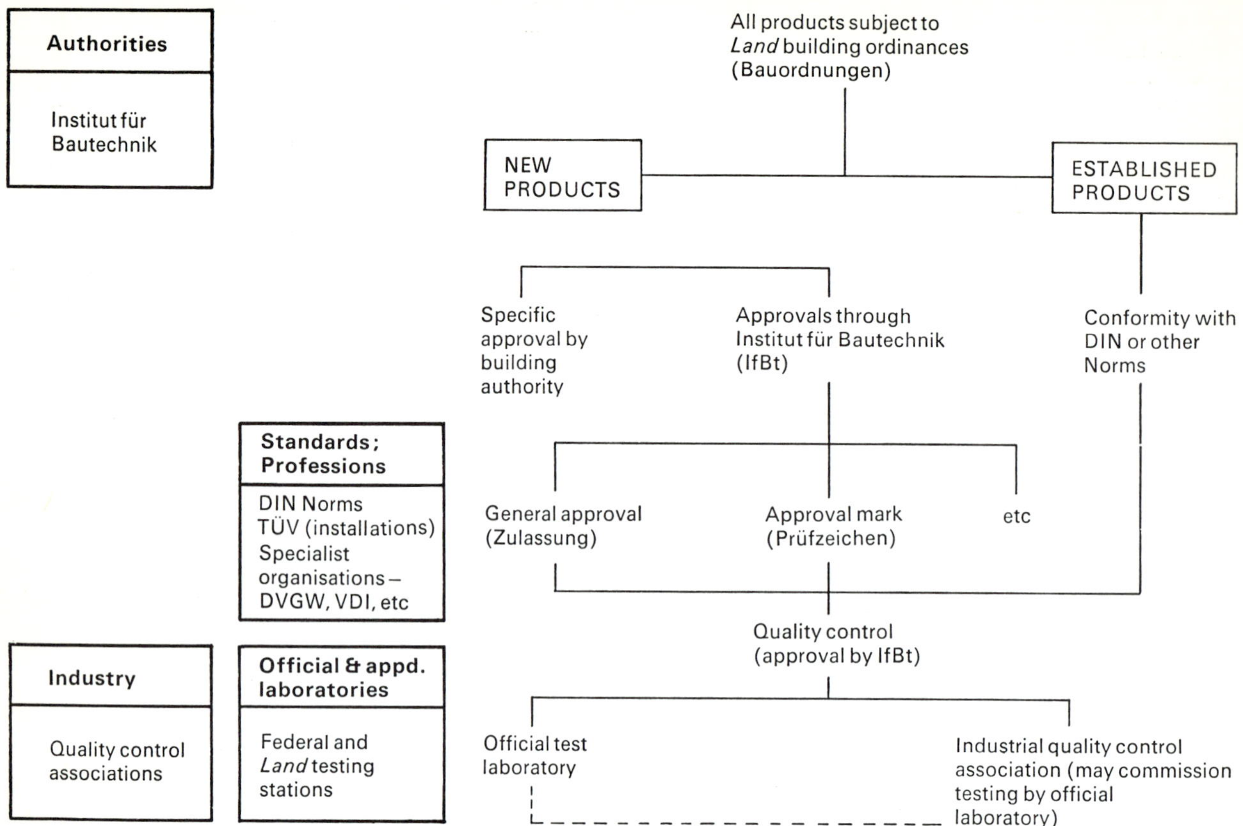

Fig 5

German Federal Republic

Since Germany has a federal constitution, building control is a responsibility of the individual Länder (federal states) but the technical work of approval is centralised through the Institut für Bautechnik (IfBt), in Berlin, which acts as the technical agency for approvals on behalf of the *Land* building ministries. The *Land* building ordinances demand that all building materials and products, so far as they affect safety and health, are of proved suitability, and a building permit is conditional on proof of this.

Traditional products must conform to recognised standards, generally DIN Norms, and must be certified and marked accordingly. Approval is required only exceptionally, for example, for certain loadbearing elements.

New products for which no standards exist must be approved for use. There are various procedures for this but the most usual is to apply to IfBt, which acts as the central agency, for a general approval (Zulassung). Applications are considered by the appropriate advisory committees, whose membership includes representatives of the federal ministries and of the *Land* building ministries. In the absence of a standard test, IfBt will devise suitable test procedures. Zulassungen are concerned only with products and aspects of performance affected by the building ordinances; approved products, subject to any conditions set out in the approval certificate and with the necessary identification and quality control marks, require no further inspection by local building authorities.

A simplified procedure, the approval mark (Prüfzeichen), is used for products listed in 'legal orders' of the *Land* building ministries for use only if they have an approval mark or are certified and marked as conforming with a DIN Norm.

Arrangements for the quality control of all products to which the building ordinances apply must be approved by IfBt and may be carried out either by an official laboratory or by one of the industrial quality control associations. Some of these do their own testing but most are administrative organisations which delegate testing to an approved laboratory, as shown by the broken link line in Fig 5.

The inspection of certain mechanical installations and services, such as lifts, electrical equipment, heating installations and fuel storage tanks, as required by various legal orders is done by inspectors of the technical supervision associations, Technische Überwachungs-Vereine (TÜV).

DIN	*Deutsche Industrienorm* German Industrial Standard
DVGW	*Deutscher Verein von Gas und Wassfachmännern* German Association of Gas and Water Engineers.
IfBt	*Institut für Bautechnik* Institute of Building Technology
TÜV	*Technischer Überwachungs-Verein* Technical Supervision Association
VDI	*Verein deutscher Ingenieure* German Association of Engineers

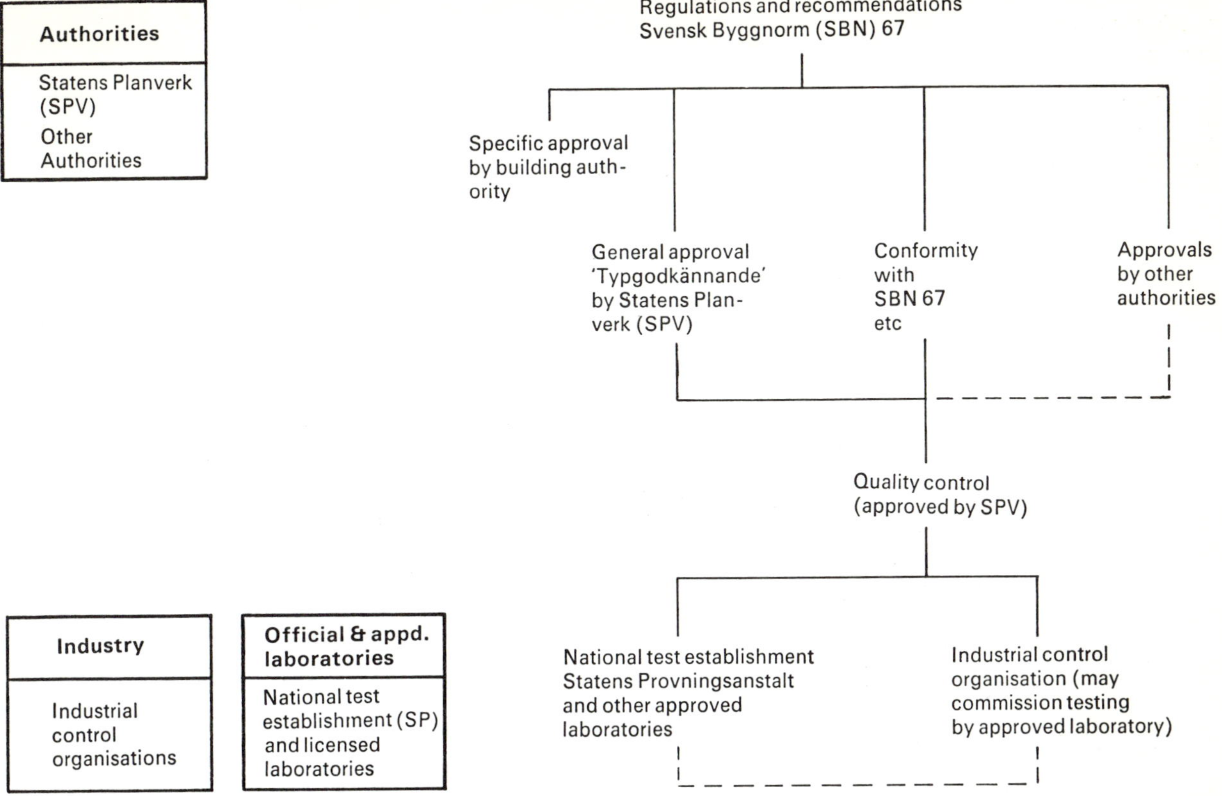

Fig 6

Sweden

The basic regulatory document in Sweden is the Swedish Building Norm, Svensk Byggnorm (SBN) 67, which brings together all mandatory requirements and recommendations, together with references to the requirements of other authorities.

The authority responsible for building regulations is the National Board of Urban Planning, Statens Planverk (SPV), which is also the central authority responsible for building product approvals. Approval in relation to regulation requirements may be given by the local building authority or through a centrally issued general approval, which local authorities must accept. Thus, although general approvals are not obligatory for manufacturers, they are usually an advantage and may be a practical necessity. Certain regulations in SBN 67, for instance those on fire classification, imply that SPV approval is needed in order to show that requirements have been met.

General approvals are normally conditional on quality control, requirements for which are set out in the approval certificate. There is also an extensive system of quality control for various product groups, co-ordinated by Statens Planverk. SBN 67 stipulates that building products must be checked in relation to the SBN requirements either by site inspection or by limited site inspection combined with control in the factory. Rules for control are broadly similar for 'generally approved' products and for the other product groups. Manufacturers must maintain 'internal'

control and must arrange for 'external' control, by an authorised 'control organisation', covering inspection and sampling, product testing by an official laboratory and periodic checking of the factory's own test equipment.

For most 'generally approved' products, the national test establishment, Statens Provningsanstalt, acts as control organisation. For other product groups, industrial control organisations have been set up to approve products in relation to requirements and to organise quality control. Several of these pre-date Statens Planverk, others have been set up under its direct sponsorship, but all are now subject to its authorisation. Their administration always includes roughly equal representation of industry and of the authorities (central and local) together with appropriate representation of professional and research interests. Their role in the building control system is illustrated by the fact that several are referred to in relevant parts of SBN 67. They normally delegate the work of inspection, sampling and testing to an official laboratory.

Affiliation to a control association is voluntary, but is usually considered to be a practical necessity.

SBN *Svensk Byggnorm* Swedish Building Norm

SP *Statens Provningsanstalt* National Institute for Materials Testing

SPV *Statens Planverk* National Board of Urban Planning

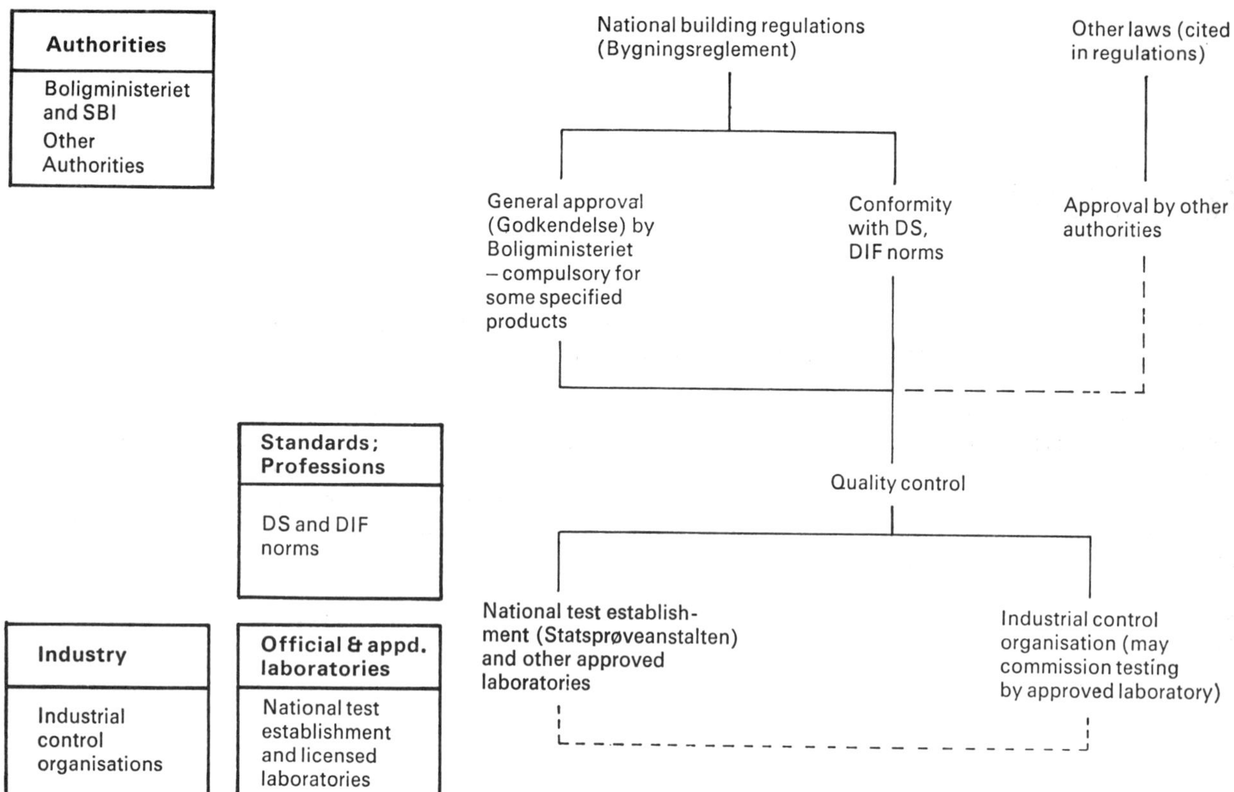

Fig 7

Denmark

The National Building Regulations (Bygningsreglement) specify certain products for which approvals are compulsory, eg, components and fittings for water supply and drainage installations. Approvals for these, and for innovations, are given by the Housing Ministry (Boligministeriet) in relation to functional requirements such as the structural characteristics of loadbearing elements, fire gradings and other safety requirements. Applications are forwarded by the Housing Ministry to SBI. Assessment is on the basis of existing standards, as far as possible; if additional test procedures need to be devised, this is done by the national testing establishment (Statsprøveanstalten) in collaboration with SBI. General approvals are nationally valid and are binding on local building authorities. The Regulations quote several other laws, for materials for gas installations, electrical equipment and underground oil-storage tanks, under which the approval of some other authorities is compulsory. Approvals are also mandatory for lifts, which are certified by a Lift Inspectorate in Copenhagen, or by local labour inspectorates. Conformity with Norms is an alternative pathway for building permit approval.

Quality control is normally a condition of all of these approvals, including certification by the standards organisation. Control organisations have been set up for some product groups, though they are not so well developed as those in Sweden—perhaps because a large proportion of building products in Denmark is imported. The Danish Society of Engineers (DIF) has taken a lead in attempting to co-ordinate procedures by publishing standard or model rules for control organisations. So far, these have been adopted by the organisations for glued laminated timber and for stress-graded timber; their adoption by the concrete element association is under consideration.

DIF   *Dansk Ingeniørforening* Danish Society of Engineers

DS   *Dansk Standardiseringsråd* Danish Standards Council

SBI   *Statens Byggeforskningsinstitut* Building Research Institute of Denmark

Further information

Technical Help to Exporters, British Standards Institution, Maylands Avenue, Hemel Hempstead HP2 4SQ

E Cibula *Product approvals for building: an international review* BRE CP 15/74 (includes addresses of organisations)

Index

This two-part index consists of a *numerical* index of BRE Digest current titles followed by an *alphabetical* subject index. The numerical section shows the location of each digest by volume and page number, whilst the alphabetical section gives the relevant digest numbers for each subject.

It will be seen that some digests appear in more than one volume, thus ensuring that the subject area of each volume is fully covered and repeating a feature which proved its value to users of the first edition of this series. The index is common to all four volumes, Building Materials, Building Construction, Services and Environmental Engineering, and Building Defects and Maintenance. When a reader is seeking a particular volume which is temporarily unavailable, as, for instance, in an office library, consulting the index in one of the other volumes will show if any of the required digests is to be found elsewhere.

Key M = Building Materials, C = Building Construction, S = Services and Environmental Engineering, D = Building Defects and Maintenance.

Numerical list of current titles

Alphabetical subject index